Jan-Peter Hildebrandt, Eberhard Teuscher, Ulrike Lindequist
Natural Poisons and Venoms

Also in this Series

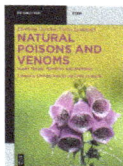

Natural Poisons and Venoms.
Volume 1: Plant Toxins: Terpenes and Steroids
Eberhard Teuscher, Ulrike Lindequist, 2023
ISBN 978-3-11-072472-1, e-ISBN (PDF) 978-3-11-072473-8,
e-ISBN (EPUB) 978-3-11-072486-8

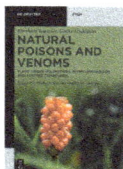

Natural Poisons and Venoms.
Volume 2: Plant Toxins: Polyketides, Phenylpropanoids and Further Compounds
Eberhard Teuscher, Ulrike Lindequist, 2024
ISBN 978-3-11-072851-4, e-ISBN (PDF) 978-3-11-072853-8,
e-ISBN (EPUB) 978-3-11-072864-4

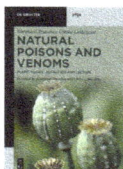

Natural Poisons and Venoms.
Volume 3: Plant Toxins: Alkaloids and Lectins
Eberhard Teuscher, Ulrike Lindequist, 2025
ISBN 978-3-11-112740-8, e-ISBN (PDF) 978-3-11-113621-9,
e-ISBN (EPUB) 978-3-11-113692-9

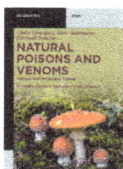

Natural Poisons and Venoms.
Volume 5: Fungal and Microbial Toxins
Ulrike Lindequist, Timo Niedermeyer, Eberhard Teuscher, 2025
ISBN 978-3-11-072856-9, e-ISBN (PDF) 978-3-11-072857-6,
e-ISBN (EPUB) 978-3-11-072865-1

Jan-Peter Hildebrandt, Eberhard Teuscher,
Ulrike Lindequist

Natural Poisons and Venoms

Volume 4
Animal Toxins

Founded by
Eberhard Teuscher and Ulrike Lindequist

DE GRUYTER

Authors

Prof. Dr. Jan-Peter Hildebrandt
Zoological Institute
Animal Physiology and Biochemistry
University of Greifswald
Loitzer Strasse 26
17489 Greifswald
Germany
jph@uni-greifswald.de

Prof. Dr. Eberhard Teuscher
Goethestr. 9
07950 Zeulenroda-Triebes
Germany
teuscher-triebes@t-online.de

Prof. Dr. Ulrike Lindequist
Lise-Meitner-Str. 6a
17491 Greifswald
Germany
lindequi@uni-greifswald.de

ISBN 978-3-11-072854-5
e-ISBN (PDF) 978-3-11-072855-2
e-ISBN (EPUB) 978-3-11-072862-0

Library of Congress Control Number: 2023933035

Bibliographic information published by the Deutsche Nationalbibliothek
The Deutsche Nationalbibliothek lists this publication in the Deutsche Nationalbibliografie;
detailed bibliographic data are available on the internet at http://dnb.dnb.de.

© 2023 Walter de Gruyter GmbH, Berlin/Boston
Cover image: Willem Van Zyl/iStock/Getty Images Plus
Typesetting: Integra Software Services Pvt. Ltd.
Printing and binding: CPI books GmbH, Leck

www.degruyter.com

Preface

Living beings produce pharmacologically active substances to ensure their survival. In many cases, such substances can harm or kill other living creatures or can trigger allergies. Thus, some of these compounds may pose danger to human beings as well as to domestic or wild animals.

Our aim is to acquaint readers, especially physicians, veterinarians, pharmacists, biologists, chemists, food chemists, biochemists, students of these disciplines, and interested laymen, with poisonous and venomous animals and their complex toxin compositions. The other volumes of this book series will focus on toxins produced by plants, microorganisms, and fungi.

The focus of this volume is on the toxic ingredients of animal poisons and venoms, their chemical structures, their effects and modes of actions in target organisms, and on the biology of poisonous or venomous animals. We highlight the symptoms of poisoning or envenoming in humans or animals, cite case reports that are available, and we try to give hints on poisoning prevention and on potential treatments. However, the book is not intended as a guide to medical practice. Local physicians should be contacted in case of poisoning or envenoming!

We try to create an understanding that animal toxins are not primarily threats to humans and domestic animals, but rather admirable evolutionary adaptations allowing animals to deter potential predators, protect themselves from infectious agents, or to efficiently obtain prey for their own survival.

This book series is based on the book *Biogene Gifte*, third edition, published in 2010 by Wissenschaftliche Verlagsgesellschaft Stuttgart in German language. Because of the enormous increase in knowledge in the field of natural poisons and venoms, to reach a larger readership, and to fit into the STEM series of the De Gruyter publishing house, the topic is now covered in five volumes in English language.

After a fundamental chapter on toxins (history, chemistry, biology, first aid, and general and clinical toxicology) Volumes 1 to 3 are devoted to poisonous plants and their active substances. Volume 4 summarizes the current knowledge about poisonous and venomous animals. Volume 5 deals with poisonous mushrooms and microalgae and their ingredients and selected microbial toxins. The volumes will appear in certain intervals, so that the highest possible timeliness of the content of each volume is guaranteed.

Despite the extensive bibliographies at the end of each chapter, the cited literature represents only a selection. We excuse ourselves for any important papers that we may have missed during our literature searches. For reasons of space, the citation format is as short as possible.

We would like to thank everyone who helped us with the preparations of the books. We would especially like to thank Dipl.-Phys. Karl-Heinz Lichtnow, Greifswald (Germany), for drawing the chemical structures of venom and poison components and his great help with computer problems of any kind. Thanks go also to Elvira Lutjanov

https://doi.org/10.1515/9783110728552-202

for her help with the literature search. We would like to thank all image authors and Getty images for the provision of photos. Our thanks go to De Gruyter publishing house, especially to Mrs. Karin Sora, for making these books possible, and to Dr. Bettina Noto for her always pleasant and helpful cooperation in the completion of the volumes and her careful editing work. We thank Mr. David Jüngst (Integra Software Services) for smoothly managing the production process of this book.

We are grateful for critical comments, reporting of errors, and suggestions for improvements!

Greifswald and Triebes, February 2023

Jan-Peter Hildebrandt, Eberhard Teuscher, Ulrike Lindequist

Contents

Abbreviations and Icons

AA	arachidonic acid
ACh	acetylcholine
AChE	acetylcholine esterase
ADP	adenosine diphosphate
AMP	antimicrobial peptide
ARF	acute renal failure
ATP	adenosine triphosphate
BTX	bungarotoxin
BW	body weight
CNS	central nervous system
CoA	coenzyme A
CYP	cytochrome P
DSP	diarrhetic shellfish poisoning
DW	dry weight
EDTA	ethylenediaminetetraacetic acid
ER	endoplasmic reticulum
3FTx	three-finger toxin
FW	fresh weight
HA	hyaluronic acid
HCN	hydrogen cyanide
i.m.	intramuscular
i.p.	intraperitoneal
i.v.	intravenous
LD	lethal dose
mAChR	muscarinic acetylcholine receptor
mRNA	messenger ribonucleic acid
nAChR	nicotinic acetylcholine receptor
$NADP^+$	nicotinamide adenine dinucleotide phosphate (oxidized form)
$NADPH + H^+$	nicotinamide adenine dinucleotide phosphate (reduced form)
NO	nitric oxide
NSP	nitrile-specifier protein
OA	okadaic acid
PFT	pore-forming toxin
PLA	phospholipase A
PNS	peripheral nervous system
p.o.	per os, oral uptake
PSP	paralytic shellfish poisoning
ROS	reactive oxygen species
s.c.	subcutaneous
TRP channel	transient receptor potential cation channel
TTX	tetrodotoxin
WHO	World Health Organization

https://doi.org/10.1515/9783110728552-204

♟ Symptoms and consequences of human exposure to animal toxins
🚑 First aid and potential treatment
⚲ Case description
✻ Allergy
🐕 Effects of animal toxins on other animals

1 Introduction to Animal Toxins

1.1 What Are Toxins?

Toxins are molecules which are synthesized by living organisms and used to parasitize or to predate on other creatures or to avoid being attacked by predators or parasites. Poisonous animals make use of one or several toxic compounds that they secrete onto the body surface (skin, feathers, fur) or accumulate internally in body tissues to repel, to harm or to kill predators or to avoid being colonized by microorganisms. In some cases, parent animals provide protection to their young by impregnating eggs or newborns with poisonous substances.

Venoms are generally mixtures of several substances produced in specialized glands and applied to target organisms by special applicators like stings or fangs. They may be used by animals as means to efficiently obtain prey or as chemical weapons for defense against attackers. In predatory animals, venoms are used to immediately immobilize or kill prey which requires that they cause rapid onsets of their paralyzing effects. Thus, venoms used by predators mostly target neuronal or muscle cell systems with the aim of interrupting signal transmission from motor neurons to muscle cells and impairing voluntary motor functions in the target organism.

The scientific field studying the generation, the application, the molecular interactions with endogenous targets, the metabolism and excretion or elimination of natural poisons, venoms or related molecules is called 'toxinology'. It is a subdiscipline of 'toxicology'.

The extent to which toxins harm their target organisms depend on
- the kind of toxin,
- the amount of toxin,
- the pathway of application to or introduction into the target organism, e.g.,
    - topical – superficial application to parts of the body surface
    - *per os* or peroral (p.o.) – through the mouth
    - intraperitoneal (i.p.) – injection through the peritoneum into the body cavity
    - intramuscular (i.m.) – injection into the body wall muscle
    - intravenous (i.v.) – injection into the blood stream
    - subcutaneous (s.c.) – injection under the skin into the subcutis
- whether the toxin is acutely or chronically applied,
- the sensitivity of the target organism against the toxin,
- the metabolism of the toxin within the target organism, or
- the rate of elimination of the toxin from the target organism.

This illustrates that the toxin character of a given substance is relative and depends on many variables. This has been recognized very early in history, e.g., by Famoso Doctor Paracelsus (Philippus Theophrastus Aureolus Bombastus von Hohenheim,

https://doi.org/10.1515/9783110728552-001

1493–1541) who is famous for his proverbial phrase: 'Alle Ding' sind Gift und nichts ohn' Gift; allein die Dosis macht, das ein Ding' kein Gift ist' (everything is a poison and nothing is no poison; just the dose is relevant for its toxicity).

Also, the administration route may be relevant whether a substance acts as a toxin or not: A protein toxin that may be highly toxic when injected into an animal may be completely harmless if ingested because it is denatured by the acidic environment in the stomach and readily digested by proteases in the intestines.

The scientific discipline dealing with questions on the uptake rates, metabolism, and elimination of toxins in or from target organisms is called 'toxicokinetics'. The mechanism of action of a toxin within a target organism is studied in 'toxicodynamics'. The term 'toxicography' describes studies of complex effects of toxins on target organisms and the reasoning about potential antidotes or therapies.

In this book, we focus on biogenic substances used by animals which are directed against other animals or humans. However, antimicrobial compounds that are generated and used by animals to combat potential pathogenic microorganisms will also be considered.

1.2 Organisms as Sources for Toxic Compounds

Animals that are able to produce their own toxins display 'primary toxicity'. 'Secondary toxicity', however, is a feature of animals which are not able to originally generate their own toxins but acquire their toxins or at least precursors of such toxins from food organisms or from commensal or symbiontic microorganisms. Evolutionary adaptations in animal species with secondary toxicity to tolerate their own toxins in the body are common. An interesting example is the impregnation of tissues of inner organs in Indo-Pacific puffer fish (Tetraodontidae) of the genus *Takifugu* with tetrodotoxin (see Section 2.9.7), a toxin that is originally produced by bacteria (*Aeromonas*, *Vibrio*) [10, 13]. Tetrodotoxin is a highly effective blocker of voltage-gated sodium channels in animals and inhibits the generation of action potentials in neurons and skeletal muscle cells resulting in paralysis. Mutations in one of the subunits of the voltage-gated sodium channel in *Takifugu* which resulted in the exchange of just a few amino acids in this protein rendered the channel tetrodotoxin-resistant in these fishes [14]. Another example of evolutionary development of resistance against specific alkaloid toxins of food organisms are the poison frogs of Central and South America (see Section 3.2.12). In these animals, mutations in the $Na_V1.4$ voltage-gated sodium channel that are conserved among the different species of frogs in the genus *Dendrobates* are associated with alkaloid toxin resistance [18]. Being resistant against these toxins but making the own body toxic for other organisms protects these animals from becoming victims of predators. Caterpillars of the monarch butterfly (*Danaus plexippus*) (see Section 3.2.9) tolerate high concentrations of cardenolides in their bodies because the α-subunit of the Na^+/K^+-ATPase carries a mutation (N122H, asparagine to histidine at amino acid position 122) which renders the enzyme insensitive to such steroid glycosides which, in turn, harm other organisms trying to feed on these caterpillars or the adult butterflies [7, 9].

Some animals store their toxins in specialized gland reservoirs and mobilize these compounds only to the external space. This is the case in many snakes where the toxins are contained in specialized salivary glands in the upper jaws [3, 8] and released through ducts connected to teeth that work as injection needles. Honeybees, e.g., *Apis mellifera*, carry their toxins in venom glands and associated storage compartments while all other tissues of such animals are entirely free of any toxic material. Such a separation of toxin compartments from other tissues in the body (sequestration) avoids being poisoned by the own toxins.

Differences in toxin profiles between individuals of the same species are common features in animals [17]. Temporal or regional differences in toxin content of animals have been observed in primarily as well as in secondarily toxic animals. Toxin profiles may also change during ontogenic development so that juveniles may express toxin compositions that differ in quality or quantity from those expressed in adults [12]. Venoms of honeybee workers vary between summer and winter and are different from that of queens [1]. Juveniles of rattlesnakes (see Section 3.2.13) may express different compositions of toxins than adult animals from the same geographical region. On the other hand, adult rattle snakes from different locations may produce venoms with different toxin compositions [6]. Similar findings have been made for the venom compositions of regional populations of the monocled cobra, *Naja kaouthia* [15]. Some of these differences are based on genetic differences (local adaptation) between subpopulations [5]. The marine box jellyfishes from Northern Australia (*Chironex fleckeri*) (see Section 3.2.2) show geographical differences in venom composition as well but it remains unclear whether these differences are due to genetic or environmental factors [16].

Especially secondarily toxic animals differ in their toxin composition depending on environmental factors. In some cases, the differences result from interaction of individuals with different types of toxin-producing microorganisms; in other cases, toxin profiles depend on the choice of food organisms. Secondary toxicity may fade away if animals are reared without access to toxin-producing microorganisms. This is the case with crust anemones (*Palythoa* sp., Cnidaria) in the Pacific Ocean which carry the highly toxic palytoxin (see Section 2.3.7) only when toxin-producing dinoflagellates, *Ostreopsis* sp., are present in the surrounding water [2].

The tissue distribution of toxins in animals that impregnate their tissues with toxic compounds to deter predators is more generalized. Certain tissues in such animals, however, may contain higher toxin concentrations than others as in the case of the *Takifugu*-puffer fish whose skin and inner organs may be highly toxic while skeletal muscle tissue contains only low amounts of tetrodotoxin [10]. In fact, thin slices of puffer fish muscle tissue (fugu) are considered a delicacy by traditional Japanese gourmets (see Section 3.2.11).

Generally, we differentiate actively toxic animals or 'venomous animals' from passively toxic animals which are also called 'poisonous animals'. Venomous animals produce highly effective toxins in specialized glands that generally comprise secretory cells and storage compartments. The entire complement or portions of the stored venom may be conditionally applied to other organisms using specialized tools like stings or biting mouthparts. This may occur in attempts to apply disabling or deadly venom to prey or to defend oneself against an attack by a predator. Examples of venomous animals are spiders and snakes. Poisonous animals, however, use their toxic compounds to impregnate the external body surface or tissues of the integument or of internal organs and become toxic as a whole. This may help to deter potential predators from preying on such animals because the poisonous tissue has bad taste or smell, may exert inflammatory effects on the predator's mucous membranes, or may harm the internal organs of a predator after consuming tissue of a poisonous animal [4]. Examples of poisonous animals are the above-mentioned puffer fish *Takifugu* or the slow loris of the genus *Nycticebus* (see Section 3.2.14) who impregnate their fur with toxic salivary compounds.

Some researchers add to these two a third category of animal toxins, the so-called 'toxungens' [11]. These are animal-borne toxic substances which are delivered to the body surface of a target organism without creating a wound. This is illustrated by the defensive action of a spitting cobra (*Naja* sp.) which can project venom from its fangs to a target that is up to 2 m away. The venom stream is directed toward the eyes of an attacker (Photo 1.1). When hit, the venom is absorbed by the moist surfaces around the eyes of the target animal and may readily unfold its cytotoxic or neurotoxic actions.

Ph. 1.1: *Naja pallida* spitting venom from its fangs toward an attacker (Source: Guido Westhoff 2007, with permission).

References

[1]   Danneels EL et al. (2015) Toxins 7(11): 4468
[2]   Deeds JR et al. (2011) PLoS ONE 6(4): e18235
[3]   Fry BG et al. (2003) Rapid Commun Mass Spectrom 17(18): 2047
[4]   Garson MJ (2010) Marine natural products as antifeedants. In: Liu H-W, Mander L (eds.) Comprehensive Natural Products II. Elsevier, Oxford, p. 503
[5]   Gren ECK et al. (2017) Geographic variation of venom composition and neurotoxicity in the rattlesnakes Crotalus oreganus and C. helleri: Assessing the potential roles of selection and neutral evolutionary processes in shaping venom variation. In: Dreslik MJ, Hayes WK, Beaupre SJ, Mackessy SP (eds.) The Biology of Rattlesnakes II. ECO Herpetological Publishing and Distribution, Rodeo, New Mexico, p. 228
[6]   Holding ML et al. (2021) Proc Natl Acad Sci U S A 118(17): e2015579118
[7]   Holzinger F et al. (1992) FEBS Lett 314(3): 477
[8]   Kardong KV (2002) J Toxicol: Toxin Rev 21(1–2): 1
[9]   Mebs D et al. (2000) Chemoecology 10(4): 201
[10]   Melnikova DI, Magarlamov TY (2022) Toxins 14(8): 576
[11]   Nelsen DR et al. (2014) Biol Rev 89(2): 450
[12]   Schonour RB et al. (2020) Toxins 12(10): 659
[13]   Simidu U et al. (1987) Appl Environ Microbiol 53(7): 1714
[14]   Soong TW, Venkatesh B (2006) Trends Genet 22(11): 621
[15]   Tan KY et al. (2017) PeerJ 5: e3142
[16]   Winter KL et al. (2010) Toxicol Lett 192(3): 419
[17]   Yu C et al. (2020) Int J Biol Macromol 165(Pt B): 2994
[18]   Yuan ML, Wang IJ (2018) PLoS ONE 13(3): e0194265

1.3 Structure and Function of Animal Toxins

Proteins and peptides are constituents of most animal poisons or venoms and may function as cytotoxins, as blockers of ion channels, as inhibitors of transmitter receptors, as enzymes, or as enzyme inhibitors. These compounds need to be injected into target organisms because oral application would result in loss of function due to protein denaturation (loss of tertiary and quaternary structure) and enzymatic digestion of the toxic molecules in the gastrointestinal tract of the target organism.

Proteins are macromolecules composed of one or more chains of amino acid residues bound to each other by the peptide bond (Fig. 1.1). Small proteins up to 100 amino acid residues are called 'peptides', longer ones or those composed of several amino acid strands are called 'proteins' in the narrow sense. There are generally 20 different amino acids that are used by organisms to build proteins. The amino acid sequence within a given protein is dictated by the nucleotide sequence of the respective gene. The chemical nature of the amino acid side chains at specific locations in the protein determines the way of local protein folding into specific 3D structures called 'protein domains'. The combination of specific domains in the overall structure determines protein function. An example is the highly specific interaction of a

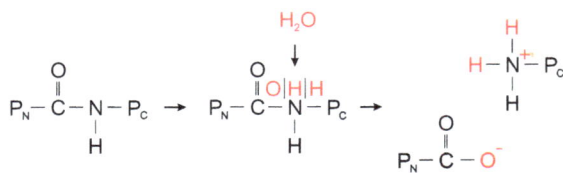

Fig. 1.1: The peptide bond (left) between two consecutive amino acids in a protein. Cleavage of the peptide bond (middle) via hydrolysis (red) is mediated by 'proteinases' (or shorter 'proteases') yielding two independent peptide strands (P) of which one has a new C-terminus (C) and the other a new N-terminus (N).

scorpion peptide, charybdotoxin (see Section 3.2.7), with homotetrameric, voltage-gated potassium channels in plasma membranes of neurons in the central nervous system. The toxin blocks this channel and elongates action potentials resulting in neuronal overexcitation in the brain [3–5].

Most animal toxins, however, are secondary metabolites generated via diverse metabolic pathways. Just to mention a few examples, formic acid is used by ants to deter attackers or to hinder microbes from overgrowing the ant brood. Amino compounds are used for the same purpose by cnidarians, molluscs, and echinoderms. Polyethers may be present in mussels or in corals. This class of molecules encompasses some of the most toxic substances in nature. Badly smelling molecules like mercaptans and sulfides are used by ants and martens to deter predators. Macrocyclic ketones are constituents of coral toxins. Benzoic acid, benzoquinones, and cyanides are used by beetles and diplopods to defend themselves against attackers. Terpenes are used for the same purpose by mites, bees, ants, butterflies and beetles, and *N*-heterocyclic compounds by fire ants or spiders. Plant-borne polycyclic hydrocarbons, especially steroids, are used by toads or salamanders or alkaloids like coccinellin by ladybirds to avert predators. Bacterial toxins as tetrodotoxin (TTX) in puffer fish or batrachotoxin (BTX) in frogs [2] and birds [1] render these animals inedible for carnivores.

References

[1]   Bartram S, Boland W (2001) Chem Bio Chem 2(11): 809
[2]   Daly JW et al. (2005) J Nat Prod 68(10): 1556
[3]   Miller C (1995) Neuron 15(1): 5
[4]   Novoseletsky VN et al. (2016) Acta Naturae 8(2): 35
[5]   Thompson J, Begenisich T (2000) Biophys J 78(5): 2382

1.4 Evolutionary Origins of Animal Toxins

Although the toxin character of proteins, peptides, or secondary metabolites produced by animals is well known in many cases, researchers often do not know the exact mechanisms of action of these substances in potential target organisms. This is mostly due to the fact that the functional domains of such molecules that actually make a molecule a toxin are hidden among a lot of other chemical groups within the respective molecule that do not have any significance in this context. This is a consequence of biological evolution that does not evolve specific toxins in a targeted manner but, instead, generates genetic diversity by accumulation of mutations in the DNA which result in the production of variants of already existing molecules that originally may have served entirely other functions than being toxins [3]. Processes like gene duplication [15], genetic recombination like exon shuffling [14], and diversification may contribute to such diversity within a given organism [4]. These processes may accidentally result in some molecules that turn out to have toxic properties against certain target organisms. Only in cases in which the interaction of a toxic animal and such a target organism is biologically relevant and gives a selective advantage to the toxin producer, the pathway of toxin synthesis and all additional traits necessary for the toxin producer to deal with this compound will be stabilized and inherited through subsequent generations. Otherwise, the trait will be lost over several generations because it does not provide the carriers of this trait with a selective advantage, i.e., production of more offspring per generation in trait carriers compared with the noncarriers.

Evolutionary genesis of a toxic compound may occur directly (by mutation of genes that code for the toxic proteins or peptides) or indirectly (by mutations in the genes of enzymes involved in generation of intermediate or secondary metabolic products with toxin properties) [7]. This evolutionary process is illustrated by the fact that many snake venoms contain protein toxins that are structural homologs of phospholipase A_2 [2], an enzyme that is important in all animal cells as a component of an essential signal transduction pathway. The toxic character of the secreted neurotoxic phospholipase A_2 in animal venoms is related to its enzymatic activity [13] but lies at least to the same degree in its ability of leaving the producing cell which was not an original feature of the signal transduction molecule or bind to target molecules that were not the original ones of the signal transduction molecule [11]. This example shows that the toxin character of a given protein molecule may not only evolve by changing its primary sequence but may be related to evolutionary changes in other molecules that result in previously impossible molecular interactions, changes in cellular trafficking or alterations in localization of a given molecule. This way, a harmless molecule may become a toxin without any changes in the molecule itself.

When this rationale of generating molecular diversity and selection of the most suitable variants would strictly apply to all animal toxins, we would expect that each individual of a given animal species expresses just one highly efficient toxin that optimally fulfills a specific function in the respective target organism. However, nature is not that strict but allows for some variability, which may turn out to be meaningful when analyzed in detail [3]. A toxic animal has to fight off not just one but several different species of predators. Vice versa, it may accept several different animal species as potential prey. Having just one toxin at hand for this variety of target organisms may not be enough to optimally match its needs. These considerations correspond well with observations that toxic animals generally produce different toxic substances at a time, either secondary metabolites or proteins and peptides, in some cases even several isoforms of proteins or peptides with more or less sequence variation [6]. Snake venoms (see Section 3.2.13), for instance, are mixtures of compounds belonging to different protein families, and each of these families contains many different toxin isoforms. Some compounds of the venoms are effective against neuronal cells (neurotoxins), others against muscle cell functions (myotoxins). One may be optimally suppressing a certain mechanism in a given prey species, another isoform may actually be better suited to target the same mechanism in another prey species. Blood-feeding animals like leeches make use of salivary proteins that are injected into the wound which they inflict to the body surface of host animals. Some of the injected substances inhibit blood clotting during feeding and thereafter and are thus called 'anticoagulants' (see Section 2.12.7). The most famous anticoagulant in leech saliva is hirudin [9], a small protein (7 kD) that is present in medicinal leeches like *Hirudo medicinalis* in three isoforms which are all effective inhibitors of thrombin, a protease that mediates an essential step in the blood coagulation cascade, the formation of insoluble fibrin from fibrinogen. Moreover, recent studies revealed the presence of several hirudin-like factors in the same species [8]. The reason for this diversity is not clear but we may hypothesize that leeches that are opportunistic ectoparasites using every opportunity to feed on potential hosts (amphibians, birds, mammals, humans) cannot know in advance which host species will come along next. Thus, it may be advantegeous to have several forms of very similar molecules at hand that may not all be highly efficient in inhibiting thrombin in the current host animal but contain variants that work better in other host species. This example illustrates that it may be evolutionarily rewarding to maintain the coding genes for different molecular variants of toxins or toxin-producing enzyme systems over many generations and to actually express them in every single individual despite the metabolic costs that are associated with this effort. This strategy prepares animals to deal with any potential nutrient resource that may come along.

The individual components of toxin compositions may also work at different time scales in target organisms. Toxins for disabling prey may work in the range of seconds to minutes. This is quick enough to hinder the prey animal to run away after being hit. Toxins that are used for defensive purposes, however, have to work in milliseconds to seconds because otherwise the defender would be severely harmed or dead before the attacker would react to the toxin. This is the reason why many animal poisons or venoms contain substances that taste or smell badly or immediately induce severe pain in the target animal. An example for the latter is serotonin, a biogenic amine that is a constituent of many animal venoms and induces immediate and intense pain when injected into peripheral tissues [10].

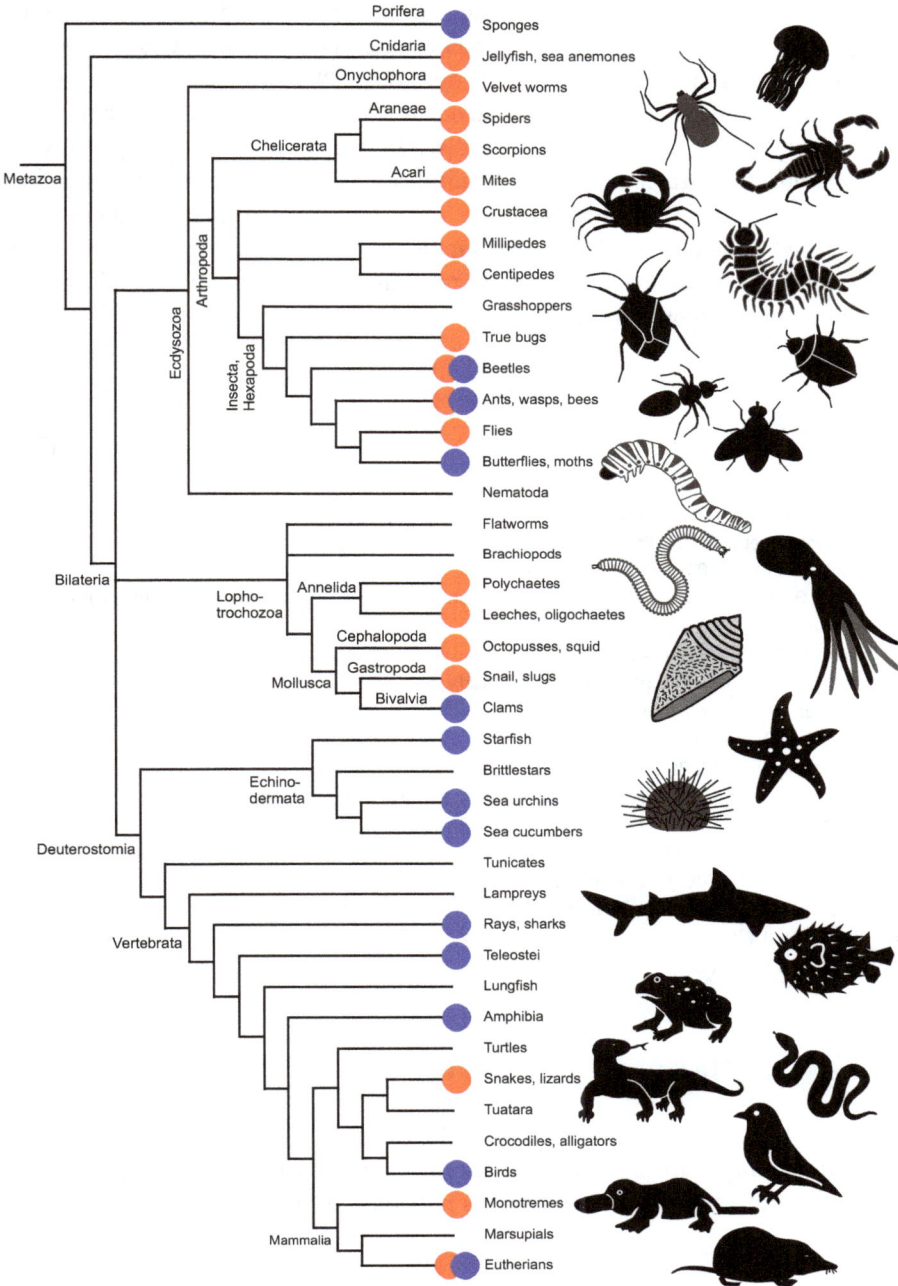

Porifera	Sponges
Cnidaria	Jellyfish, sea anemones
Onychophora	Velvet worms
Araneae	Spiders
	Scorpions
Acari	Mites
	Crustacea
	Millipedes
	Centipedes
	Grasshoppers
	True bugs
	Beetles
	Ants, wasps, bees
	Flies
	Butterflies, moths
	Nematoda
	Flatworms
	Brachiopods
Annelida	Polychaetes
	Leeches, oligochaetes
Cephalopoda	Octopusses, squid
Gastropoda	Snail, slugs
Bivalvia	Clams
	Starfish
	Brittlestars
	Sea urchins
	Sea cucumbers
	Tunicates
	Lampreys
	Rays, sharks
	Teleostei
	Lungfish
	Amphibia
	Turtles
	Snakes, lizards
	Tuatara
	Crocodiles, alligators
	Birds
	Monotremes
	Marsupials
	Eutherians

Fig. 1.2: Simplified schematic phylogenetic tree of the animal kingdom specifying those taxa in which venomous or poisonous animals have developed. Branches highlighted in red contain venomous animal

Considering all this it is not surprising that the trait of being 'toxic' has been developed many times independently during evolution of the animal kingdom [3] and that closely related animal species may be toxic (Fig. 1.2) or not and may express variants of molecules (even isomers) of which one is toxic while the others are not.

Researchers have established a new discipline to study the diversity of toxins, their different structural features, the chemical compositions of poisons or venoms as well as the diversity of potential targets in other organisms. This new science discipline is called 'venomics' [1, 12]. The term indicates that researchers nowadays try to implement high-throughput methods as 'genomics', 'transcriptomics', and 'proteomics' to extend our knowledge in this field.

References

[1]   Abd El-Aziz TM et al. (2020) J Chromatogr B 1160: 122352
[2]   Bickler EP (2020) Toxins 12(2): 68
[3]   Casewell NR et al. (2013) Trends Ecol Evol 28(4): 219
[4]   Chothia C et al. (2003) Science 300(5626): 1701
[5]   Giribet G, Edgecombe GD (2020) The Invertebrate Tree of Life. Princeton University Press, Princeton, NJ, USA
[6]   Malhotra A (2017) Mutation, duplication, and more in the evolution of venomous animals and their toxins. In: Malhotra A, Gopalakrishnakone P (eds.) Evolution of Venomous Animals and Their Toxins. Springer, Dordrecht, Netherlands, p. 33
[7]   Mebs D (2001) Toxicon 39(1): 87
[8]   Müller C et al. (2017) Parasitol Res 116(1): 313
[9]   Nowak G, Schrör K (2007) Thrombosis Haemostasis 98(1): 116
[10]   Sommer C (2010) Handb Behav Neurosci 21: 457
[11]   Šribar J et al. (2019) Sci Rep 9(1): 283
[12]   Sunagar K et al. (2016) J Proteom 135: 62
[13]   Vardjan N et al. (2013) Commun Integr Biol 6(3): e23600
[14]   Wang X et al. (2017) Toxins 9(1): 10
[15]   Wong ES, Belov K (2012) Gene 496(1): 1

1.5 Toxicokinetics of Animal Toxins

Toxins may have very complicated chemical structures, although they are generally derived from simple precursor molecules. Thus, researchers studying animal toxins are trying to understand the biosynthesis pathways of toxic compounds. A second

Fig. 1.2 (continued)
species and branches in blue contain poisonous animal species. The phylogenetic tree has been drawn according to current knowledge [5]. Note that most animal taxa contain venomous or poisonous species besides nontoxic ones. This indicates that toxicity in animals has been developed multiple times in evolution.

goal is to understand what the target mechanisms of these toxins in other living organisms are and how specific compositions of such toxins are used to reach specific goals related to the ecology of the respective species (biological significance). Furthermore, researchers in toxicokinetics wish to understand and deal with the impacts of toxic animals and their toxins on other organisms, especially humans.

Discrimination of acute and chronic toxic effects is less important when dealing with animal toxins than when studying toxins of bacteria, plants, or fungi. This is due to the fact that most animals use their toxins for rather immediate effects like stunning or paralyzing prey or deterring predators. Chronic effects of animal toxins are relevant only in scenarios in which a consumer feeds for longer periods on animals that contain toxic compounds which are not immediately harmful but exert their effects only after reaching certain minimal concentrations or have to be present over an extended period of time in the target organism to get harmful. The latter may apply to substances that are genotoxic and interfere with the integrity, the replication, or the processing of DNA. However, there are as yet only rare reports on such compounds in animal toxins. Exceptions are the dolastatins, a family of potent cyanotoxins accumulating in the marine sea hare *Dolabella auricularia* [6]. Their mechanisms of action include inhibition of tubulin polymerization and microtubule formation in animal cells [1], processes that indirectly affect genomic stability of such cells and results in inhibition of cell proliferation.

A potentially toxic compound needs to reach its molecular interaction partner in the target organism. This requires that the toxin passes the body surface by diffusion or secondarily active transport. Uptake may occur through the integument (skin), through the airways upon inhalation of gaseous or aerosolized toxins or through the gastrointestinal surfaces after ingestion of the substance. Alternatively, and this is actually the main route for toxins of venomous animals, the toxin gets injected into the target organism through the skin.

Upon injection, the toxin is locally present at very high concentrations and may exert special effects depending on the amount of toxin that has been injected. Large portions of the toxin, however, are subsequently distributed within the body by the circulating body fluid which results in dilution of the toxin but enables the toxin to exert systemic effects, i.e., to interfere with target mechanisms in every tissue. Since the circulation time of blood through the entire body in humans is less than a minute [7], the toxins are theoretically able to quickly reach any cell in the body.

Yet, this is only the case for small molecules that can easily pass biological barriers (like epithelial cell layers) by diffusion. In vertebrates, however, proteins beyond a molecular mass of 50 kDa do generally not diffuse out of the circulatory system into the extracellular space due to the fact that the endothelial cell layer with its underlying basal membrane provides an efficient permeability barrier for such molecules. Especially tight in this respect is the blood-brain barrier in mammals [3] so that protein toxins of this size stay in the blood plasma and do not reach the neurons or glia cells of the central nervous system (CNS). This is the reason why most snake toxins (see Section 3.2.13) exert their effects on blood and endothelial cells, peripheral neurons, and the muscular system, but do not affect the brain straight away. Just certain peptides in snake venom may be able to penetrate the blood-brain barrier [5]. Thus, animals or humans affected by snake bites are often fully conscious about what is happening to them, at least initially.

The distribution of an injected toxin in the target organism also depends on its chemical nature. Molecular size plays an important role as small molecules generally have higher diffusive mobilities in body fluids than large molecules like proteins. The net charge of a molecule is also important, especially if it needs to pass cellular membranes to reach its molecular target. If the toxin is transported by an endogenous carrier, it may more easily reach the intracellular space than molecules of the same size that are not subject to facilitated diffusion. Lipophilic toxins may be transported within the blood stream in association with nonspecific carrier proteins like albumins but may easily pass the lipid environment of the plasma membrane to reach intracellular targets when getting in contact with cell surfaces. Water-soluble toxins, however, will not be able to pass the plasma membranes by diffusion and are likely to target extracellular interaction partners like endogenous plasma membrane receptors.

The concentration level of a given toxin in the target organism is not only dependent upon its absorption and distribution kinetics but also on its metabolic fate and excretion rate. Proteins and peptides are generally degraded by proteases which occur in the extracellular space as well as in internal compartments of cells. Biotransformation of small molecules in the body occurs by nonspecific monooxygenase systems (e.g., the cytochrome P450 monooxygenase in the liver) which introduce polar functional groups (hydroxyl or epoxy groups) into the molecules making them better soluble in water so that they can easily be disposed off via renal excretion [2]. It is a somewhat ironic feature of this biotransformation mechanism that some toxin molecules get even more toxic when oxidized compared with the native form. Such toxins may actually harm liver cells and are considered hepatotoxins. Some lipophilic molecules are, especially in liver cells, conjugated with more hydrophilic endogenous molecules like monosaccharides, glucuronic acid, sulfate moieties, or amino acids to make them more polar and facilitate renal excretion.

Excretion of such derivatized toxin molecules from the body of the target organism may occur by renal excretion because this pathway does not require any specific transporters for these molecules. In some cases, however, liver cells transport toxin molecules directly into the bile fluid. Most of these molecules subsequently pass the intestines and get excreted along with the faeces.

Considering all these differences in uptake pathways and rates, in the processing and in the disposal of toxins in target organisms, it is not easy to define an objective measure of toxicity of a given substance. Toxicologists use the 'lethal dose 50' (LD_{50}) as an approximation for the relative toxicity of a toxic substance. This is the dose of a toxic substance that each experimental animal has to receive using a specific application technique at which 50% of all treated animals will not survive. For one of the most toxic compounds that occurs in animals, maitotoxin (see Section 2.3.10), the LD_{50} (mouse, i.p.) is approximately 80 ng/kg BW [4].

References

[1]   Bai RL et al. (1990) Biochem Pharmacol 40(8): 1859
[2]   Hasler JA et al. (1999) Mol Aspects Med 20(1): 1
[3]   Hawkins BT, Davis TP (2005) Pharmacol Rev 57(2): 173
[4]   Holmes MJ, Lewis RJ (1994) Nat Toxins 2(2): 64
[5]   Osipov A, Utkin Y (2012) Central Nervous System Agents in Medicinal Chemistry 12(4): 315
[6]   Wang E et al. (2020) Biomolecules 10(2): 248
[7]   Wolff CB et al. (2005) Circulation time in man from lung to periphery as an indirect index of cardiac output. In: Okunieff P, Williams J, Chen Y (eds.) Oxygen Transport to Tissue XXVI. Springer, Boston, MA, USA, p. 311

1.6 Toxicodynamics of Animal Toxins

Toxicodynamics is a subdiscipline of toxicology investigating the systemic, cellular, or molecular targets of toxins, the toxin–target interactions, and the resulting physiological effects in the target organism. As we have discussed above, there is a high degree of diversity of natural substances that have toxic potencies. At least as diverse as those are the molecular targets that may be termed 'receptors' to pinpoint the fact that they are the initial interaction partners for toxins in the target organism.

Due to the purpose of animal toxins to immediately immobilize or kill prey or to deter potential predators, a typical animal toxin shows a rapid onset of its effects. Thus, toxins used by predators mostly target neuronal or muscle cell systems with the aim of interrupting signal transmission from motor neurons to muscle cells and impairment of voluntary motor functions in the target organism.

Some toxins are directed against components of presynaptic endings of motor neurons. Some toxins interfere with the activities of action potential-generating ion channels in the plasma membrane. Others modify vesicular trafficking in the presynaptic ending and alter the rate of neurotransmitter release. α-Latrotoxin [10] which is a component of the venom produced by black widow spiders (*Latrodectus* sp.) is an example of such toxins (see Section 3.2.7). This toxin induces uncontrolled neurotransmitter release from motor neurons which results in muscle cramps and, in extreme cases, spastic paralysis of the ventilatory muscles [7].

Other toxins targeting the motor system are directed against components of the postsynaptic cell, the myocyte of the skeletal muscle system. Vertebrates use acetylcholine as the transmitter of motor neurons to activate nicotinic acetylcholine receptors (nAChRs) in the subsynaptic plasma membrane of skeletal muscle cells. Some snake toxins (see Section 3.2.13) bind to and block the nicotinic acetylcholine receptor (nAChR) and its sodium channel that is an integral component of the receptor molecule. An example of such a toxin is α-bungarotoxin from the Chinese krait (*Bungarus multicinctus*) which binds with extremely high affinity to the nAChR and blocks sodium influx into the muscle cell. This renders the neuromuscular junction dysfunctional and results in paralysis of the target organism [1].

Some toxins interfere with the activity of ion pumps in the cell membranes in target organisms. The sodium/potassium-ATPase (Na^+/K^+-ATPase) is present in virtually all animal cells and hydrolyzes ATP to provide energy for the translocation of sodium ions (Na^+) from the cytosol to the extracellular space and potassium ions (K^+) in the other direction. These processes result in steady-state concentrations within quiescent human cells of 10 mmol/L Na^+ and 120 mmol/L K^+ compared with 140 mmol/L Na^+ and 4–6 mmol/L K^+ in the extracellular space. These ion gradients provide the driving forces for secondary transport processes like nutrient uptake or transepithelial ion and fluid transport. Palytoxin (see Section 2.3.7), a product of dinoflagellates like *Ostreopsis siamensis*, accumulates in crust anemones of the genus *Palythoa*. When animals ingest or inhale this toxin it interacts with Na^+/K^+-ATPase and converts this ion pump into an open cation channel [8]. This results in breakdown of ion gradients and membrane potential across the cellular membranes and cell death. The LD_{50} (mouse) of palytoxin is 100 ng/kg BW.

Animal toxins usually contain enzymes (see Section 2.12.2) like phospholipase A_2 [9], sphingomyelinases [3], or hyaluronidases [4] which support the action of other toxin components by destroying cells or loosening the extracellular matrix, thereby improving the distribution of other toxin components in tissues of the target animal. On the other hand, toxins may be inhibitors of enzymes in the target organism. We have already mentioned the cardenolides in poisons of insects that are inhibitors of the Na^+/K^+-ATPase and the anticoagulant hirudin which is used by leeches to suppress the activity of the protease thrombin in host blood. In addition, many antimicrobial peptides or antimicrobially active protein domains in animal toxins are effective inhibitors of secreted microbial proteases. These inhibitors protect biological material of the toxin producers from degradation and consumption by microorganisms. Among others, components in the skin secretions of amphibians (see Section 3.2.12) fulfill similar functions [2].

Animal toxins may contain proinflammatory or skin blistering substances like cantharidin (see Section 2.4.3) that is a poisonous substance in the body fluids and glandular secretions of melyrid beetles. When orally administered cantharidin is highly toxic to most animals. Reports of LD_{50} values for humans vary between 1 and 60 mg/kg BW. Contact of cantharidin with moist body surfaces induces burning sensations and blistering, i.e., shedding of the outermost layers of epithelial cells. These effects of cantharidin usually keep potential predators away from feeding on melyrid beetles. Therapeutically, physicians use cantharidin topically to remove warts from the skin [5]. Skin blisters heal within 1 week without forming scars.

Some proteins and peptides in animal toxins may function as allergens in humans when applied repeatedly. One substance that has been implicated in this context is melittin which is highly abundant (up to 50% of the venom dry mass) in honeybee venom (*Apis mellifera*) (see Section 3.2.9). Bee stings result in intense pain sensations and local edema which is generally enough to deter any potential attacker of the bee hive. However, in sensitized persons it can result in life-threatening immunological responses such as anaphylactic shock [6].

References

[1]   Albuquerque EX et al. (2009) Physiol Rev 89(1): 73
[2]   Ali MF et al. (2002) Biochim Biophys Acta (BBA) Proteins Proteom 1601(1): 55
[3]   Fry BG et al. (2009) Ann Rev Genomics Hum Genet 10: 483
[4]   Kuhn-Nentwig L et al. (2019) Toxins 11(3): 167
[5]   Moed L et al. (2001) Arch Dermatol 137(10): 1357
[6]   Perez-Riverol A et al. (2015) Toxins 7(7): 2551
[7]   Südhof TC (2001) Annu Rev Neurosci 24: 933
[8]   Takeuchi A et al. (2008) Nature 456(7220): 413
[9]   Vardjan N et al. (2013) Commun Integr Biol 6(3): e23600
[10] Yan S, Wang X (2015) Toxins 7(12): 5055

1.7 Animal Toxins and Human Health

Most cases of intoxication with animal toxins in humans occur by accident. Being outdoors for professional or recreational purposes raises the chance of getting in contact with toxic animals. As humans are generally not considered prey by toxic animals the usual scenario of intoxication is that humans intrude the habitats of toxic animals and get bitten or stung in a situation that the animal feels threatened and tries to deter the intruder or the alleged attacker.

Snake bites occur when humans unexpectedly come very close to such a reptile, e.g., when working in the fields or during child play. Life-threatening snake bites occur in many tropical and subtropical countries in Africa, Asia, and the Americas. More than 5 million snake bites per year are registered by the World Health Organization. At least 20,000 of these accidents are fatal [5], and in another estimated 300,000 cases the victims suffer from permanent disabilities.

In many cases the victims of snake bites do not even see the snake which has bitten them because things occur very quickly. The snake will rapidly disappear after having stunned the alleged attacker. Snake bites, however, are generally easy to diagnose. Typical initial symptoms of a snake bite are the presence of two puncture wounds in the skin, swelling and redness around the wounds, pain at the site, difficulty breathing, nausea and vomiting, blurred vision, sweating, and salivation as well as numbness in face and limbs (see Section 3.2.13). These symptoms may occur individually or in combination and in different intensities. Every snake bite should be treated as if it is venomous and has to receive medical attention. As the relevant components of snake venom are proteins there are effective antivenoms available. This is the reason why relatively few of the many snake bites in humans result in fatalities or disabilities.

Insect bites or stings may just be a nuisance for humans. However, apart from the chance of parasite or pathogen transmission there are some cases in which the immune system reacts very heavily to the venom proteins that have been injected into the wound. Allergic reactions to bee venoms occur in sensitized people, i.e., those that have been stung repeatedly [1]. Immediate symptoms of bee or wasp stings are intense local pain, erythema, or edema that may vary in extent or size. Generalized or systemic reactions may vary greatly and set in approximately 10 min after the sting. They comprise urticaria (hives) and facial or generalized angio-edema (blood blisters), feelings of extreme illness, hypotension, or loss of consciousness. Dyspnea may result from laryngeal edema which is a life-threatening complication.

In some instances, accidental intoxications in humans occur by consuming meat or other products of toxic animals. People are usually not aware of the toxic nature of their food. Examples are certain types of fish poisoning in humans who consumed tropical fish that may have become poisonous (ciguatera toxin, maitotoxin (see Section 2.3.10), or brevetoxin (see Section 2.3.4)) upon periods of algal blooms in tropical marine ecosystems [2].

In rare cases, however, humans use animal toxins voluntarily as drugs and may suffer from the effects of overdosing. This has been reported for skin extracts of toads which are used as hallucinogens and for 'Spanish fly', a preparation of meloid beetles (*Lytta vesicatoria*) containing the terpenoid cantharidin (see Section 2.4.3), that is used as an aphrodisiac [6].

Because of their exceptional specificities, some animal toxins have found their ways in medical or research uses. Recombinant variants of the anticoagulant hirudin which is found in saliva of blood-sucking leeches is used in the clinic for thrombosis prophylaxis in cases in which the patients do not tolerate heparin [3]. Conotoxins of predatory marine snails (see Section 3.2.3) are of interest to clinicians as well as to neuroscientists due to their high selectivities and potencies for binding to and interfering with functions of specific signal transduction modules, primarily in prey but also in mammalian cells. This compound class provides a rich source for molecular probes and therapeutic leads [4]. A peptide in the venom of the Brazilian viper *Bothrops jararaca* was used as a template to develop inhibitors of angiotensin-converting enzyme (ACE) which cleaves angiotensin I and produces angiotensin II. These inhibitors, e.g., ramipril, are used as antihypertensive drugs in humans [7].

References

[1]   Ewan PW (1998) Br Med J (Clin Res Ed) 316(7141): 1365
[2]   Friedman MA et al. (2017) Marine Drugs 15(3): 72
[3]   Greinacher A, Warkentin TE (2008) Thrombosis Haemostasis 99(5): 819
[4]   Jin AH et al. (2019) Chem Rev 119: 11510
[5]   Kasturiratne A et al. (2008) PLoS Med 5(11): e218
[6]   Prischmann DA, Sheppard CA (2002) Am Entomol 48(4): 208
[7]   Teetz V et al. (1984) Arzneimittel-Forschung/Drug Res 34–2(10b): 1399

1.8 Therapeutic Approaches upon Intoxication in Humans

Effective therapy of poisoning with animal toxins is only possible if professional help is immediately available (contact a physician, rush the person to the next hospital, and contact the local or national Poison Control Center as soon as possible) and depends on several points of information that should be communicated when alerting such a center:

1. Toxic animal species that caused the poisoning
2. Circumstances of toxin application to judge intensity of transmission or dose of transmitted toxin (accident, drug abuse/overdose, suicide, etc.)
3. Pathway of toxin administration (bite, sting, oral ingestion, etc.)
4. Age, sex, height, and body weight of the poisoned person
5. Day and time of the event
6. Exact location
7. Description of symptoms (pain, paralysis, unconsciousness, etc.)
8. Medical history (previous illness, drug prescriptions)
9. First aid measures already taken

As most animal toxins work quickly every incidence should be taken seriously and attention of medical professionals should be sought immediately. If venom contains tissue destructive enzymes, the wound may initially seem harmless. The severity of tissue damage becomes obvious only after hours or days. Such a delayed onset of tissue destruction is observed upon bites from recluse spiders of the genus *Loxosceles* which result in pain sensation, erythema, and edema during the first 3 days, but may end with ulceration down to the bone 1 week later (necrotic cutaneous loxoscelism) (see Section 3.2.7). Early surgical intervention may limit the damage [6].

If the toxin just came into contact with the body surface, clothes should be removed and the skin rinsed with warm water that may contain a mild detergent (e.g., soap). When toxic substances have got in contact with the eyes, a sodium chloride solution (9 g NaCl/L of water, body temperature) should be used to rinse the eyes continuously for at least 10 min. An ophthalmologist should be contacted.

Upon oral ingestion of toxins, a physician may decide to give the patient a suspension of activated carbon in water to drink (approximately 1 g/kg BW). Absorption of toxins to the activated carbon may prevent transfer of toxins from the intestines into the body. This may be combined with a mild laxative to support rapid excretion. Physicians may decide from case to case whether vomiting should be induced to remove toxins from the stomach. This may be considered when the intoxication occurred within the last 1–2 h; otherwise the stomach content may be already transferred to the small intestine.

Application of antisera against the toxins of certain animals is very efficient if the animal species that caused the intoxication is known. Antisera are available against the toxins of certain species of spiders [3, 4], scorpions [1], or snakes [2, 5]. The local medical personnel or the local authorities generally know where to obtain these antisera and may have stored suitable material for immediate use.

References

[1]   Abroug F et al. (2020) Intensive Care Med 46(3): 401
[2]   Cocchio C et al. (2020) Am J Health – Syst Pharm 77(3): 175
[3]   Isbister GK (2010) Toxicology 268(3): 148
[4]   Isbister GK et al. (2003) J Toxicol – Clin Toxicol 41(3): 291
[5]   Silva A, Isbister GK (2020) Biochem Soc Trans 48(2): 537
[6]   Swanson DL, Vetter RS (2006) Clin Dermatol 24(3): 213

2 Chemistry, Pharmacology, and Toxicology of Animal Toxins

2.1 Aliphatic Acids

Aliphatic acids are used as defense substances by a large number of insect species. Formic acid (HCOOH) appears to be the most effective one. It has a pungent smell, is painful when in contact with mucous parts of the body surface, and irritates even normal skin. Insects generate and store formic acid in specialized glands and spray it upon potential attackers or inject it using stings associated with the venom apparatus.

When used as a defense substance, formic acid is usually accompanied by lipophilic substances that promote skin penetration. In ground beetles (Carabidae) of the genus *Platynus*, formic acid is stored as a highly concentrated aqueous solution (30–80% formic acid) in the paired pygidial glands. The lipophilic compounds in this solution are mostly alkanes (*n*-undecane, *n*-tridecane, and *n*-pentadecane) and ketones (2-dodecanone, 2-tridecanone, 2-pentadecanone, and 2-heptadecanone). These substances facilitate the permeation of formic acid through lipophilic body surfaces of target organisms. In other species of ground beetles (e.g., in *Calathus ruficollis*), formic acid is mainly accompanied by fatty acids. All species are able to spray up to 0.5 mg of fluid in one round onto a potential attacker [17]. In another carabid species, the false bombardier beetle (*Galerita lecontei*), the pygidial gland secretions contain 80% formic acid. The lipophilic additions are long-chain hydrocarbons and esters [13].

Many species in the order of Hymenoptera (sawflies, wasps, bees, and ants; see Section 3.2.9) have specialized toxin glands and application tools to transmit the stored material to target organisms. Venoms composed of several different substances are used to deter attackers or competitors, to kill prey, or to suppress the growth of microorganisms [3]. Most of these mixtures contain aliphatic acids (e.g., formic acid) besides other toxins of different chemical structure [16].

Among the ants (Formicidae), formic acid occurs only in members of the subfamily Formicinae. The females of these species have an acidopore, a nozzle-shaped structure at the apex of the seventh abdominal sternite used to spray formic acid [8]. Formic acid is produced in a venom gland and stored in a reservoir from which it is released through the acidopore. Stingless weaver ants (e.g., *Oecophylla longinoda*) usually bite prey organisms using their powerful jaws, bend their abdomen forward, and inject formic acid into the wound [1]. In some formicine ants, however, formic acid is used as a defensive tool to protonate and inactivate alkaloid toxins [4] that have been applied by fire ant species ('acidopore grooming' [10]).

The biosynthesis of formic acid in the ants probably takes place from serine or glycine as precursors via tetrahydrofolate intermediates [7]. Formic acid in the poison glands of formicine ants is often accompanied by hydrocarbons, *n*-undecane, terpenes,

https://doi.org/10.1515/9783110728552-002

acyclic ketones, or aromatic compounds [6] which facilitate the diffusive uptake of formic acid through the chitinous integument in target organisms.

Remarkable is the suicidal defense of their foraging territory by workers of *Camponotus* species (carpenter ants of the *Camponotus cylindricus* complex) that live in the rainforests of Malaysia. These animals have hypertrophied mandibular glands that extend from the base of their jaws to the back of their bodies. When attacked, they contract their abdominal muscles so intensely that the epithelia of the mandibular glands as well as the intersegmental membranes of the abdomen (gaster) rupture. Among the substances that are released from the mandibular glands are several corrosive and irritating compounds like *m*-cresol (3-methylphenol) and resorcinol (benzene-1,3-diol) besides hydroxyacetophenones, aliphatic hydrocarbons, formic acid and alcohols like octadecanol that make the mixture sticky [9].

Many bird species (e.g., the common blackbird, *Turdus merula*) make use of the ability of ants to spray formic acid. The birds sit directly on top of an ant's nest, spread their wings, and let the ants spray their defensive secretions onto the feathers (see Section 3.2.13). Some birds have been observed to grab ants using their beaks and slide them over their feathers. The procedure called 'anting' brings formic acid and other ingredients of the ant's toxin mixture on the skin and the feathers of the bird where they unfold their insecticidal, miticidal, fungicidal, or bactericidal effects [12].

Some caterpillars (larvae of butterflies, Lepidoptera) also use formic acid as an ingredient of defense secretions. When larvae of the puss moth, *Cerura vinula*, feel threatened, they spray 1–3 mg of a 30% solution of formic acid from glandular organs that are located in a skin fold in front of the first pair of legs [15].

The caterpillars of the swallowtail, *Papilio machaon*, and relatives carry an evertible neck fork, the osmeterium. It is an often intensely colored defensive organ which, along with eye-like spots at the sides of the body surface, deters potential predators visually. While the osmeterial organ is usually invisible in its inverted position, the animal may evert and display it when threatened. In addition, gland cells in the osmeterial organ secrete a mixture of volatile organic acids, such as isobutyric acid and 2-methylbutyric acid [5] that serve to repel birds, reptiles, ants, spiders, or mantids by their unpleasant odor. Mono- and sequiterpenes are constituents of such secretions in some species of Papilionidae depending on their food plants.

Defensive secretions of ground beetles (Carabidae) contain formic acid, acetic acid, isobutyric acid, isovaleric acid, methacrylic acid, or tiglic acid ((2*E*)-2-methylbut -2-enoic acid). These organic acids are often accompanied by C_{10} to C_{13} hydrocarbons. The gold ground beetle, *Carabus auratus*, releases almost pure methacrylic acid with traces of tiglic acid from its defensive glands [14].

The predatory larvae of some fungus gnat (Diptera, Mycetophilidae) species (e.g., the glowworm, *Arachnocampa* sp.) use their saliva to build small nets or long strings from protein and mucus and deposit droplets of oxalic acid solution on the surface to kill prey organisms that get in contact with these traps [2, 11].

🪲 Although long-term exposure of the skin to formic acid may lead to severe burns with blistering, the symptoms that occur in humans or mammals upon contact with formic acid-containing defense secretions of invertebrates, such as reddening, swelling, or slight inflammation, are not medically relevant due to the very small amounts released by the producers and short exposure times.

References

[1]   Bradshaw JWS et al. (1979) Physiol Entomol 4(1): 39
[2]   Broadley RA, Stringer IAN (2001) Invertebrate Biol 120(2): 170
[3]   Brütsch T et al. (2017) Ecol Evol 7(7): 2249
[4]   Chen L et al. (2014) Angew Chem Int Ed Engl 53(44): 11762
[5]   Chow YS, Tsai RS (1989) Experientia 45(4): 390
[6]   Hayashi N, Komae H (1980) Biochem Syst Ecol 8(3): 293
[7]   Hefetz A, Blum MS (1978) Science 201(4354): 454
[8]   Hung ACF, Brown WL (1966) J New York Entomol Soc 74(4): 198
[9]   Jones TH et al. (2004) J Chem Ecol 30(8): 1479
[10]   LeBrun EG et al. (2015) J Chem Ecol 41(10): 884
[11]   Mansbridge GH, Bruston HW (1933) Trans Royal Entomol Soc London 81: 75
[12]   Revis HC, Waller DA (2004) The Auk 121(4): 1262
[13]   Rossini C et al. (1997) Proc Natl Acad Sci U S A 94(13): 6792
[14]   Schildknecht H (1970) Angewandte Chemie 82(1): 17
[15]   Schildknecht H, Schmidt H (1963) Zeitschrift für Naturforschung 18 b(7): 585
[16]   Touchard A et al. (2016) Toxins 8(1): 30
[17]   Will KW et al. (2010) J Insect Sci 10(1): 12

2.2 Polyynes

2.2.1 General

The polyynes are aliphatic, non-branched hydrocarbons containing at least one triple bond between neighboring carbon atoms (C≡C). Derivatives may occur in the form of ring structures, molecules containing oxygen (epoxy-, hydroxyl-, or oxo-groups), sulfur (e.g., in methylthiopolyynes), or halogens. Most polyynes are found in plants, but sponges (Porifera) use polyynes as defensive substances or antimicrobials as well. The number of carbon atoms in sponge polyynes varies from 14 up to 48. Polyynes may also occur in defense secretions of marine mollusks, e.g., in the sea hare, *Aplysia oculifera* [3]. Polyynes with chain lengths of 12–17 carbon atoms were also found in reef-forming hard corals of the genus *Montipora* [1]. In plants, polyynes seem to be synthesized from fatty acids, e.g., oleic acid [2, 9], by desaturase-mediated reactions. The synthesis pathways of animal polyynes are still unknown.

2.2.2 Polyynes of Sponges (Porifera)

There are several short-chain (C_{12}–C_{17}) polyynes and numerous long-chain (C_{22}–C_{48}) unbranched polyynes in sponges (see Section 3.2.1). Some of these polyynes have hydroxyl-, keto-, or carboxyl groups. Brominated polyynes as well as those that are sulfated, esterified with sterols, or linked with amines in an amide-like manner occur. Examples include petrotetrayndiol, dihomo-(3S, 14S) petrocortyn A (Fig. 2.1), or petrosianyn A and B from *Petrosia* species like the stony sponge *Petrosia ficiformis* that occurs in the Mediterranean Sea and in the North Atlantic [4, 7, 8]. Such agents are able to inhibit ion transport ATPases and may function as immune suppressants or as cytotoxins in target animals. In addition, they have antifungal and antimicrobial effects. Other examples for long-chain polyynes with cytolytic activities are the nepheliosynes A and B from sponges of the genera *Xestospongia* or *Niphates*, respectively [5, 6]. Sponges of the genus *Callyspongia* contain polyynes that are linked to phenylethylamine with an amide-like bond or are esterified with sulfuric acid [10, 11].

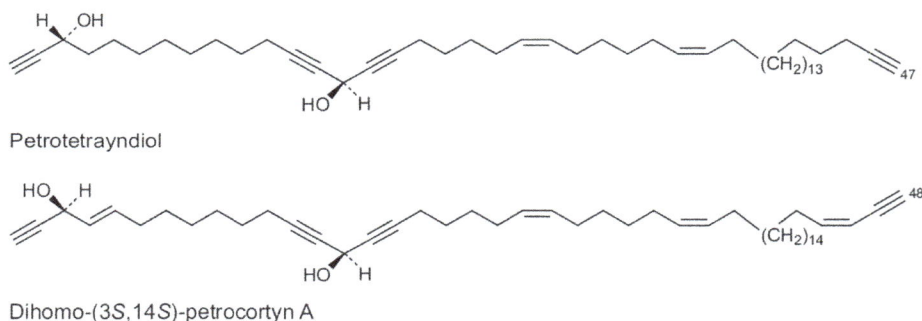

Petrotetrayndiol

Dihomo-(3S,14S)-petrocortyn A

Fig. 2.1: Long-chain polyynes of sponges of the genus *Petrosia*.

References

[1]   Alam N et al. (2001) J Nat Prod 64(8): 1059
[2]   Bohlmann F, Burkhardt T (1969) Chem Ber 102(5): 1702
[3]   de Silva ED et al. (1983) J Org Chem 48(3): 395
[4]   Kim JS et al. (1999) J Nat Prod 62(4): 554
[5]   Kobayashi J et al. (1994) J Nat Prod 57(9): 1300
[6]   Legrave N et al. (2013) Marine Drugs 11(7): 2282
[7]   Lim YJ et al. (2001) J Nat Prod 64(12): 1565
[8]   Lim YJ et al. (2001) J Nat Prod 64(1): 46
[9]   Minto RE, Blacklock BJ (2008) Prog Lipid Res 47(4): 233
[10] Youssef DTA et al. (2003) J Nat Prod 66(6): 861
[11] Youssef DTA et al. (2003) J Nat Prod 66(5): 679

2.3 Polyketides

2.3.1 General

Polyketides are a large group of natural substances that are extremely heterogeneous in terms of their chemical structures. The common biosynthesis route is the polyketide pathway starting with the polymerization of acetyl-CoA or malonyl-CoA to form poly-β-ketoesters (Fig. 2.2). These may form complex ring systems (including macrocyclic or heterocyclic rings) that are rich in keto or hydroxyl groups [25]. Polyketides that are built from uniform acyl residues are called 'simple polyketides' (e.g., polyacetates or polypropionates). When polyketides are synthesized from different types of starter acyl residues, the products are termed 'mixed polyketides'.

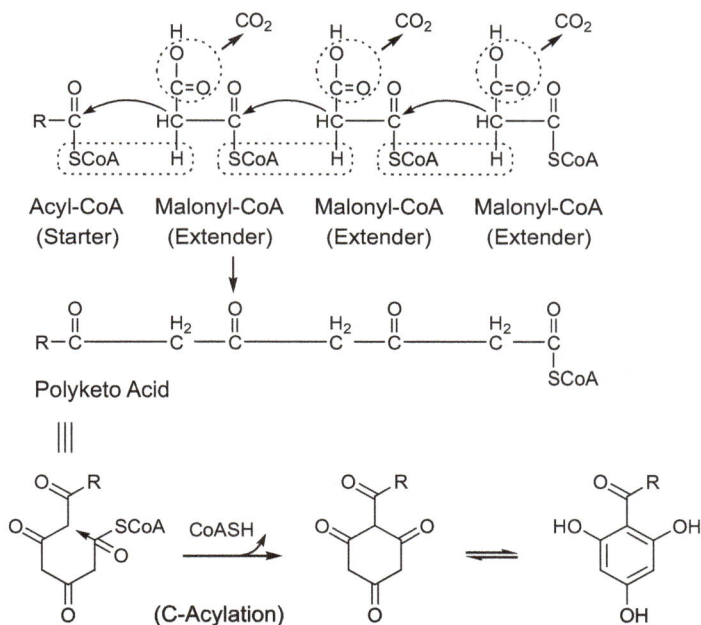

Fig. 2.2: Polyketide synthesis pathway.

Most of the natural polyketides are produced in (cyano)bacteria, dinoflagellates, fungi, or plants [33, 49]. The polyketides present in animals originate from microbial symbionts or food organisms [14, 32, 43], or their origin is uncertain [42]. Only recently, enzymes of the polyketide synthesis pathway, fatty acid synthase-like polyketide synthase proteins, have been identified in tissues of sacoglossan sea slugs [56]. Occurrence of polyketide, however, is widespread in marine animals. Representatives of this substance class were isolated from sponges (Porifera), corals (Cnidaria), moss animals (Bryozoa), mussels and

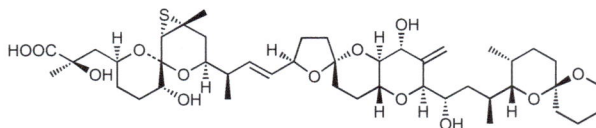

Acanthifolicin

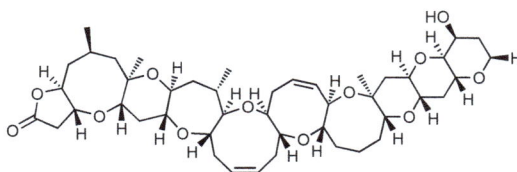

Brevetoxin A

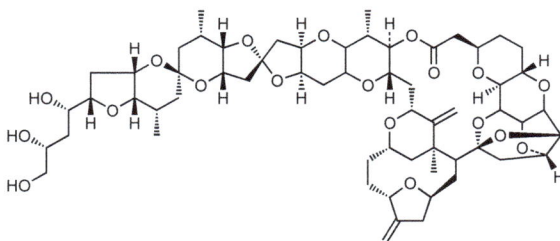

Halichondrin B

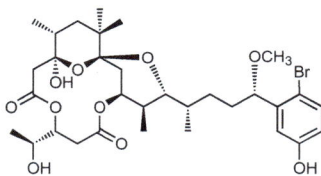

Aplysiatoxin

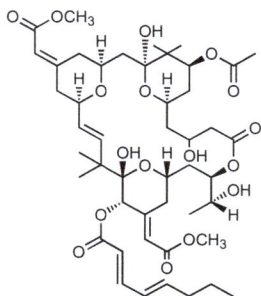

Bryostatin I

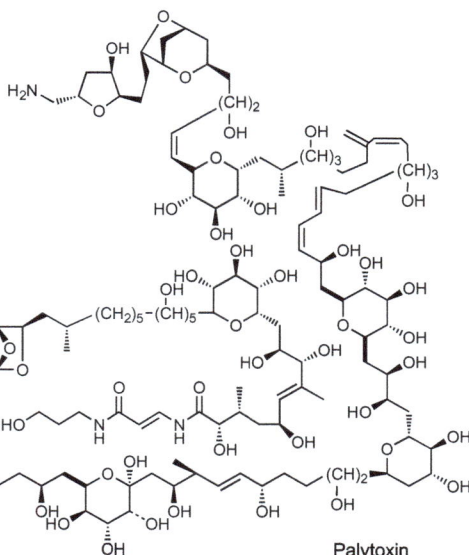

Palytoxin

Fig. 2.3: Polyketide toxins in animals I. Molecular structures of some examples of polyketide toxins in marine animals.

Toxin	R^1	R^2	R^3
Okadaic acid	H	H	CH$_3$
Dinophysistoxin 1 (DTX-1)	H	CH$_3$	CH$_3$
Dinophysistoxin 2 (DTX-2)	H	CH$_3$	H
Dinophysistoxin 3 (DTX-3)	Acyl	H	CH$_3$

Pederin   R=CH$_3$
Pseudopederin R=H

Maitotoxin

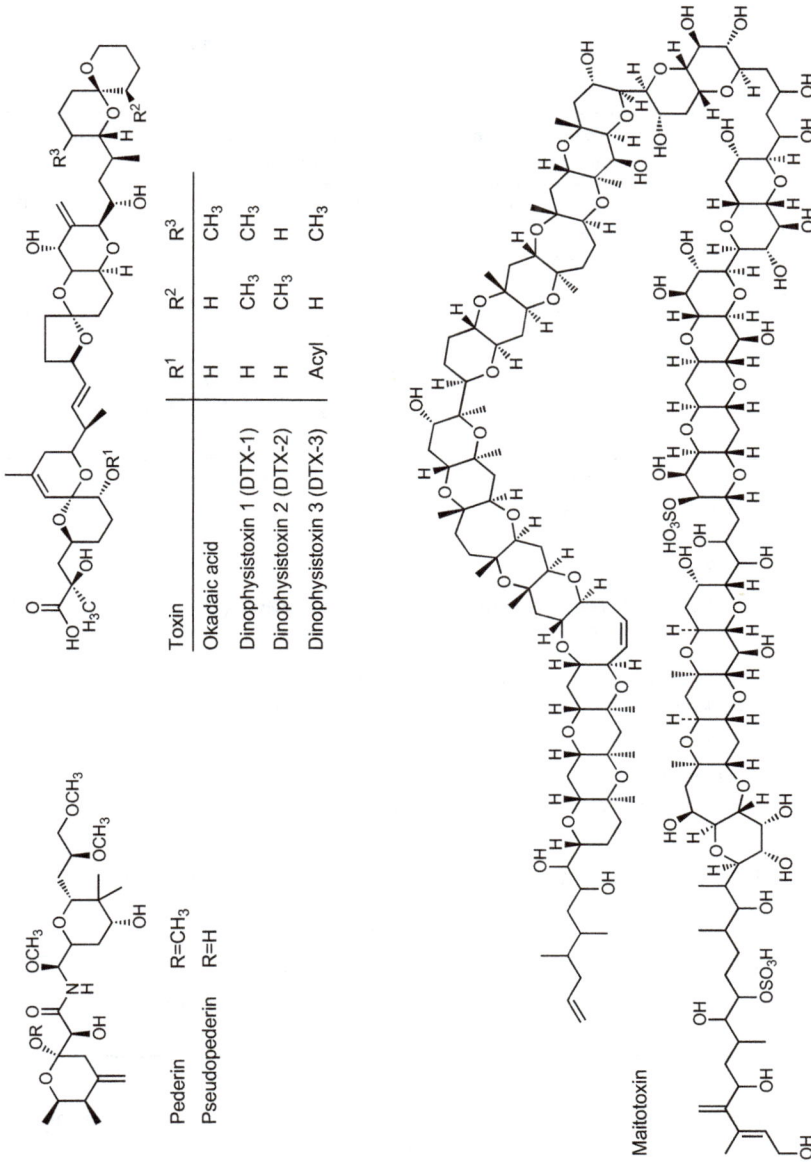

Fig. 2.4: Polyketide toxins in animals II. Molecular structures of some examples of polyketide toxins in marine animals.

Tab. 2.1: Polyketide toxins, their potential producers, and animals that accumulate such toxins for purposes of defense against potential predators with preliminary data on their effects on cells in target organisms.

Toxin	Potential producers	Toxin-carrying animals	Major physiological effects on target organisms	References
Acanthifolicin B	Bacteria	Marine sponge, *Pandaros acanthifolium* C	Cytotoxicity	[48]
Aplysiatoxin C	Cyanobacteria	Marine slugs, *Aplysia* sp.	Tumor promoter, activation of protein kinase C	[3]
Brevetoxin D	Dinoflagellates, *Karenia brevis*	Bivalve mollusks, fish	Activation of Na$^+$ influx through voltage-gated Na$^+$ channels in nerve and muscle cells	[7]
Bryostatin B	Bacteria, Candidatus *Endobugula sertula*	Bryozoa, *Bugula neritina*	Activation of protein kinase C	[47]
Halichondrin B D	Dinoflagellates?	Marine sponges, *Halichondria okadai*	Tubulin-targeted mitosis inhibitor	[8]
Maitotoxin D	Dinoflagellates, *Gambierdiscus toxicus*	Tropical fish	Activation of nonselective calcium channels in cell membranes	[66]
Okadaic acid D	Dinoflagellates, *Prorocentrum* sp., *Dinophysis* sp.	Marine sponges, *Halichondria okadai*	Inhibition of protein phosphatases, neurotoxicity	[13, 40]
Palytoxin D	Dinoflagellates, *Ostreopis siamensis*	Crust anemones, *Palythoa toxica*	Renders the sodium–potassium pump passively permeable to sodium and potassium ions	[54]
Pederin B	Bacteria, *Pseudomonas* sp.	Staphylinid beetles, *Paederus* sp.	Inhibition of DNA synthesis, cell cycle arrest, erythema, skin blisters	[18]

slugs (Mollusca), or tunicates (Tunicata) [45]. The molecular structures of some important polyketide toxins listed in Tab. 2.1 are presented in Figs. 2.3 and 2.4.

2.3.2 Acanthifolicin

Most of or even all of the polyketides in sponges are metabolic end products of symbiontic bacteria which can make up to 40% of the spongy tissue mass. Bacterial polyketides in sponges are generally polyethers or macrocyclic compounds like acanthifolicin from *Pandaros acanthifolium* [48] whose molecular structure is shown in Fig. 2.3. Acanthifolicin, which is a product of dinoflagellates, carries an episulfide group which is rare in natural substances. It is able to inhibit protein phosphatases with similar potency as okadaic acid [37].

2.3.3 Aplysiatoxin

Aplysiatoxin (Fig. 2.3) is synthesized by cyanobacteria (e.g., genera *Lyngbya*, *Schizothrix*, and *Planktothrix*) and may accumulate in marine slugs (sea hares, Aplysiomorpha) as they graze on biofilms containing these microorganisms in shallow coastal water. The mollusks compartmentalize aplysiatoxins in their tissues and secrete these substances in their skin, mucus, and genital secretions to impregnate their egg masses. This renders the animals as well as their egg masses inedible and untouchable for other marine animals and humans [30].

> ♣ Aplysiatoxins bind to and activate protein kinase C (PKC). They may function as tumor promoters in animal or human cells [68]. Handling sea hares with bare hands may induce contact dermatitis and skin blistering [62].

2.3.4 Brevetoxins

Brevetoxins are cyclic polyether compounds produced by dinoflagellates such as *Karenia brevis*. Brevetoxins of the subgroup A have 10 ring systems, and those of the subgroup B have 11 [29]. These substances may occur in high concentrations in marine environments during periods of mass production of dinoflagellates (red tides) which happen on a regular basis along the coastline of the Gulf of Mexico and elsewhere during summer. Brevetoxins (Fig. 2.3) are metabolically stable and are transferred to animals (e.g., mussels, crustaceans, or fish) via the food chain. Brevetoxins may be poisonous to consumers of contaminated food like some fish species, shorebirds, or marine mammals [16, 61]. Some fish species as well as invertebrates, however, accumulate substantial amounts of brevetoxins in their tissues. Humans consuming such contaminated fish or shellfish may develop 'neurotoxic shellfish poisoning' when certain amounts of brevetoxins have been ingested [63].

> ♣ Brevetoxins bind to voltage-gated sodium channels in neuronal cells of animals and prevent their intrinsic inactivation. Sodium overload of such cells with sodium ions depolarizes the membrane potential and exaggerates the action potential formation and transmitter release. LD_{50} for brevetoxins in mice ranges from 0.05 mg/kg BW for i.v. administration to 0.5 mg/kg BW for oral or i.p. administration [7]. Besides the occurrence of intestinal problems, paresthesia (tingling) of lips and skin, erratic temperature sensations, myalgia (muscle pain), cramps, vertigo, ataxia (loss of coordination), or bradycardia (lowering of heart rate) may result upon poisoning. Human fatalities due to acute brevetoxin poisoning are rare.

However, brevetoxin has been shown to form DNA adducts in lymphocytes which may induce cell cycle problems and alter functions of the immune system [46].

🔬 Mild brevetoxin poisonings in humans do not require medical attention. In severe cases, a physician should be consulted. Brevenal, another polyether from *Karenia brevis*, functions as an antagonist of brevetoxin. It also binds to the voltage-activated sodium channel, but at a different site. Nevertheless, it effectively inhibits the action of brevetoxin on the channel [10].

2.3.5 Bryostatins

Bryostatins are macrocyclic lactones (Fig. 2.3) produced by symbionts (*Endobugula sertula*) of the marine bryozoan *Bugula neritina*. The bryozoan host accumulates the substance and utilizes it as a mildly cytotoxic deterrent against potential predators. Bryostatin I is a member of a group of approximately 20 bioactive natural products with antineoplastic activities [15] that are under investigation for their therapeutic potentials for cancer therapy, treatment of Alzheimer's disease, or HIV eradication from AIDS patients. Bryostatin I may reverse synaptic loss in animal models with several neurological disorders [50, 51]. It seems to play a role in protecting cell tight junctions (TJs) in epithelial cells [67]. It is more potent than phorbol myristate acetate in activating the protein kinase C isoforms δ and ε whose activation do not have any tumor-promoting functions [20].

2.3.6 Halichondrins

Halichondrin B (Fig. 2.3) is a polyether macrolide originally isolated from the marine sponge *Halichondria okadai* [27]. It functions as a cytotoxin in the affected animal or human cells by inhibiting the formation of microtubules [31]. A synthetic analogon of halichondrin B, eribulin mesilate, is an approved drug for patients with locally advanced or metastatic breast cancer [12]. Another sponge polyketide, (+)-discodermolide, from the deep-sea sponge *Discodermia dissoluta,* however, was recognized as one of the most potent natural promoters of tubulin assembly [44]. Other sponge polyketides, e.g., the reidispongiolides A and B as well as the sphinxolides, exert their cytotoxic effects through their interaction with the actin cytoskeleton of human or animal target cells [2].

2.3.7 Palytoxin

One of the most effective poisons (LD_{50} (mouse, i.v.) of approximately 150 ng/kg BW [60]) occurring in marine animals is palytoxin (Fig. 2.3). It was isolated from crust anemones (Zoanthidae), e.g., *Palythoa tuberculosa, P. toxica, P. caribaeorum,* and *P. mamillosa* [39, 58] inhabiting the intertidal zones of the tropical parts of the Atlantic and Pacific Oceans.

If the animals are disturbed, they excrete a slimy secretion containing palytoxin. It is accompanied by homopalytoxin (includes butanolamine) and bishomopalytoxin (includes pentanolamine) [57]. The native Hawaiians used *Palythoa toxica* secretions as arrow poison. Palytoxin has also been found in other marine animals including some crustaceans (e.g., the xanthid crabs *Demania alcalai* and *Lophozozymus pictor* [65], polychaetes [22], starfish (*Acanthaster planci*), or fish (*Chaetodon* sp.)) which feed on crust anemones [21].

These animals do not produce the poison themselves but obtain it through the food web. Dinoflagellata of the genus *Ostreopsis* seem to produce palytoxin and other related compounds like ostreocine D (42-hydroxy-3,26-didesmethyl-19,44-dideoxypalytoxin) [59, 60]. Especially during periods of algal blooms, they transfer their toxins to consumers like crust anemones, other filter feeders, or predators that prey on such organisms.

Palytoxin targets the Na^+/K^+-ATPase in animal and human cells and converts it into a nonspecific ion channel [24, 26]. The resulting transmembrane pore is permeable for both types of ions [5, 6]. Thus, exposure of human or animal cells to palytoxin results in a loss of ion gradients across the plasma membrane, cessation of secondary transport processes depending on these gradients, a decrease in plasma membrane potential, and, ultimately, cell death. Sublethal concentrations of palytoxin may have tumor-promoting effects or mediate losses in actin cytoskeletal filaments in affected cells [4].

🐁 Human poisoning may occur during consumption of marine animals that contain palytoxin or similar compounds [41, 55]. The typical symptoms of palytoxin poisoning are severe muscle pain, low back pain, and discharge of black urine. Serum creatine phosphokinase (CPK) levels are higher than usual [55]. The lethality of the toxin is mainly due to a rapid disruption of cardiac function together with severe vasoconstriction [38]. Skin irritation occurs upon contact with the mucus of poisonous animals like crust anemones. Respiratory symptoms (flu-like) were observed upon inhalation of aerosols containing palytoxin [11].

💊 Poisoning is generally treated symptomatically as removal of the toxin from the body is difficult. Gastric lavage with activated charcoal and forced mannitol-alkaline diuresis therapy may save lives in mild cases of palytoxin poisoning [41]. In animal experiments, application of a serum obtained by immunizing rabbits with a conjugate of palytoxin and bovine albumin protected mice from the lethal effects of palytoxin [35].

🕯 A 49-year-old man living in the Philippines ate a crab, which was later identified as *Demania reynaudii*, cooked over a charcoal fire. A few minutes after eating about one fourth of the crab, the man felt dizzy, nauseated, tired, and he was in cold sweat. When the man saw that a dog that had been eating the remains of the crab died an hour later, he had himself taken to a hospital. His hands and lower extremities became numb. Vomiting, muscle cramps, and restlessness were noted on arrival at the hospital. Periods of normal heart rate and bradycardia (30 beats/min) followed each other. Breathing was quick and shallow. Eventually, his kidneys failed and he died the next day from palytoxin intoxication [1].

2.3.8 Pederin

Compounds that are presumably formed from polyketo acids have also been found in insects. These include the defense secretions of some beetles. Pederin (Fig. 2.4) is the only complex polyketide that has been isolated from terrestrial animals. Pederin, a toxic amide with two tetrahydropyran rings, is produced by the bacterium *Pseudomonas paederi*, which settles in the oviducts of the females of staphylinid beetles, e.g., *Paederus fuscipes* (see Section 3.2.9). The bacterial symbiont is vertically transmitted to the offspring via the eggs. Pederin is found in almost all female animals of *Paederus fuscipes*, but not in male beetles. Hemolymph, tissues, and skin secretions of females as well as egg packages contain the skin-irritating substance which is accompanied by pseudopederin. Both substances act as growth inhibitors of tumor cells in vitro [44]. In vivo, they may protect the beetles against spiders and fungi [18].

2.3.9 Okadaic Acid

Okadaic acid (OA) and its derivatives (Fig. 2.4) are produced by marine dinoflagellates (e.g., genera *Prorocentrum* and *Dinophysis*). These cytotoxic polyethers accumulate in filter feeders like sponges and bivalves. These animals use OA and its derivatives as defensive molecules. They occur in sponges of the genera *Halichondria* [52] and *Suberites* [64]. They also accumulate in organisms considered seafood and may cause 'diarrhetic shellfish poisoning' (DSP) if humans consume seafood during seasons of dinoflagellate blooms. OA is a potent inhibitor of protein phosphatases and disrupts signaling and other cell physiological processes associated with protein phosphorylation. Moreover, it has tumor-promoting activity and neuronal toxicity [40].

2.3.10 Maitotoxin

Maitotoxin (Fig. 2.4) is considered the most potent marine toxin discovered to date. The LD_{50} (i.p., mouse) is 50–130 ng/kg BW [66]. It is one of the largest natural substances (molecular mass of 3,425.86 g/mol) not having polymeric structures. Maitotoxin is produced by dinoflagellates, e.g., *Gambierdiscus toxicus*, which are distributed in tropical and subtropical areas of the Pacific Ocean [36]. Filter-feeding animals ingest such dinoflagellates and, due to being resistant against the toxin, accumulate it. The toxin is present in some tropical fish species (e.g., in the striated surgeonfish *Ctenochaetus striatus*). If contaminated fish is eaten by humans, the symptoms of intoxication with maitotoxin or similar compounds (ciguatoxins, CTX) are subsumed as 'ciguatera poisoning'.

In case of maitotoxin, the toxic effect is based on the activation of calcium-permeable, nonselective cation channels in animal or human cell membranes [38]. This leads to a massive influx of calcium ions into the cytosol of the cells. In smooth muscle cells this results in contractions [23]. Uncontrolled exocytosis of neurotransmitters may occur in maitotoxin-exposed neuronal cells [53]. Maitotoxin may also affect ion channels of the TRP family [17]. The resulting fatal neurological damage is typical of ciguatera poisoning.

The related ciguatoxins [34] are also products of dinoflagellates (e.g., *Gambierdiscus toxicus*, *Prorocentrum* sp., *Ostreopsis* sp., *Coolia monotis*, *Thecadinium* sp., or *Amphidinium carterae*) and accumulate in fish (e.g., in the giant moray, *Gymnothorax javanicus*) through the food chain [28]. They act by shifting the voltage dependence of voltage-activated sodium channels which activates these channels at normal membrane potential [9]. The resulting uncontrolled transmitter release in neuronal cells elicits similar effects as those observed with maitotoxin.

🦠 Ciguatera poisoning in mammals and humans goes along with diarrhea, vomiting, numbness, itchiness, or sensitivity to hot and cold of the skin, dizziness, and general weakness [19]. These symptoms may last from weeks to years, sometimes leading to long-term disability.

💊 There is no effective treatment or antidote for ciguatera poisoning. Calcium channel blockers like nifedipine or verapamil may be suitable for alleviating some of the symptoms.

References

[1]   Alcala AC et al. (1988) Toxicon 26(1): 105
[2]   Allingham JS et al. (2005) Proc Natl Acad Sci U S A 102(41), 14527
[3]   Arcoleo JP, Weinstein IB (1985) Carcinogenesis 6(2): 213.
[4]   Ares IR et al. (2005) J Exp Biol 208(Pt 22), 4345
[5]   Artigas P, Gadsby DC (2003) Proc Natl Acad Sci U S A 100(2): 501.
[6]   Artigas P, Gadsby DC (2004) J Gen Physiol 123(4): 357.
[7]   Baden DG (1989) FASEB J 3(7): 1807.
[8]   Bai RL et al. (1991) J Biol Chem 266(24), 15882
[9]   Bidard JN et al. (1984) J Biol Chem 259(13), 8353
[10] Bourdelais AJ et al. (2004) Cell Mol Neurobiol 24(4), 553
[11] Ciminiello P et al. (2006) Anal Chem 78(17), 6153
[12] Cortes J et al. (2011) Lancet 377(9769), 914
[13] Dounay AB, Forsyth CJ (2002) Curr Med Chem 9(22): 1939.
[14] Faulkner DJ (2000) Nat Prod Rep 17(1): 7.
[15] Figuerola B, Avila C (2019) Marine Drugs 17(8): 477.
[16] Flewelling LJ et al. (2005) Nature 435(7043), 755
[17] Flockerzi V (2007) Handbook Exp Pharmacol 179: 1
[18] Frank JH, Kanamitsu K (1987) J Med Entomol 24(2): 155.
[19] Friedman MA et al. (2017) Marine Drugs 15(3), 72
[20] Geiges D et al. (1997) Biochem Pharmacol 53(6), 865
[21] Gleibs S, Mebs D (1999) Toxicon 37(11): 1521.
[22] Gleibs S et al. (1995) Toxicon 33(11), 1531
[23] Gusovsky F, Daly JW (1990) Biochem Pharmacol 39(11): 1633.

[24] Habermann E (1989) Toxicon 27(11): 1171.
[25] Hertweck C et al. (2007) Nat Prod Rep 24(1), 162
[26] Hilgemann DW (2003) Proc Natl Acad Sci U S A 100(2): 386.
[27] Hirata Y, Uemura D (1986) Pure Appl Chem 58(5): 701.
[28] Holmes MJ, Lewis RJ (2022) Toxins 14(8): 534.
[29] Hort V et al. (2021) Marine Drugs 19(12), 656
[30] Johnson PM, Willows AOD (1999) Mar Freshw Behav Physiol 32(2–3): 147.
[31] Jordan MA et al. (2005) Mol Cancer Ther 4(7), 1086
[32] König GM et al. (2006) ChembioChem 7(2), 229
[33] Lane AL, Moore BS (2011) Nat Prod Rep 28(2): 411.
[34] Legrand AM et al. (1989) J Appl Phycol 1(2), 183
[35] Levine L et al. (1987) Toxicon 25(12), 1273
[36] Litaker RW et al. (2010) Toxicon 56(5), 711
[37] MacKintosh C et al. (1994) Plant J 5(1), 137
[38] Meunier FA et al. (2009) Marine toxins potently affecting neurotransmitter release. In: Fusetani M, Kem W (eds.) Marine Toxins as Research Tools. Springer Verlag, Berlin, Heidelberg, p.. 159
[39] Moore RE, Scheuer PJ (1971) Science 172(3982): 495.
[40] Munday R (2013) Toxins 5(2): 267.
[41] Okano H et al. (1998) Int Med 37(3), 330
[42] Pankewitz F, Hilker M (2008) Biol Rev 83(2): 209.
[43] Piel J (2004) Nat Prod Rep 21(4): 519.
[44] Piel J (2006) Curr Med Chem 13(1): 39.
[45] Proksch P et al. (2002) Appl Microbiol Biotechnol 59(2–3), 125
[46] Sayer A et al. (2005) Arch Toxicol 79(11), 683
[47] Schaufelberger DE et al. (1991) J Nat Prod 54(5), 1265
[48] Schmitz FJ et al. (1981) J Am Chem Soc 103(9), 2467
[49] Shimizu Y (2003) Curr Opin Microbiol 6(3): 236.
[50] Silva M et al. (2021) Marine Drugs 19(7), 373
[51] Sun M-K, Alkon DL (2006) CNS Drug Rev 12(1): 1.
[52] Tachibana K et al. (1981) J Am Chem Soc 103(9), 2469
[53] Takahashi M et al. (1982) J Biol Chem 257(13), 7287
[54] Takeuchi A et al. (2008) Nature 456(7220), 413
[55] Taniyama S et al. (2002) J Nat Toxins 11(4), 277
[56] Torres JP et al. (2020) Nat Commun 11(1), 2882
[57] Uemura D et al. (1985) Tetrahedron 41(6), 1007
[58] Uemura D et al. (1981) Tetrahedron Lett 22(29), 2781
[59] Ukena T et al. (2001) Biosci Biotechnol Biochem 65(11), 2585
[60] Usami M et al. (1995) J Am Chem Soc 117(19), 5389
[61] van Deventer M et al. (2012) Botanica Marina 55(1), 31
[62] Vranješ N, Jovanović M (2011) Arc Oncol 19(3–4): 64.
[63] Watkins SM et al. (2008) Marine Drugs 6(3), 431
[64] Wiens M et al. (2003) Mar Biol 142: 213.
[65] Yasumoto T et al. (1986) Agric Biol Chem 50(1), 163
[66] Yokoyama A et al. (1988) J Biochem 104(2), 184
[67] Yoo J et al. (2003) Am J Physiol Cell Physiol 285(2), C300
[68] Zanchett G, Oliveira-Filho EC (2013) Toxins 5(10): 1896.

2.4 Terpenes

2.4.1 General

Terpenes are natural products occurring in all groups of organisms. They are composed of isoprene building blocks containing five carbon atoms (Fig. 2.5). Biosynthesis of isoprene in humans and animals occurs through the acetate–mevalonate pathway and starts with three acetate molecules. Several enzymatically mediated intermediate steps (Fig. 2.6) yield isopentyl-5-pyrophosphate, the so-called activated isoprene, which is then used for the generation of higher order isoprenes (terpenes) by adding such C_5 units to other C_5, C_{10}, C_{15} units, and so forth.

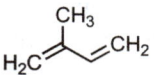

Fig. 2.5: Basic structure of isoprene.

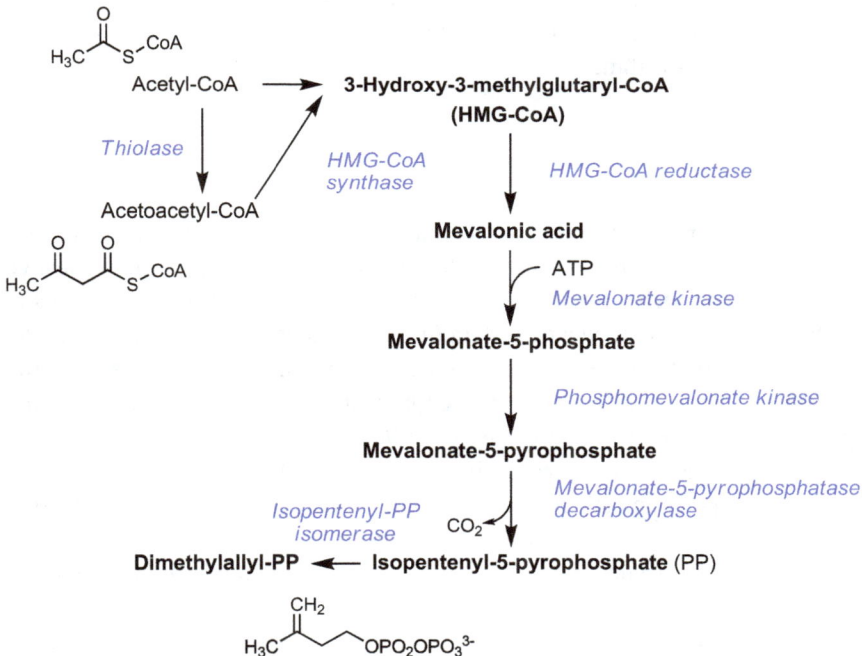

Fig. 2.6: Acetate–mevalonate pathway of isoprene biosynthesis in animals and humans.

According to the number of isoprene units in the final terpene molecules, the following molecule classes can be distinguished:

Hemiterpenes – one isoprene residue (C_5)
Monoterpenes – two isoprene residues (C_{10})
Sesquiterpenes – three isoprene residues (C_{15})
Diterpenes – four isoprene residues (C_{20})
Sesterterpenes – five isoprene residues (C_{25})
Triterpenes – six isoprene residues (C_{30})
Tetraterpenes – eight isoprene residues (C_{40})
. . .
Polyterpenes – more than eight isoprene residues

Derivatization of these units by shifts or deletions of methyl groups, ring formation, secondary ring openings, ring extensions or elimination of elements from rings, oxidation, or introduction of halogens may yield molecules that are hardly recognizable as terpene derivatives anymore and highly diverse (Fig. 2.7) [16]. Introduction of nitrogen into terpenes results in 'terpene alkaloids'. The steroids are generally derived from squalene, a triterpenoid (Fig. 2.7).

Meroterpenoids are compounds that combine a terpene and a polyketide or a compound of other biosynthetic origin. The terpenophenolics methoxyconidiol, epiconicol, and didehydroconicol, isolated from the tunicate *Aplidium densum*, are examples of such defensive substances.

2.4.2 Monoterpenes

The biosynthesis of monoterpenes used as building blocks for toxins may occur directly in the animals [7] or may occur in other organisms and get into the respective animals via the food web. Monoterpenes in mixtures of animal toxins are mostly those containing a cyclopentane ring system (iridoids). They are generally used as defensive substances that are either components of skin gland secretions (e.g., in the dorsal glands of Chrysomelidae) or may even be actively applied to attackers using mouthparts (e.g., in soldier termites of the genus *Nasutitermes*) [58].

Iridoids are defensive substances in insects (Hexapoda) (see Section 3.2.9). Cyclopentanoid monoterpenes, mostly iridoids, occur in the defense secretions of representatives of the ant subfamily Dolichoderinae (Formicidae), the ghost or stick insects (Phasmatidae), the leaf beetles (Chrysomelidae), rove beetles (Staphylinidae), and the longhorn beetles (Cerambycidae).

Iridodial, dolichodial, iridomyrmecin, isoiridomyrmecin (Fig. 2.8), iridolactone, dihydronepetalactone, and the monoterpene alkaloid actinidine were found in ants, e.g., in the Australian cocktail ant, *Anonychomyrma nitidiceps* [25]. When ants spray their defensive secretions toward an attacker, they coat the mouthparts and the legs of the attacker. Iridodial polymerizes rapidly when getting in contact with air, forms a resin-like layer, and thus glues mouthparts and legs together making the attacker unable to move.

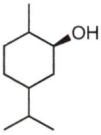

Monoterpenes: Menthol   Sesquiterpenes: Farnesol

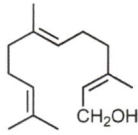

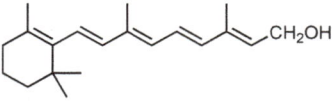

Diterpenes: Retinol (Vitamin A)

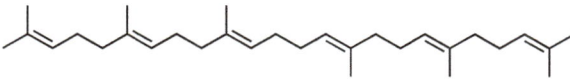

Squalene

Triterpenes: Triterpenoid precursor (squalene) of cholesterol and steroids

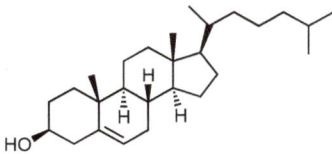

Cholesterol

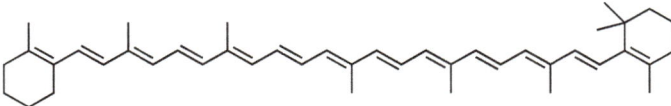

Tetraterpenes: β-Carotene (Provitamin A)

Fig. 2.7: Examples of terpenes.

When disturbed, larvae of the red pine sawfly *Neodiprion sertifer* (Hymenoptera) release an oily droplet from the mouth opening and transfer this droplet aiming at an attacker. This deterrent is chemically identical with the terpenoid resin of the host plants (*Pinus sylvestris*) and contains monoterpenes: α-pinene and β-pinene. The material is stored in two compressible pouches of the foregut from where it is regurgitated in case of danger [34].

The stick insect *Anisomorpha buprestoides*, which is native to Florida, sprays a poisonous mixture of substances from defensive glands in the thorax. The most effective component is anisomorphal (Fig. 2.8). It repels ants, bugs, and even mice. In larger animals, e.g., dogs or humans, the poison irritates the eyes and bronchi [75].

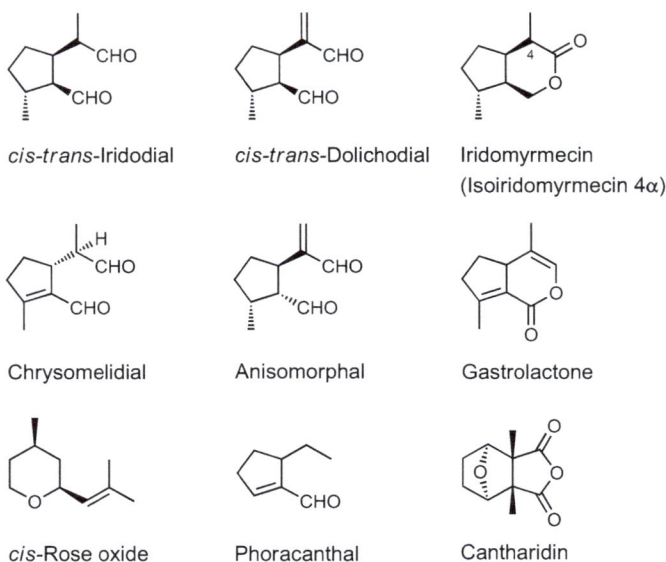

cis-trans-Iridodial   cis-trans-Dolichodial   Iridomyrmecin
                                                 (Isoiridomyrmecin 4α)

Chrysomelidial   Anisomorphal   Gastrolactone

cis-Rose oxide   Phoracanthal   Cantharidin

Fig. 2.8: Monoterpenes in defense secretions of insects.

Among leaf beetles (Chrysomelidae), it is the larvae of the subfamily of the Chrysomelinae (e.g., *Chrysomela*, *Hydrothassa*, *Chrysolina*, or *Phaedon*) which have nine pairs (rarely only one pair) of dorsal glands in the meso- and metathorax that can be everted when the animal is irritated. The glands then secrete an opaque and sticky fluid that deters attackers. The secretions contain, besides other substances, chrysomelidial (Fig. 2.8), plagiodial, plagiolactone, and gastrolactone [13, 73].

Citronellal, iridodial, limonene, isopulegol, as well as the monoterpene alkaloids actinidine and *N*-ethyl-3-(2-methylbutyl) piperidine could be detected in the defensive secretions of rove beetles (Staphylinidae) [10, 32, 91].

Among the longhorn beetles (Cerambycidae) are species that produce monoterpenes as important ingredients of their defense secretions. The repellent of the musk beetle *Aromia moschata* contains *cis*- and *trans*-rose oxides as well as γ- and δ-iridodial (Fig. 2.8) [100]. Threatened adults of the longhorn beetle *Chloridolum loochooanum* release a frothy secretion from their metasternal glands which contain cyclopentanoid monoterpenoids (iridodials) [72]. Some butterfly caterpillars (Lepidoptera) store iridoid glycosides in their tissues and are protected from predators, especially birds, by their bitter taste [15]. Iridoids do not have any direct toxicological relevance for humans.

Defensive secretions of termites contain monoterpenes, sesquiterpenes, diterpenes, polyketides, e.g., macrocyclic lactones [80], nitroalkenes (e.g., 1-nitro-1(*E*)-pentadecene, long-chain vinyl- or divinyl-ketones, and ketoaldehydes [81]), proteins, or mucopolysaccharides [64]. The secretions act as irritants and repel attackers. Some of the fatty-acid-derived substances function as contact insecticides [94]. After the volatile components have evaporated, they can also act as adhesives to immobilize opponents.

The monoterpene ingredients of termite defense secretions comprise α-thujene, sabinene, α-pinene, limonene, α-phellandrene, and terpinolene [36]. These monoterpenes not only have toxic effects on the attacking insects but also serve as alarm or recognition pheromones and as solvents for the cembrane-type diterpenes that are contained in the same secretions and act as adhesives [79], as well as pheromones. The diterpenes in turn act as fixatives for monoterpenes [36].

2.4.3 Cantharidin

Cantharidin (Fig. 2.8) occurs in the testes and in the accessory glands of the genital tract of the males in some representatives of the blister beetles (Meloidae) belonging to the genera *Meloe, Cyaneolyta, Cysteodesmus, Epicauta, Lytta, Megetra, Mylabris, Nemognatha, Pseudomeloe, Pyrota*, and *Zonitis* [23]. During copulation with females, the stored cantharidin together with the sperm is transferred from the genital tract of the males to the genital tract of the female. Portions of the toxin are absorbed into the hemolymph. Other portions are used to impregnate the fertilized eggs before deposition [71].

Predatory beetles avoid cantharidin-containing food in lab experiments and, in the wild, do not attack blister beetles. Blister beetles that come close to an ant nest are carried away and released in safe distance from the nest without being stung by the ants. When blister beetles feel seriously threatened, they release droplets of fluids containing cantharidin from the intersegmental membranes of their legs ('reflex bleeding'). The oily fluid works effectively as a feeding deterrent on predatory insects.

From a formal point of view, cantharidin is a monoterpene, but it is actually derived from the aliphatic sesquiterpene farnesol by elimination of five carbon atoms [76]. Cantharidin is a potent inhibitor of protein phosphatases, especially phosphoprotein phosphatase 2A [45, 56, 69]. The result of protein phosphatase inactivation is the accumulation of phosphorylated proteins in cells followed by epithelial inflammation, loosening of the epithelial cell layers from basal membranes (blister formation), as well as inhibition of cell division [14]. The skin contact with cantharidin may result in the disintegration of the epidermis due to the cellular release of endogenous proteases [11]. This is extremely irritating to the skin, especially in the moist parts of the body surface of animals and humans.

Humans may be affected by cantharidin as a medication [63] or as an alleged aphrodisiac [82]. Cantharidin-containing natural products, mostly dried materials prepared from blister beetles (*Lytta vesicatoria, Mylabris cichorii*, and *Mylabris phalerata*), are used in traditional medicine. The cantharidin content of *Lytta vesicatoria* and *Mylabris* species is between 0.3% and 2.5% of dry BW [20]. Solutions prepared with pure cantharidin in the lab are used to remove skin warts [63, 70].

Reports exist about accidental cantharidin poisoning in humans due to blister beetle ingestion [97], but in most cases poisoning is due to cantharidin overdosing during consumption as an aphrodisiac called 'Spanish fly' [48]. Cantharidin has been used since ancient times in attempts to enhance sexual potency. The origin of this drug abuse was the observation that oral ingestion of high doses of cantharidin induces priapism in men and pelvic organ engorgement in women due to its ability to cause vascular congestion and inflammation of the urinary tract. While these effects are just painful to most consumers, some others seem to interpret them as sexually stimulating.

☠ Exposure of human skin to high concentrations of cantharidin results in blister formation that heals without scarring or, in severe cases, in necrosis. In case of oral intake, severe damage to the mucous surfaces in the entire gastrointestinal tract may occur. Cantharidin is eliminated from the body via the kidneys which results in inflammation in the urinary tract. Acute cantharidin poisoning through ingestion manifests itself in salivation, thirst, swallowing difficulties, nausea, vomiting, severe pain in the kidneys, bladder and urethra, priapism as well as in hematuria, dysuria, or anuria that may result in fatal uremia.

🚑 The most important therapeutic measure after oral intake is drinking of copious amounts of fluid which accelerate renal excretion of cantharidin. Gastric lavage and the administration of activated charcoal may also be indicated if ingestion occurred recently. Lipophilic laxatives (e.g., castor oil) are contraindicated as they would favor absorption of cantharidin. Further treatment is symptomatic, mainly with analgesics. The skin damage may be treated with dry powder.

🐎 Cantharidin is potentially dangerous to domestic animals. Fatalities have been reported in grazing animals (i.e., in horses) in North America feeding on alfalfa hay which was infested with blister beetles of the genus *Epicauta*. Clinical signs of cantharidin toxicosis in horses vary according to the dose of beetles that have been consumed from mild lethargy and anorexia to acute severe colic, hypovolemic shock, and death. The toxin is excreted in the urine which may result in urinary tract irritation. The LD_{50} of cantharidin in the mouse (i.p.) is approximately 1 mg/kg BW. Lethal doses for humans and horses are 0.5 mg/kg BW (*Epicauta immaculata* contains up to 4.8 mg per beetle [20]), for cats and dogs 1.0–1.5 mg/kg, and for rabbits 20 mg/kg BW [85].

2.4.4 Sesquiterpenes

Formally, sesquiterpenes are composed of three isoprene units (15 C atoms). This class of molecules comprise more than 11,000 different natural substances that have been found in different kingdoms of organisms. Sesquiterpenes are ingredients of defensive toxin mixtures in sea anemones (e.g., 1,10-epoxy-14-hydroperoxy-4-lepidozene in *Anthopleura pacifica* [103]) and in leather corals and sea hares [6]. They occur also in terrestrial animals like ants (e.g., dendrolasin or 3-(4,8-dimethylnona-3,7-dienyl)-furan

[84]), caterpillars of butterflies feeding on celery (e.g., α-selinene or (3*R*,4α*R*,8α*R*)-5,8α-dimethyl-3-prop-1-en-2-yl-2,3,4,4α,7,8-hexahydro-1*H*-naphthalene [89]) and in termites (e.g., ancistrodial, amiteol, or helminthogermacrene) [5, 68, 90].

Dendrolasin-containing secretions are released from mandibular glands of Eurasian jet ants (*Lasius fuliginosus*) in case of disturbances of their nests by other organisms [44]. This sweet-tasting volatile compound serves as an alarm pheromone for nest mates and functions as a deterrent with some degree of toxicity on other ant species of the genera *Formica* and *Lasius*.

Especially enriched are sesquiterpenes in marine sponges, sessile marine animals that feed on planktonic microorganisms. They accumulate sesquiterpenes, diterpenes, and sesterterpenes provided by food or endosymbiontic microorganisms within their tissues to render themselves unpalatable for potential predators.

Terpenes in sponges (see Section 3.2.1) show a great structural diversity [21]. The sesquiterpenes detected in sponges (Fig. 2.9) are often terpenophenols like those found in brown algae. In addition, furosesquiterpenes and sesquiterpenes occur with similar structures as those from green algae. These molecules in sponges may contain NH_2-CO-NH, NH_2-C(= NH)-NH, CN, OCN, or SCN moieties. The basic sesquiterpene bodies, however, may have different structures, e.g., in bisabolene derivatives or in decalin derivatives. Such compounds may originally be produced by fungi that are associated with sponges [96, 101].

Terpenoid quinones and hydroquinones are widespread among sessile animals as sponges (see Section 3.2.1) [93]. These substances have great structural diversity, different biosynthetic pathways, and pleiotropic biological activities. Some of them are formed via the shikimate–mevalonate pathway, resulting in the formation of bicyclic sesquiterpene skeletons attached to a (hydro)quinone moiety. Examples are avarol and avarone of *Dysidea avara*, ilimaquinone of *Petrosaspongia metachromia*, and isospongiaquinone of *Fasciospongia turgida* (Fig. 2.9). Terpenophenols in sponges, avarone, and the corresponding hydroquinone derivative avarol deserve special attention as these sesquiterpenes have been found to function as cytostatics in eukaryotic cells [66] likely by inhibiting tubulin polymerization [65]. The dactylospongenones A to D (Fig. 2.9), which have been detected in *Dactylospongia* species, result from the contraction of the benzene ring from ilimaquinone derivatives [53]. Ilimaquinone and similar compounds are supposed to have antiviral activity against HIV by inhibiting the reverse transcriptase [83] and may, in addition, function as antimicrobial substances [53].

Other notable meroterpenoids are the metachromins A to C (Fig. 2.9) from *Hippospongia metachromia* [50], which have been found to inhibit hepatitis B virus production in the infected liver cells [102] as well as panicein B_2 from *Halichondria panicea*, the breadcrumb sponge, or the Mediterranean sponge *Haliclona mucosa*. Panicein B_2 (Fig. 2.9) and similar compounds inhibit certain types of multidrug resistance proteins and may potentially be used to enhance the efficiency of chemotherapy [38]. Aromatic sequiterpenes, which may also occur in gorgonian corals, may have antimicrobial effects [61].

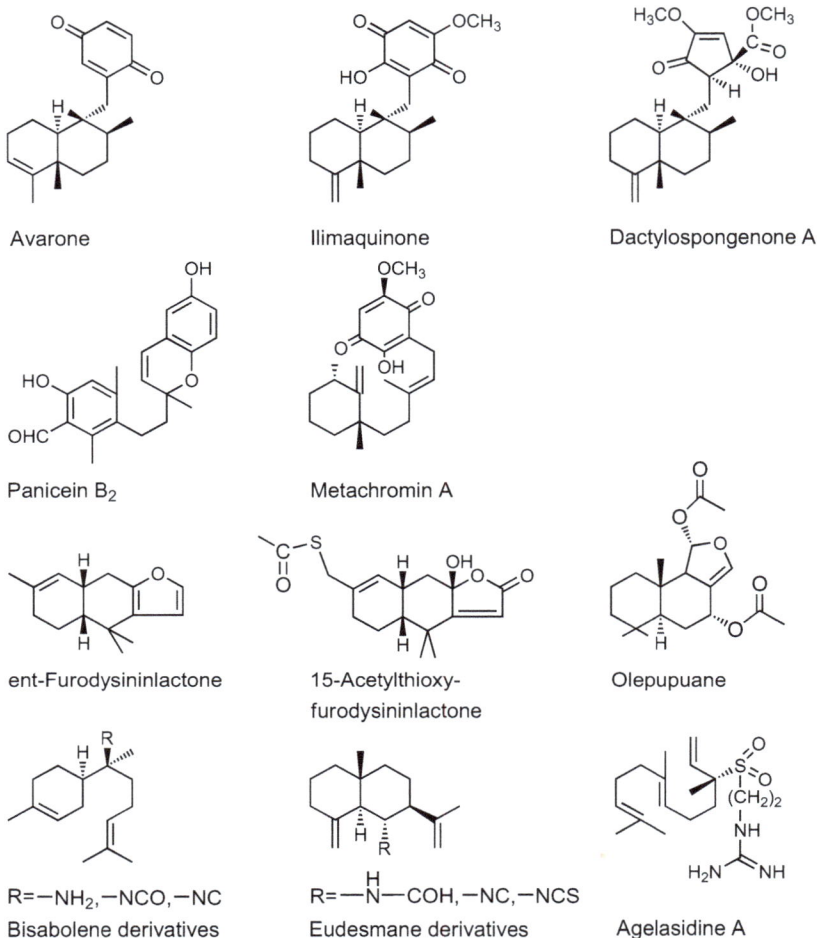

Fig. 2.9: Sesquiterpenes in sponges.

Furosesquiterpenes (Fig. 2.9) have been found in *Halichondria panicea*, *Fasciospongia rimosa*, *Dysidea* sp., and *Spongia mycofijensis* among others. One member of this molecule family, 9-hydroxyfurodysinin-*O*-ethyl lactone, was isolated from the keratose sponge *Dysidea arenaria* from New Caledonia [77]. Another furosesquiterpene, olepupuane (Fig. 2.9), does not only occur in sponges but also in a nudibranch gastropod, *Dendrodoris limbata*, which feeds on spongy tissues [18].

Sponge sesquiterpenes may also have functional groups containing nitrogen (Fig. 2.9). Examples are bisabolene derivatives of *Ciocalypta* sp. [42] or *Epipolasis* sp., or eudesmane of *Axinella* sp. [29].

An unconventional sesquiterpene is agelasidine A (Fig. 2.9), which is produced by sponges of the genus *Agelas* from the Japanese Sea. It has antimicrobial effects and inhibits the growth in cultures of *Staphylococcus aureus* [62]. In addition, agelasidines reversibly inhibit the sodium pump (Na^+/K^+-ATPase) in animal cells by interacting with the potassium-binding site [52].

2.4.5 Diterpenes

Diterpenes are formally composed of four isoprene units and comprise 20 carbon atoms. Aliphatic as well as cyclic diterpenes are known from animals. Besides being important structural or functional compounds in animal cells (e.g., retinol, in the form of light-sensitive retinal in the visual pigment rhodopsin, Fig. 2.7), diterpenes function as nonvolatile defense molecules to deter predators or inhibit microbes.

The furospongins are aliphatic terpenes with one or more furan units occurring in sponges (see Section 3.2.1). The furospongins 1 and 2 (Fig. 2.10) were first isolated from the marine sponges *Spongia officinalis* and *Hippospongia communis* from the Gulf of Naples [30, 51]. Both furospongins are partially degraded C_{21} sesterterpenoids. Micromolar concentrations of furospongin 1 inhibit mitochondrial ATP synthesis and inhibit bacterial growth [3]. It also acts as an inhibitor of protein tyrosine phosphatase 1B (PTB1B) in eukaryotic cells [2].

Spongialactone A is a polycyclic furosesquiterpene (Fig. 2.10) originally isolated from *Spongia arabica* [43]. This and similar compounds from sponges may have antimicrobial properties [49].

2-Tetraprenylhydroquinone, a terpenophenol (Fig. 2.10), was isolated from the sponge *Sarcotragus foetidus* and may act as a cytotoxin on animal cells [9]. The highly oxygenated gracilin B was isolated from *Spongionella* sp. [60] and may have inhibitory effects on integrin-mediated cell–cell adhesion [88] while gracilin A from related species may exert immunosuppressive and neuroprotective effects in animals [1].

Agelasines and similar molecules are 7,9-dialkylpurinium salts (Fig. 2.10) isolated from tissues of marine sponges (e.g., *Agelas* sp.) [40, 62]. These substances have cytotoxic properties in human and animal cells, inhibit Na^+/K^+-ATPase [52, 67], and may act as antimycobacterial agents [4].

Cembranes and other diterpenoid secondary natural products consist of a ring system containing 14 carbon atoms that is substituted with 3 methyl groups and 1 isopropyl group. There may be one or more ether, furan, or lactone rings. Cembranes and unsaturated variants, the cembrenes, occur in soft corals (Alcyonacea) (Fig. 2.11) [99], in ants of the genus *Crematogaster* [55], as well as in termites of the genus *Nasutitermes* (Fig. 2.12). Examples of such molecules are the cembranoids (nanolobols A–C; Fig. 2.11) from the soft coral *Sinularia nanolobata* [26] that are cytotoxic for eukaryotic cells. Cembrene A (Fig. 2.11) occurs in soft corals of the genus *Nephthea* as well as in secretions of termites which use it as a predator deterrent as well as a trail pheromone [12]. Other examples of cembrane derivatives (Fig. 2.11) are asperdiol [98] from gorgonian-type octocorals of the genus *Eunicea* which functions as an acetylcholine esterase inhibitor [24], the

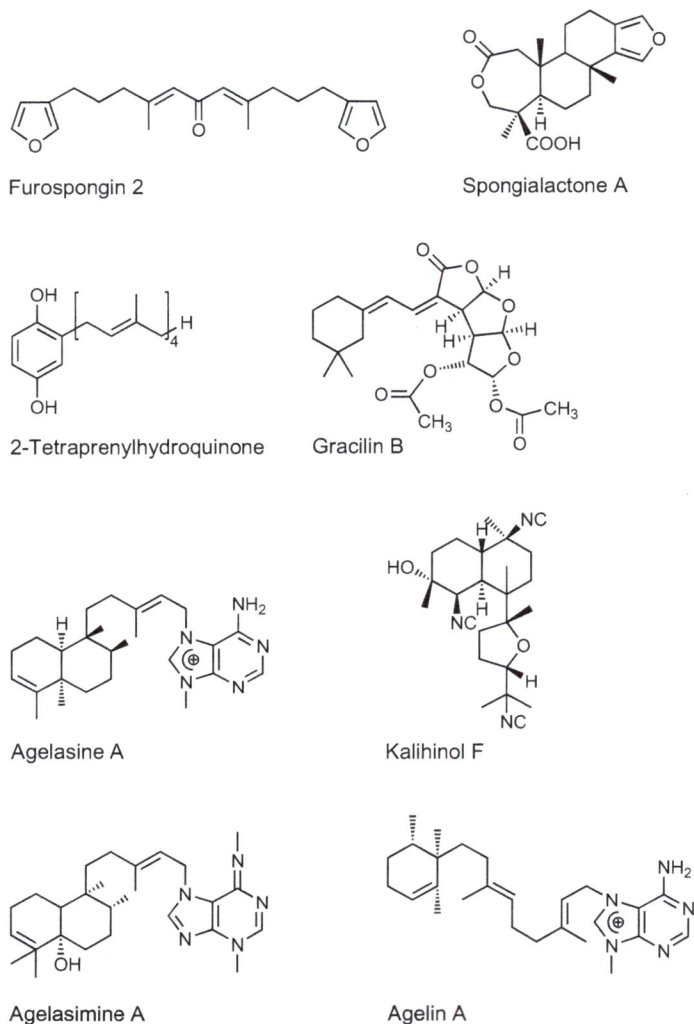

Furospongin 2

Spongialactone A

2-Tetraprenylhydroquinone

Gracilin B

Agelasine A

Kalihinol F

Agelasimine A

Agelin A

Fig. 2.10: Diterpenes in sponges.

solenolides from *Solenopodium* sp., which have anti-inflammatory and antiviral properties [41, 54] or the junceellolides A to D from the fragile sea whip *Junceella fragilis* [92]
which also function as anti-inflammatory agents in animal tissues.

Remarkable diterpenes with anti-inflammatory properties are the pseudopterosines
(Fig. 2.11) of *Antillogorgia elisabethae,* a soft coral from the tropical parts of the Atlantic
Ocean. These glycosides carry the monosaccharides L-fucose or D-arabinose and an aglycon comprising a hexahydrophenalene ring system [86]. The anti-inflammatory effects
of these substances in animal cells result from their ability to inhibit enzymes converting

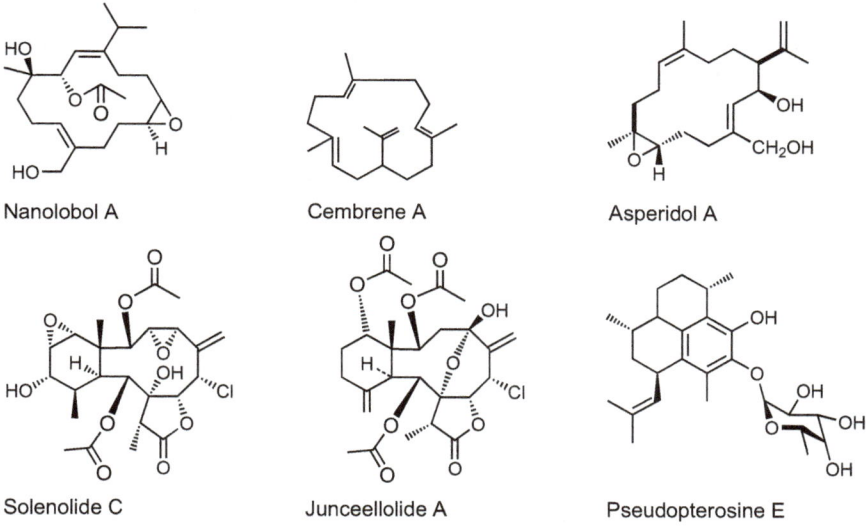

Nanolobol A   Cembrene A   Asperidol A

Solenolide C   Junceellolide A   Pseudopterosine E

Fig. 2.11: Diterpenes in soft corals.

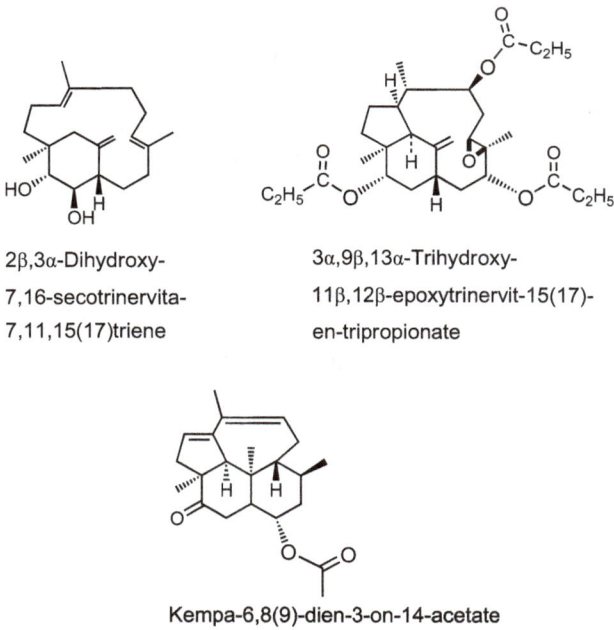

2β,3α-Dihydroxy-
7,16-secotrinervita-
7,11,15(17)triene

3α,9β,13α-Trihydroxy-
11β,12β-epoxytrinervit-15(17)-
en-tripropionate

Kempa-6,8(9)-dien-3-on-14-acetate

Fig. 2.12: Diterpenes in termites.

arachidonic acid to inflammatory mediators. Solenolide A and the pseudopterosines inhibit the 5-lipoxygenase. Solenolide E inhibits the cyclooxygenase, while the junceellolides attenuate arachidonic acid generation by inhibiting phospholipase A_2 [41, 86].

Kempanes are diterpenes that are derived from cembranes in which the outer ring system forms a combination of three or four internal rings (Fig. 2.12). Unsaturated variants, the kempenes, are used as defense substances by termites (see Section 3.2.9) [27, 28]. The diterpenes in defensive secretions of individuals of termites of the taxa *Nasutitermes* or *Bulbitermes* are generally dissolved in volatile monoterpenes. Upon secretion and transfer of these secretions to an attacker, the monoterpenes evaporate leaving the sticky diterpenes behind. This may immobilize the predator completely.

2.4.6 Sesterterpenes

Sesterterpenes are made up of five isoprene units (total of 25 C atoms). They form a relatively small group of substances which is mainly found in spongy tissues (Porifera; see Section 3.2.1), in sponge-eating mollusks, and rarely in insects [57]. Secretions from the Dufour glands of stingless bees (*Frieseomelitta silvestrii*) contain linear sesterterpenes (i.e., geranylfarnesol) of unknown functions [74].

Spongy tissues are rich in linear sesterterpenes which are probably used as deterrents against potential predators and as antimicrobials. The terminals of the aliphatic terpenes may carry furan, butenolide, or pyran ring systems or phenol groups (Fig. 2.13). Variabilin is such a linear sesterterpene in marine sponges of the genus *Sarcotragus* [8]. Manoalide is a calcium channel blocker and has antibiotic, analgesic, and anti-inflammatory effects by inhibiting phospholipases A_2 and C. It has been isolated from the West Pacific sponge *Luffariella variabilis* [35]. N-heterocyclic, e.g., molliorin A from *Cacospongia mollior* which induces analgesia when ingested by animals [19, 31], or sulfated compounds, e.g., halisulfate 1 from the marine sponge *Coscinoderma* sp. with antiplasmodial activity [47], occur frequently. Unusual sponge sesterterpenes are the cheilanthanes (terpenes containing unsaturated γ-hydroxybutyrolactone), which are able to inhibit protein kinases involved in mitogenic and stress signaling in animal cells [17]. Hyrtiosal, a sesterterpenoid possessing a rearranged tricarbocyclic skeleton, has been isolated from the Okinawan sponge *Hyrtios erectus* [46]. Cyclic sesterterpenes, the ircinianin lactones B and C, have been isolated from the marine sponge *Ircinia wistarii*. They display toxicity against protists [59].

2.4.7 Triterpenes

Triterpenes are a class of chemical compounds composed of six isoprene units (total of 30 C atoms). Animals produce triterpenes for various purposes. Some are used as components of defense secretions, e.g., the triterpene saponin 3-*O*-β-D-glucopyranosyl-(1→4)-β-D-glucuronopyranosyl-hederagenin of the leaf beetle *Platyphora kollari* contains a

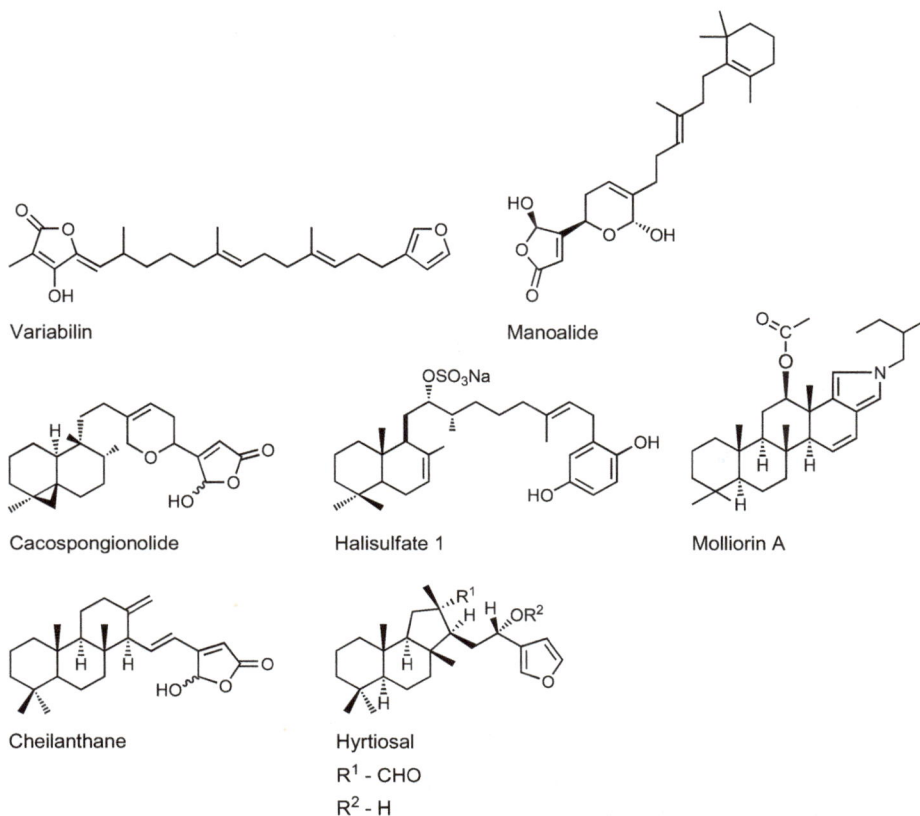

Variabilin

Manoalide

Cacospongionolide

OSO₃Na

Halisulfate 1

Molliorin A

Cheilanthane

Hyrtiosal
R¹ - CHO
R² - H

Fig. 2.13: Sesterterpenes in sponges.

triterpene as aglycon [39, 78]. The triterpene squalene is the precursor to cholesterol, and all steroids or steroidal compounds of which some are used as toxins or deterrents. Marine sponges (see Section 3.2.1) use triterpenes among other terpenoid compounds as deterrents against predators [33]. Important nonsteroidal triterpenes isolated from marine sponges and other marine organisms are the polyether triterpenes. Most of these metabolites consist of two separate ring systems with a typical linker which may be an ethylene bridge, a modified linker, or a butylene bridge [37]. An example of a sponge substance containing an ethylene bridge is sipholenone B (Fig. 2.14) from *Callyspongia siphonella* [22]. A compound with a modified linker is muzitone, a triterpene isolated from the Red Sea sponge *Ptilocaulis spiculifer* [87]. Finally, testudinariol A was isolated from the skin and the mucus of the marine gastropod *Pleurobranchus testudinarius* [95]. This lipophilic compound carrying a butylene bridge between the ring systems is an important lipophilic component of the defensive secretion of this opisthobranch. When applied to animal tissues, polyether triterpenes generally display cytotoxic properties. Testudinariol A was found to be highly toxic for fish (ichthyotoxicity).

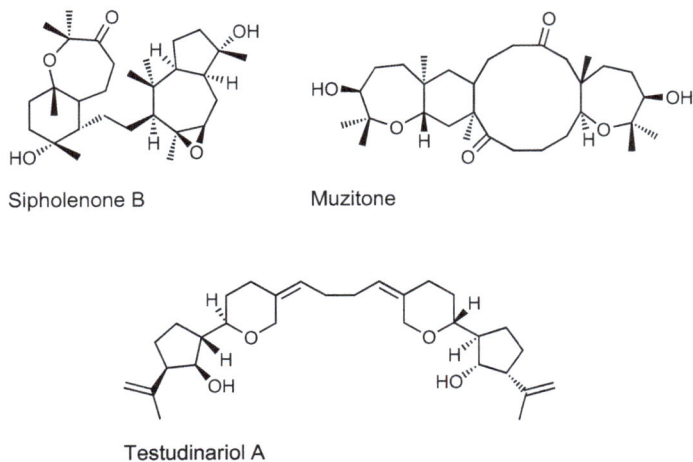

Sipholenone B

Muzitone

Testudinariol A

Fig. 2.14: Polyether triterpenes.

References

[1]   Abbasov ME et al. (2019) Nat Chem 11(4): 342
[2]   Abdjul DB et al. (2017) Bioorganic Med Chem Lett 27(5): 1159
[3]   Anderson AP et al. (1994) Clin Exp Pharmacol Physiol 21(12): 945
[4]   Arai M et al. (2014) ChemBioChem 15(1): 117
[5]   Baker R et al. (1978) J Chem Soc Chem Commun 9: 410.
[6]   Baker B et al. (1988) Tetrahedron 44(15): 4695
[7]   Banthorpe DV et al. (1972) Chem Rev 72(2): 115
[8]   Barrow CJ et al. (1988) J Nat Prod 51(2): 275
[9]   Baz JP et al. (1996) J Nat Prod 59(10): 960
[10] Bellas TE et al. (1974) J Insect Physiol 20(2): 277
[11] Bertaux B et al. (1988) Br J Dermatol 118(2): 157
[12] Birch AJ et al. (1972) J Chem Soc, Perkin Trans 1: 2653.
[13] Blum MS et al. (1978) J Chem Ecol 4(1): 47
[14] Bonness K et al. (2006) Mol Cancer Ther 5(11): 2727
[15] Bowers MD, Puttick GM (1986) J Chem Ecol 12(1): 169
[16] Brock NL, Dickschat JS (2013) Biosynthesis of terpenoids. In: Ramawat KG, Mérillon J-M (eds..)
      Natural Products – Phytochemistry, Botany and Metabolism of Alkaloids, Phenolics and Terpenes.
      Springer-Verlag, Berlin, Heidelberg, p.. 2693
[17] Buchanan MS et al. (2001) J Nat Prod 64(3): 300
[18] Butler MS, Capon RJ (1993) Aus J Chem 46(8): 1255
[19] Cafieri F et al. (1977) Tetrahedron Lett 18(5): 477
[20] Capinera JL et al. (1985) J Econ Entomol 78(5): 1052
[21] Capon RJ (2001) Eur J Org Chem 2001(4): 633
[22] Carmely S, Kashman Y (1983) J Org Chem 48(20): 3517
[23] Carrel JE, Eisner T (1974) Science 183(4126): 755
[24] Castellanos F et al. (2019) Nat Prod Res 33(24): 3533

[25]   Cavill GWK et al. (1982) Tetrahedron 38(13): 1931

[26]   Chao CH et al. (2016) Marine Drugs 14(8): 150

[27]   Chuah C-H (2005) J Chem Ecol 31(4): 819

[28]   Chuah CH et al. (1989) J Chem Ecol 15(2): 549

[29]   Ciminiello P et al. (1987) J Nat Prod 50(2): 217

[30]   Cimino G et al. (1972) Tetrahedron 28(2): 267

[31]   De Pasquale R et al. (1988) Pharmacol Res Commun 20(Suppl 5): 23

[32]   Dettner K, Schwinger G (1986) Zeitschrift für Naturforschung C – J Biosci 41(3): 366

[33]   Ebada SS et al. (2010) Marine Drugs 8(2): 313

[34]   Eisner T et al. (1974) Science 184(4140): 996

[35]   Ettinger-Epstein P et al. (2008) Mar Biotechnol (NY) 10(1): 64

[36]   Everaerts C et al. (1988) Biochem Syst Ecol 16(4): 437

[37]   Fernandez JJ et al. (2000) Nat Prod Rep 17(3): 235

[38]   Fiorini L et al. (2015) Oncotarget 6(26): 22282

[39]   Ghostin J et al. (2007) Naturwissenschaften 94(7): 601

[40]   Gordaliza M (2009) Marine Drugs 7(4): 833

[41]   Groweiss A et al. (1988) J Org Chem 53(11): 2401

[42]   Gulavita NK et al. (1986) J Org Chem 51(26): 5136

[43]   Hirsch S, Kashman Y (1988) J Nat Prod 51(6): 1243

[44]   Hölldobler B, Wilson EO (1990) The Ants. Harvard University Press, Cambridge, Mass., USA

[45]   Honkanen RE (1993) FEBS Lett 330(3): 283

[46]   Iguchi K et al. (2001) J Org Chem 57(2): 522

[47]   Jeong H et al. (2019) Zeitschrift für Naturforschung C 74(11–12): 313

[48]   Karras DJ et al. (1996) Am J Emerg Med 14(5): 478

[49]   Keyzers RA et al. (2006) Nat Prod Rep 23(2): 321

[50]   Kobayashi J et al. (1989) J Nat Prod 52(5): 1173

[51]   Kobayashi M et al. (1992) J Chem Res-S 11: 366.

[52]   Kobayashi M et al. (1987) Arch Biochem Biophys 259(1): 179

[53]   Kushlan DM et al. (1989) Tetrahedron 45(11): 3307

[54]   Kwak JH et al. (2001) J Nat Prod 64(6): 754

[55]   Leclercq S et al. (2000) Tetrahedron Lett 41(5): 633

[56]   Li YM, Casida JE (1992) Proc Natl Acad Sci U S A 89(24): 11867

[57]   Li K, Gustafson KR (2021) Nat Prod Rep 38(7): 1251

[58]   Lubin YD, Montogomery GG (1981) Biotropica 13(1): 66

[59]   Majer T et al. (2022) Marine Drugs 20(8): 532

[60]   Mayol L et al. (1985) Tetrahedron Lett 26(10): 1357

[61]   McEnroe FJ, Fenical W (1978) Tetrahedron 34(11): 1661

[62]   Medeiros MA et al. (2006) Zeitschrift für Naturforschung C 61(7–8): 472

[63]   Moed L et al. (2001) Arch Dermatol 137(10): 1357

[64]   Moore BP (1968) J Insect Physiol 14(1): 33

[65]   Müller WEG et al. (1985) Basic Appl Histochem 29: 321.

[66]   Müller WEG et al. (1985) Comp Biochem Physiol C: Comp Pharmacol 80(1): 47

[67]   Nakamura H et al. (1984) Tetrahedron Lett 25(28): 2989

[68]   Naya Y et al. (1982) Tetrahedron Lett 23(30): 3047

[69]   Neumann J (1995) J Pharmacol Exp Ther 274(1): 530

[70]   Nguyen AL et al. (2019) Dermatologic Ther 32(6): e13143

[71]   Nikbakhtzadeh MR et al. (2007) J Insect Physiol 53(9): 890

[72]   Ohmura W et al. (2009) J Chem Ecol 35(2): 250

[73]   Pasteels JM et al. (1982) Tetrahedron 38(13): 1891

[74]   Patricio EFLRA et al. (2003) Apidologie 34(4): 359
[75]   Paysse EA et al. (2001) Ophthalmology 108(1): 190
[76]   Peter MG et al. (1977) Helvetica Chimica Acta 60(8): 2756
[77]   Piggott A, Karuso P (2005) Molecules 10(10): 1292
[78]   Plasman V et al. (2001) Chemoecology 11(3): 107
[79]   Prestwich GD (1982) Tetrahedron 38(13): 1911
[80]   Prestwich GD, Collins MS (1981) Tetrahedron Lett 22(46): 4587
[81]   Prestwich GD, Collins MS (1982) J Chem Ecol 8(1): 147
[82]   Prischmann DA, Sheppard CA (2002) Am Entomol 48(4): 208
[83]   Qiu Y, Wang XM (2008) Molecules 13(6): 1275
[84]   Quilico A et al. (1957) Tetrahedron 1(3): 177
[85]   Ray AC et al. (1989) Am J Vet Res 50(2): 187
[86]   Roussis V et al. (1990) J Org Chem 55(16): 4916
[87]   Rudi A et al. (1999) Tetrahedron 55(17): 5555
[88]   Rueda A et al. (2006) Lett Drug Des Discov 3(10): 753
[89]   Ruzicka L, Stoll M (1923) Helvetica Chimica Acta 6(1): 846
[90]   Scheffrahn RH et al. (1986) J Nat Prod 49(4): 699
[91]   Schildknecht H et al. (1976) J Chem Ecol 2(1): 1
[92]   Shin J et al. (1989) Tetrahedron 45(6): 1633
[93]   Sladić D et al. (2006) Molecules 11(1): 1
[94]   Spanton SG, Prestwich GD (1982) Tetrahedron 38(13): 1921
[95]   Spinella A et al. (1997) Tetrahedron 53(49): 16891
[96]   Suzue M et al. (2016) Tetrahedron Lett 57(46): 5070
[97]   Tagwireyi D et al. (2000) Toxicon 38(12): 1865
[98]   Tello E et al. (2009) J Nat Prod 72(9): 1595
[99]   Tseng W-R et al. (2019) Marine Drugs 17(8): 461
[100]   Vidari G et al. (1973) Tetrahedron Lett 14(41): 4065
[101]   Wu Q et al. (2019) Marine Drugs 17(1): 56
[102]   Yamashita A et al. (2017) Antiviral Res 145: 136.
[103]   Zheng GC et al. (1990) J Org Chem 55(11): 3677

2.5 Steroids

2.5.1 General

Steroids are natural compounds based on the hypothetical tetracyclic hydrocarbon sterane. The steroid biosynthesis starts from the triterpene squalene (Fig. 2.15).

Steroids and steroid esters that affect the heart function in vertebrate animals have been detected in a number of insect species. For adult insects or their larvae, the steroid body (Fig. 2.16 [42]) or cholesterol, respectively, are vitamins and need to be taken up from forage organisms. Most of the other animal species are able to synthesize steroids by themselves using squalene as a precursor (Fig. 2.15). Many insect species acquire steroid esters directly from certain food plants, mainly from Asclepiadaceae or Apocynaceae. Some insect species, however, are able to derivatize steroids [32].

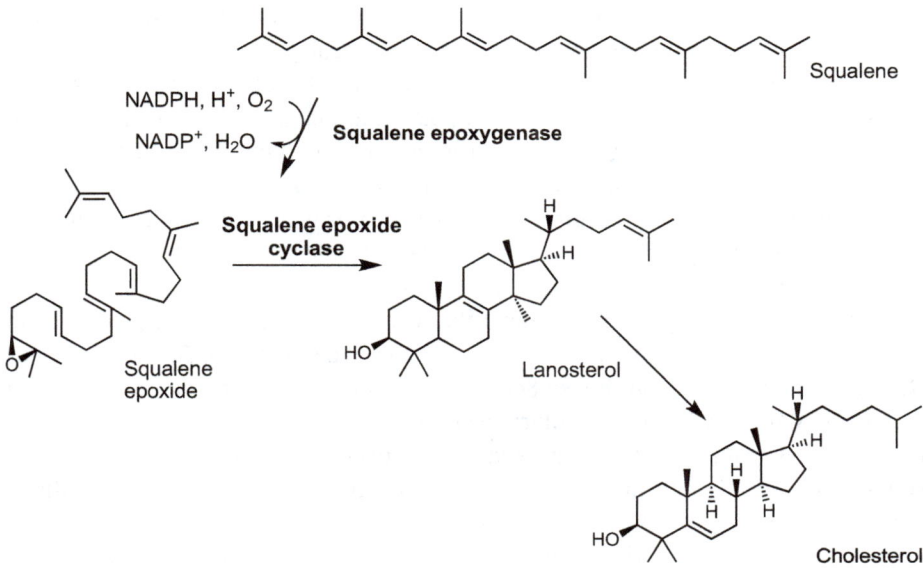

Fig. 2.15: Steroid biosynthesis from squalene.

These steroid derivatives are used to impregnate tissues or as components of defensive secretions and protect the animals from attacks by insectivores, especially birds, through their bitter taste and emetic effects [3, 4].

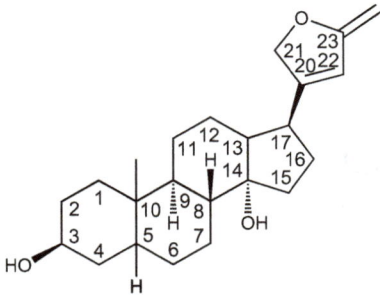

Fig. 2.16: C-atom numbering in steroids [6, 42].

2.5.2 Steroids as Defensive Molecules in Animals

Cardenolides are cardioactive steroidal glycosides derived from steroids that are secondary metabolites in plants. The term is derived from 'card-' (from Greek καρδία – heart) and the suffix '-enolide', which refers to the lactone ring at the C_{17} position of the steroid ring system. In many cases, cardenolides contain individual or chains of sugar molecules (cardenolide glycosides). These molecules are toxic mainly because they inhibit the pumping action of the Na^+/K^+-ATPase. This ubiquitous enzyme is responsible for maintaining the sodium and potassium ion gradients across cell membranes of animal cells [22]. In the vertebrate heart, inhibition of the ATPase causes elevations in cytosolic concentrations of sodium and calcium ions which results in increased contraction force in cardiac muscle cells. This is why cardenolides are used in prevention and therapy of heart failure and are also called 'cardiac glycosides' [1, 45], a term that is generally used but not fully correct.

A prominent example for an animal that uses plant-borne cardiac glycosides to protect itself from becoming prey of birds is that of the monarch, *Danaus plexippus* (Nymphalidae). The monarch (see Section 3.2.9) is a butterfly native to the United States and Canada whose caterpillars feed on milkweed (*Asclepias* sp., Asclepiadaceae). The caterpillar and the image can contain over 15 different cardenolide glycosides in the hemolymph in relatively high concentrations (approx. 700 µg/animal, approx. 0.3% of wet weight), among them digitoxin (Fig. 2.17). In animals, the plant glycosides are absorbed through the intestinal lining upon ingestion and subsequently chemically modified by splitting off the deoxyhexosuloses in positions 2 and 3 of the aglycone, and conversion to β-glucosides [27, 35] to make the molecules more hydrophilic. To avoid negative effects on the endogenous Na^+/K^+-ATPase in the insects, monarch caterpillars and images express an alpha-subunit of this transport ATPase in which an asparagine residue is replaced by a histidine residue in position 122 of the amino acid chain [21, 28]. This domain of the protein is also the cardenolide-binding site. This amino acid exchange renders Na^+/K^+-ATPase in the monarch insensitive to cardenolides.

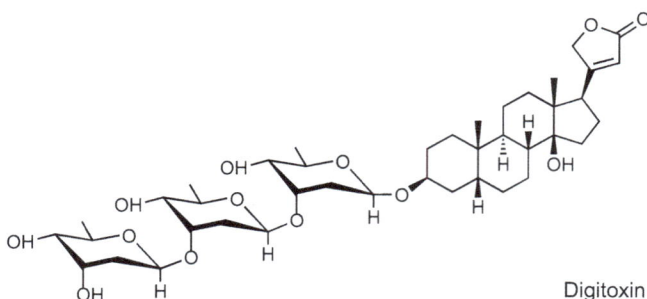

Fig. 2.17: Digitoxin.

Cardiac glycosides are utilized as defensive substances by many insect species (see Section 3.2.9). Besides in Lepidoptera [36], poisonous insect species are present in the taxa Saltatoria [13, 34], Hemiptera [26, 30, 43], and Coleoptera [8, 11, 19]. An interesting fact is that many animal species that are exposed to cardenolides one way or the other have evolved very similar mutations in their Na^+/K^+-ATPase alpha-subunits in a convergent manner [12, 13, 40].

The oleander aphid (*Aphis nerii*) takes up a variety of 17 different cardenolides from its food plant and accumulates them in its body fluids. These cardenolides or derivatives thereof are also excreted in the form of 'honeydew' through the gut [26].

Some chrysomelid beetles of the genera *Chrysolina*, *Dlochrysa*, and *Oreina* are themselves capable of producing steroid glycosides [8, 33]. In the leaf beetle *Chrysolina coerulans*, it was shown that the cardenolide aglycones sarmentogenin, periplogenin, and bipindogenin (Fig. 2.18), which occur in the form of their xylosides in these animals, can be built from radioactively labeled cholesterol fed to these beetles [41]. In *Oreina gloriosa*, males with higher toxicity are preferred as sexual partners by the females, thus ensuring that the genetic trait 'toxicity' is inherited to the offspring [25].

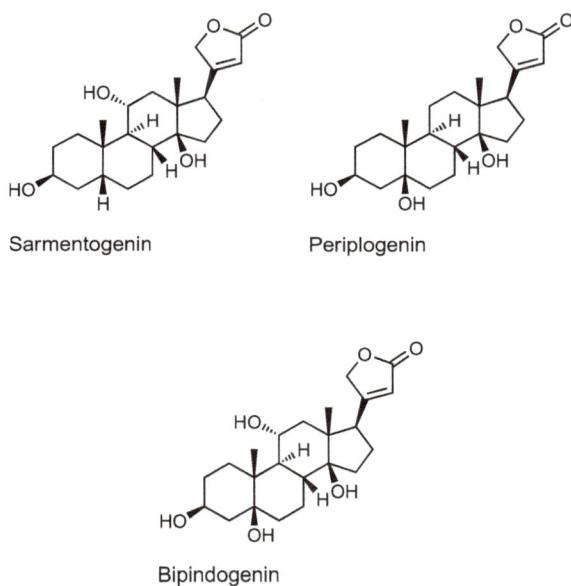

Sarmentogenin   Periplogenin

Bipindogenin

Fig. 2.18: Cardenolide aglycones in leaf beetles.

Photinus ignitius and *P. marginellus*, two fireflies (Lampyridae) that are widespread in the northwestern United States, are also able to synthesize defensive substances on their own. They contain lucibufagin (Fig. 2.19), 12-oxo-2β,5β, 1α-trihydroxy-bufaline esterified with isobutyric acid, propionic acid, or acetic acid in position 3 and partly with acetic acid in position 11 [15]. Fireflies of the genus *Photuris*, however, are not able to produce lucibufagins themselves. Thus, the females of these species imitate the flashing signals of *Photinus* females, catch the approaching *Photinus* males, and

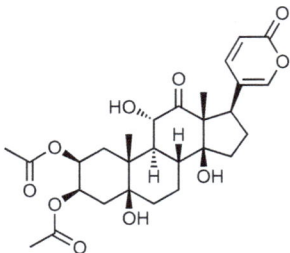

Fig. 2.19: Lucibufagin.

obtain their lucibufagins from them by devouring the toxin containing males of the other genus ('femmes fatales') [14].

Caterpillars of the small ermine moth *Yponomeuta cagnellus* (Yponomeutidae) feeding on leaves of the European spindle tree *Euonymus europaeus* are generally not attacked by birds. Investigation of their defense substances revealed that they do not make use of cardenolides of their forage plant but accumulate butenolide and isosiphonidin (3-hydroxymethyl-2-(5H)-furanone). There are doubts that this substance is solely produced by the plant and just utilized by the caterpillar since this butenolide is also found in caterpillars feeding in isosiphonidin-free plants [16].

Many beetle species accumulate steroids as defensive substances in their body fluids and in pygidial gland secretions that are identical with steroid hormones in vertebrates (e.g., glucocorticoids, pregnane derivatives, testosterone, and estradiol) [9]. Dytiscidae (e.g., *Dytiscus marginalis*) possess paired pygidial glands next to the anus as well as defense glands in the prothorax which secrete a milky fluid enriched in 11-deoxycorticosterone [37]. This steroid also occurs as a corticoid in the adrenal glands in vertebrates which may have mixed functions (mineralo- and glucocorticoid actions). An amount of 0.4 mg 11-deoxycorticosterone can be extracted from one beetle, which is the same amount that can be extracted from the adrenal glands of 1.000 cows. 11-Deoxycorticosterone has anesthetizing effects on fish [29] and emetic effects (inducing vomiting) in amphibia.

Pregnane derivatives occur as defensive substances in other swimming beetles, e.g., *Agabus bipustulatus, Acilius sulcatus, Cybister lateralimarginalis*, and *Ilybius fenestratus*. In addition to pregnane derivatives, the secretion of the pond swimmer *Ilybius fenestratus* also contains boldenone (1-dehydrotestosterone), estrone, and 17β-estradiol. The quantities of sterane derivatives formed are between 1 and 1,000 μg/ beetle [29]. Whirligig beetles of the genera *Gyrinus* and *Dineutus* (Gyrinidae) use the more toxic aliphatic or cyclopentanoid norsesquiterpene derivatives gyridinal, gyridinone, and gyrinidione for defense instead of the less toxic sterane derivatives [29]. Insects containing steroid derivatives as defensive substances like cardenolides or other butenolides have no toxicological significance for humans.

Besides catechol- and indolylalkylamines and peptides, the secretions of the skin glands of toads (Anura and Bufonidae; see Section 3.2.12) contain cardiac glycosides

(Fig. 2.20) of the bufodienolide type (bufogenin) or of the cardenolide type (cardenobu-fagin) and esters of bufadienolides (bufotoxins) or cardenolides (cardenobufotoxins). Among the studies on the composition of skin secretions in various toad species, research-ers put a focus on *Bufo gargarizans*, the Asian giant toad. Skin preparations of this species are used in traditional Chinese medicine (Ch'an Su) for the treatment of rheumatoid ar-thritis, epilepsy, and skin diseases. As in other toads, the skin glands contain more than 20 chemically different bufadienolides and approximately 10 different cardenolides [44]. These include some that are also found in plants, e.g., hellebrigenin, gamabufotalin, oleandigenin, digitoxigenin, periplogenin, sarmentogenin, telecinobufagin, and areno-bufagin, while others have only been found in toads, e.g., bufalin, 19-hydroxybufalin, bu-fotalin, resibufogenin, bufarogenin, argentinogenin, marinobufagin, cinobufagin, and desacetylcinobufagin. Noteworthy is the occurrence of epoxybufenolides, e.g., 3-formyl -14,15-epoxy-resibufogenin, including representatives that act cytostatically on tumor cells in vitro, in addition to 3-formyl-14,15-epoxyresibufogenin also 19-oxobufalin, 19-oxodesacetylcinobufagin, 6α-hydroxycinobufagin, and 1β-hydroxybufalin [31].

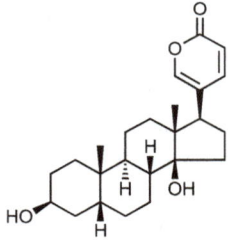

Bufogenin

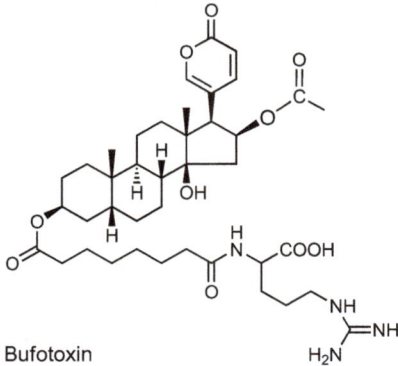

Bufotoxin

Fig. 2.20: Cardiac glycosides of toads.

In bufotoxins or cardenobufotoxins, the steroid ring systems are esterified with sulfuric acid, formic acid, succinic acid, glutaric acid, adipic acid, pimelic acid, or suberic acid on the OH groups at position 3. An L-arginine, L-glutamine, L-histidine, L-1-methylhistidine, or L-3-methylhistidine residue may be bound in an amide-like manner to the second carboxyl group of the organic dicarboxylic acids [18, 24, 31, 38, 39].

🔬 Cardioactive steroids are components of the defensive skin secretions of toads and are effective against both microorganisms and potential predators. The ingestion of cardioactive steroids by animals or humans in small doses increases the contraction force of the heart and decreases the heart rate [10]. They also have strong local anesthetic effects [46]. In large doses, they cause gastrointestinal irritation, rise in blood pressure, convulsions, and bradycardia [5]. Serious poisoning, often resulting in death, occurs from time to time in East Asia, where toad skin extracts and skin preparations are used as traditional medicines or aphrodisiacs (Ch'an Su, Yixin Wan). The consumption of toads, toad eggs, toad legs, the licking of toads, or drinking of a brew made from toad skins to generate hallucinogenic effects are reported. Several cases of accidental poisoning have been reported as well [7, 20]. Rapidly developing symptoms of intoxication from ingestion of skin secretions of toads or of toad skins are nausea, vomiting, sweating, severe bradycardia, hyperkalemia, and acidosis, followed by cardiac arrhythmia. Death may occur within a few hours. Toxicity data for several compounds isolated from European toads are shown in Tab. 2.2.

💊 Fab fragments of digoxin-specific antibodies have been successfully used to treat poisoned humans who ingested toad skin. Atropine or other antiarrhythmic drugs may be used to combat heart arrhythmia [5, 17].

🐾 The cane toad (*Rhinella marina*) is native to South America, but has been introduced to Australia to control insect, snail, and small mammal pests in sugarcane plantations. It has quickly established stable populations and has become a pest itself as it eradicates endemic animal species. Among the domestic animals, dogs are mostly affected by poisoning by toads. Symptoms include neurological abnormalities, mucosal swelling, salivation, vomiting, accelerated breathing, and collapse [2, 20].

⚠ A male Cambodian from Kampong Cham province prepared a soup from *Bufo melanostictus*, believing he had frog legs in front of him. He died after enjoying the soup. His two daughters who had only tasted suffered severe poisoning.

Tab. 2.2: Toxicity of cardioactive steroids from skin glands of toads (genus *Bufo*) (data compiled from different sources).

Compound	Toad species	LD$_{50}$ (mouse, s.c.) in mg/kg BW
Bufotalin B	*Bufo bufo*	1.13
Bufotalinin B	*Bufo bufo*	0.62
Hellebrigenin B	*Bufo bufo*	0.08
Marinobufagin B	*Bufo bufo*	0.15
Telocinobufagin B	*Bufo bufo*	0.10
Viridobufotoxin B	*Bufo viridis*	0.27

At least some southeast Asian snake species of the genus *Rhabdophis*, e.g., the red-necked keelback snake *Rhabdophis subminiatus*, are strongly venomous because of their bufadienolide content and may be dangerous to humans. Other *Rhabdophis* species, however, are less venomous but are in fact poisonous. The tiger keelback, *Rhabdophis tigrinus*, carries a pair of glands located in the neck (nuchodorsal glands) that sequester bufadienolides obtained from poisonous toads they feed on [23] and uses the glandular secretions as a defense against predation. The secretions contain 3,5,11,14-, 3,7,11,14-, or 3,5,14,16-tetrahydroxy-bufadienolide and gamabufalin, some of which have the same cardiac effects as ouabain. When humans handle these snakes, these substances may get into the eyes and cause severe corneal damage.

References

[1]   Babula P et al. (2013) Anticancer Agents Med Chem 13(7): 1069
[2]   Barbosa CM et al. (2009) J Venom Anim Toxins Incl Trop Dis 15(4): 789
[3]   Brower LP, Fink LS (1985) Ann New York Acad Sci 443: 171.
[4]   Brower LP et al. (1967) Proc Natl Acad Sci U S A 57(4): 893
[5]   Brubacher JR et al. (1996) Chest 110(5): 1282
[6]   Cahn RS et al. (1966) Angew Chem Int Ed Engl 5(4): 385
[7]   Chern MS et al. (1991) Am J Cardiol 67(5): 443
[8]   Daloze D, Pasteels JM (1979) J Chem Ecol 5(1): 63
[9]   Dettner K (1987) Annu Rev Entomol 32: 17.
[10]   Dmitrieva RI, Doris PA (2002) Exp Biol Med 227(8): 561
[11]   Dobler S et al. (2011) Phytochemistry 72(13): 1593
[12]   Dobler S et al. (2012) Proc Natl Acad Sci U S A 109(32): 13040
[13]   Dobler S et al. (2019) Proc R Soc Lond B Biol Sci 286: 20190883
[14]   Eisner T et al. (1997) Proc Natl Acad Sci U S A 94(18): 9723
[15]   Eisner T et al. (1978) Proc Natl Acad Sci U S A 75(2): 905
[16]   Fung SY et al. (1988) J Chem Ecol 14(4): 1099
[17]   Gowda RM et al. (2003) Heart 89(4): e14
[18]   Habermehl G (1969) Naturwissenschaften 56(12): 615
[19]   Hilker M et al. (1992) Experientia 48(10): 1023
[20]   Hitt M, Ettinger DD (1986) N Engl J Med 314(23): 1517
[21]   Holzinger F et al. (1992) FEBS Lett 314(3): 477
[22]   Horisberger JD (2004) Physiology 19(6): 377
[23]   Hutchinson DA et al. (2007) Proc Natl Acad Sci U S A 104(7): 2265
[24]   Kamano Y et al. (2002) J Nat Prod 65(7): 1001
[25]   Labeyrie E et al. (2003) J Chem Ecol 29(7): 1665
[26]   Malcolm SB (1990) Chemoecology 1(1): 12
[27]   Martin RA, Lynch SP (1988) J Chem Ecol 14(1): 295
[28]   Mebs D et al. (2000) Chemoecology 10(4): 201
[29]   Miller JR, Mumma RO (1976) J Chem Ecol 2(2): 115
[30]   Moore LV, Scudder GGE (1986) J Insect Physiol 32(1): 27
[31]   Nogawa T et al. (2001) J Nat Prod 64(9): 1148
[32]   Pasteels JM, Daloze D (1977) Science 197(4298): 70

[33] Pasteels JM et al. (1989) Cell Mol Life Sci 45(3): 295
[34] Pugalenthi P, Livingstone D (1995) J New York Entomol Soc 103(2): 191
[35] Reichstein T et al. (1968) Science 161(3844): 861
[36] Rothschild ML et al. (1975) Proc R Soc Lond B Biol Sci 190(1098): 1
[37] Schildknecht H et al. (1966) Angew Chem Int Ed Engl 5(4): 421
[38] Shimada K et al. (1987) Chem Pharm Bull 35(12): 4996
[39] Shimada K et al. (1987) Chem Pharm Bull 35(6): 2300
[40] Ujvari B et al. (2015) Proc Natl Acad Sci U S A 112(38): 11911
[41] Van Oycke S et al. (1987) Experientia 43: 460
[42] Verkade PE (1969) Biochemistry 8(6): 2227
[43] von Euw J et al. (1971) Insect Biochem 1(4): 373
[44] Wang D-L et al. (2011) Chem Biodivers 8(4): 559
[45] Wasserstrom JA, Aistrup GL (2005) Am J Physiol Heart Circ Physiol 289(5): H1781
[46] Yoshida S et al. (1976) Chem Pharm Bull 24(8): 1714

2.6 Saponin-Like Triterpenes and Steroid Derivatives

2.6.1 General

The saponins and saponin-like substances have lyobipolar properties, i.e., they are amphiphilic molecules that may function as surfactants or detergents. The lipophilic part of the molecule is a steroid (Fig. 2.21) or a triterpene base (Fig. 2.22). The polar part can be a mono- or an oligosaccharide. In other cases, the polar groups are sulfate groups, hydroxyl groups, and carboxyl groups, or amines. The sugar-free compounds (aglycones) are referred to as 'sapogenins'.

The efforts to isolate secondary metabolites from marine organisms and to solve their molecular structures have led to the discovery of a large number of interesting steroid or triterpene saponins as well as other sterol derivatives with similar physical and pharmacological properties. Most of these compounds have been found in aquatic animals, namely, in the taxa Porifera (sponges; see Section 3.2.1), Cnidaria (sea anemones and corals; see Section 3.2.2), and Echinodermata (starfishes, sea urchins, brittle starfish, and sea cucumbers; see Section 3.2.4). Substances similar to saponins were also found in bony (Osteichthyes) and cartilaginous fish (Chondrichthyes) (see Section 3.2.11) as well as in beetles (Coleoptera; see Section 3.2.9).

The steroid and triterpene saponins, the polyhydroxysteroids, and the salts of steroid sulfates that occur in animals are able to generate foam when dissolved in aqueous solutions. When such surfactants are brought into contact with biological membranes, they interfere with the layers of ordered lipid molecules and may destroy eukaryotic cells, but also microorganisms. Thus, such substances are considered to be cytotoxins [61] and function as antimicrobial agents as well [22].

Marine animals that have insufficient escape or mechanical defense options in avoiding predators (e.g., sponges, starfish, sea urchins, and sea cucumbers) use chemical defense strategies to protect themselves, particularly against attacks from fish. In

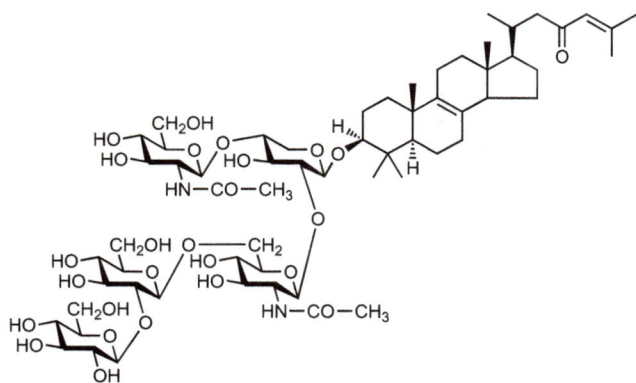

Sarasinoside A$_1$

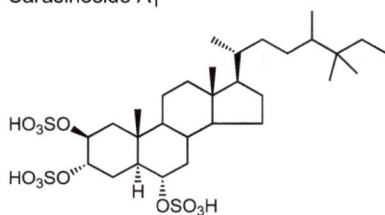

Halistanolsulfate F

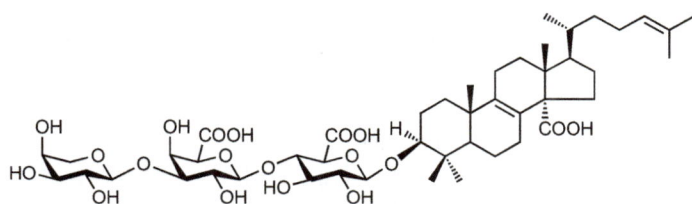

Nobiloside

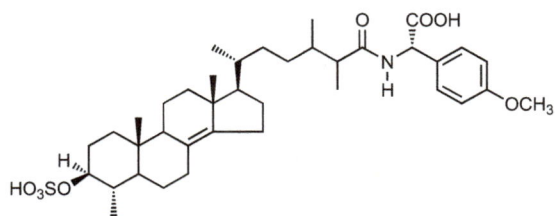

Polymastiamide A

Fig. 2.21: Saponin-like steroid derivatives of marine sponges.

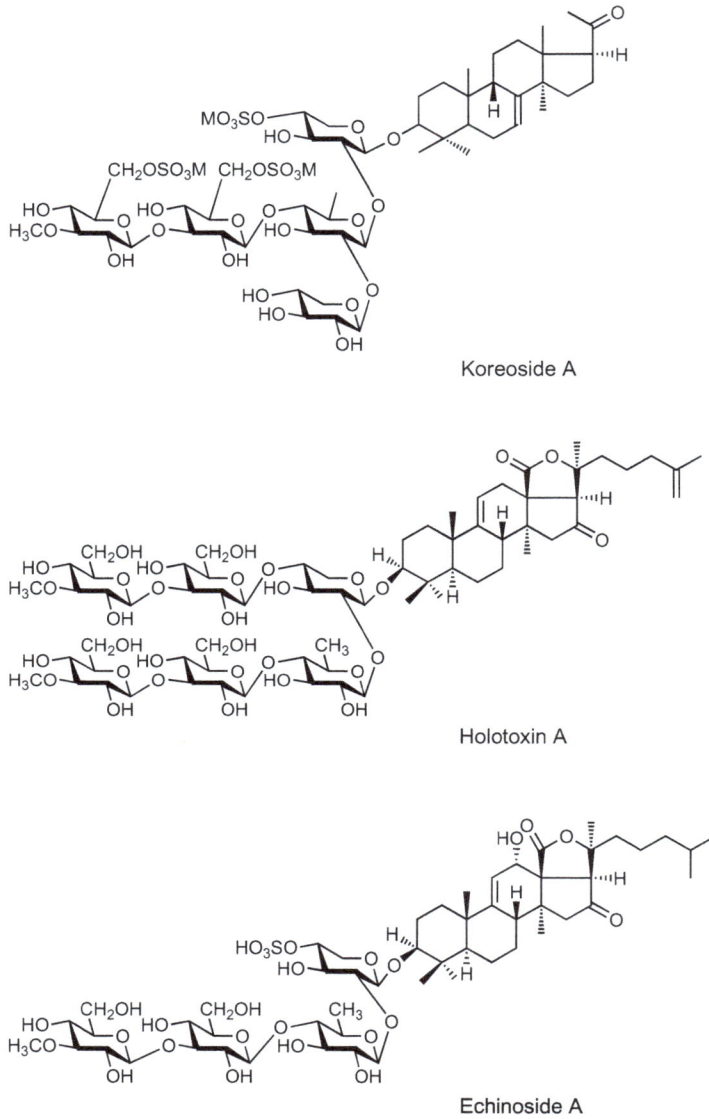

Koreoside A

Holotoxin A

Echinoside A

Fig. 2.22: Triterpene saponins of sea cucumbers. M (metal ion)–Na$^+$, K$^+$, or H$^+$.

case of an attack, these animals release surfactants into the water which may destroy the cell membranes of the fish gills. The resulting loss of electrolytes may anesthetize the fish or may even be fatal. Moreover, these substances have bacteriostatic, fungicidal, and virostatic effects and protect marine animals from fungal and algal overgrowth and from viral and bacterial infection [25, 45].

Apart from these defensive functions, saponins are also relevant and beneficial in biological processes, e.g., reproduction. In Echinodermata, the sulfated steroid saponins contained in the gelatinous egg shell facilitate the acrosome reaction which is an essential process necessary for the initiation of successful fertilization of an egg by sperm. Without the acrosome reaction, sperm is not able to penetrate the outer layers of the egg membrane [2, 31–33].

2.6.2 Saponin-Like Steroid Derivatives of Sponges (Porifera)

Saponins and saponin-like substances in sponges (see Section 3.2.1) do generally not pose a threat to humans because sponges are not part of our diet. These substances are directed against fish and may also have functions in regulating growth and assortment of bacterial symbionts.

Sarasinoside A_1 (Fig. 2.21) was the first saponin identified in sponges. It was isolated from the sponge *Melophusisis* sp. [48]. Sarasinoside A_1 contains 3β-hydroxy-4,4-dimethylcholest-8,24-dien-23-one as the aglycone and the two amino sugars *N*-acetyl-2-amino-2-deoxy-galactose and *N*-acetyl-2-amino-2-deoxy-glucose as monosaccharide components in addition to D-glucose and D-xylose. Halistanol was identified as an antiviral substance in *Halichondria moorei* [25]. Nobiloside was isolated from the marine sponge *Erylus nobilis*. It is a rare example of marine saponins that contain 14-carboxylanosterol derivatives as aglycones. Its structure was determined as a penasterol trisaccharide. It inhibits the clostridial neuraminidase with an IC_{50} value of 0.46 µg/mL and is thus considered to be an antimicrobial substance [77]. Polymastiamide A of the marine sponge *Polymastia boletiformis* was the first example of a new type of marine natural products formed by combination of steroid and amino acid components. Polymastiamide A exhibits antimicrobial activity against various bacterial pathogens in vitro [46].

Saponin-like steroid alkaloids are formed by replacing the OH group in position 3 with an acetylated or dimethylated amino group, as well as an amino group on C-4, condensation of a pyrroline or pyrrolidine ring on C-23 and C-24 and/or a piperidine ring on C-16 and C-17 [34, 81].

The knowledge of the pharmacological activities of steroidal saponines is still poor. The quantities of the substances that can be isolated from sponges, which are often in the single-digit milligram range, are mostly insufficient for extended investigations. Thus, studies are mostly limited to determination of potential cytotoxic effects. Cytotoxicity has been found with the sterols of *Phorbas amaranthus* [52], *Corticium niger* [75], *Theonella swinhoei* [64], and *Erylus nobilis* [71]. Only in some cases, we have more detailed information on the exact effects of certain steroid saponins on cellular processes in target organisms. For example, ophirapstanol trisulfate inhibits the nucleotide exchange at the small G protein p21[Ras] and silences this signal transduction pathway which, among other cellular processes, affects the eukaryotic cell cycle [30]. The immunosuppressive 4α-methyl-

5α-cholest-8-en-3β-ol from the elephant ear sponge *Agelas flabelliformis* [29] and the anti-leukemic actions of penasterol (5α-lanosta-9,24-dien-3β-ol-32-acid) from the sponge *Penares* sp. (IC$_{50}$ for the inhibition of leukemia cell proliferation: 3.6 µg/mL) [9] are further examples.

2.6.3 Saponin-Like Steroid Derivatives of Corals (Anthozoa) and Beetles (Coleoptera)

The cnidarian (see Section 3.2.2) taxon Anthozoa includes the Octocorallia (eight-pointed corals) with the major taxa Alcyonacea (soft corals and gorgonian corals), Helioporacea (colonial corals forming lobed crystalline calcareous skeletons), Pennatulacea (sea pens), and the Hexacorallia (six-pointed corals) with the major taxa Actiniaria (sea anemones), Zoanthidea (zoantids or crust anemones), Scleractinia (stony corals), and Antipatharia (black corals). All of them are sessile animals that need to use chemical defense to deter potential predators. Corals feed primarily on zooplankton. Many species harbor symbiontic algae (Zooxanthellae) to cover large parts of their energy needs by using portions of the algal photosynthesis products.

The defensive steroid saponins of corals are similar to those of the sponges in terms of structure and function. They are polyhydroxylated, may contain epoxy- or keto groups, and are partially acetylated or esterified with sulfuric acid. Rings B or D can be open. Ring closure of carbon atoms in the side chain to form cyclopropane rings or breakdown of the side chain under formation of a pregnane derivative have been described. Glycosides of such aglycones have also been found in corals.

Sterols from soft corals have been described to have antieczemic, anti-inflammatory, antipsoriatic effects, or antitumor activity, when appropriately tested or applied [20]. However, cytotoxic effects of such ingredients of corals have also been described, e.g., for the sterols of the soft corals *Nephtheae recta* [17] and *Sinularia* sp. [70]. The 9,10-secosteroids subergorgiaoles A–L from *Subergorgia rubra* [74], calicoferol A and calicoferol B from *Calicogorgia* sp. [57], and calicoferol D (Fig. 2.23) from *Muricella* sp. [69] are able to kill the larvae of the brine shrimp (*Artemia salina*) (LC$_{50}$ of the subergorgiaoles is approximately 2 µmol/L).

🔖 Injuries from contact with the sharp-edged calcium carbonate skeleton of corals while swimming, snorkeling, or diving are common. Abrasions or cuts are often followed by pruritus, erythema, urticaria, and inflammation of the subcutaneous connective tissue, sometimes also by necrosis. Wound healing takes long periods which may be attributed to defensive poisons of the corals or secondary bacterial infections.

🩹 First aid consists of rinsing the wound area with water, washing it with soap and water, removing particles with hydrogen peroxide, treating with antiseptics, and bandaging. Antibiotic prophylaxis is only necessary if the immune system of the patient is weak.

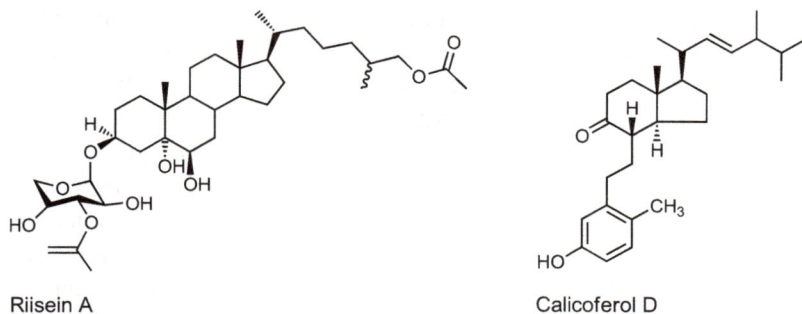

Riisein A   Calicoferol D

Fig. 2.23: Saponin-like steroid derivatives of corals.

Leaf beetles (Chrysomelidae; see Section 3.2.9) use chemical protection against predators by impregnating their tissues with toxic substances which they usually acquire from their host plants (Boraginaceae, Asteraceae, Asclepiadaceae, Convolvulaceae, Solanaceae, or Apocynaceae). An exception may be the presence of two triterpene oligosides that were isolated from the defensive glands of the leaf beetle *Platyphora ligata* (native to Central and South America). The oleanane triterpenes ligatoside A (Fig. 2.24) and ligatoside B are supposed to be synthesized by the animal itself [60].

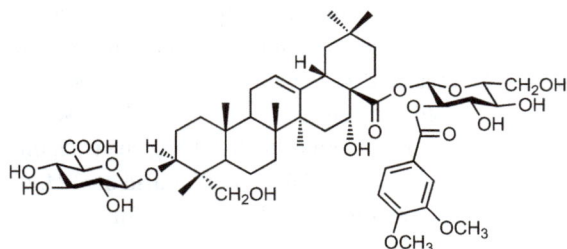

Fig. 2.24: Ligatoside A, a triterpene saponin of the chrysomelid beetle *Platyphora ligata*.

2.6.4 Saponin-Like Steroid Derivatives of Echinoderms (Echinodermata)

Echinodermata (see Section 3.2.4) have an endoskeleton that is mainly composed of $CaCO_3$, i.e., all parts of the skeleton including the spikes and thorns are covered by living tissue. This is important as these tissues may themselves be impregnated with poisonous substances or may contain poison glands with reservoirs. When predators or other organisms come into close contact with these animals, they may suffer bruises of their own body surface from the spikes and thorns of the echinoderms or poison themselves by ingesting echinoderm tissue.

Steroid saponins are present in the toxin mixtures of all echinoderms as well as in the mucus covering the body surface of the animals. The mucus restricts the rate of diffusion of the membrane-active compounds and thus prevents them from being quickly washed away by the surrounding water.

Steroid sulfates have been isolated from Pacific ophiurids. The biological roles of these sulfated polyhydroxy steroids are not yet clear. Some showed moderate cytotoxicity and antitumor activity when tested in mouse T-cell lymphoma cells [12]. In *Asterias rubens*, highly diverse mixtures of steroid glycosides bearing sulfate groups attached to the aglycone parts of the molecules are present in different body compartments, namely in the entire body wall, in the stomach, in the pyloric caeca, and in the gonads [14]. A similar distribution of the saponins asterone and iso-asterone had been previously found in another starfish species, *Leptasterias polaris* [27].

Many asterosaponins (Fig. 2.25) have aglycones with cholestane, 24-norcholestane, stigmastane, pregnane, or cholane structures. Frequently occurring aglycones are $\Delta^{9(11)}$-3β,6α-dihydroxysteroid derivatives, Δ^7-3β,6β-dihydroxysteroid derivatives, polyhydroxycholestane derivatives, or in some cases $\Delta^{8(14)}$ -3β,6β,15-trihydroxysteroid derivatives. The synthesis pathways of these substances have not been fully characterized. The monosaccharides bound to these aglycones may be D-glucose, D-xylose, D-galactose, D-fucose, D-quinovose (D-glucomethylose), 2,4-dimethyl-D-quinovose, D-glucuronic acid, L-arabinose, L-rhamnose, 2-*O*-methyl-D-xylose, 2,4-*O*-dimethyl-D-xylose, or 6-deoxy-β-D-xylo-hexos-4-ulose. Monosides and oligosides with up to six monosaccharide residues occur as coupling partners of the aglycones. Aglycones with a monosaccharide residue at C-3 and one to two monosaccharide residues at C-24 have also been detected. One example is granulatoside A from the granulated sea star *Choriaster granulatus* which activates mitogen-activated protein kinases in mammalian cells [63]. Long sugar chains are often branched and mostly bound to an OH group at C-6, less often also at C-24, or rarely at C-26, C-28, or C-29. The asterosaponins are often esterified with sulfuric acid on the OH group at C-3 of the aglycone, in some cases also on a sugar residue. Examples are thornasteroside A which occurs in the crown-of-thorns starfish *Acanthaster planci* and the glycoside B_2 from *Asterias amurensis* (Fig. 2.25 [36, 54]). However, there are also polyhydroxylated steroid aglycones without hydroxyl groups at C-24 or C-28, or those in which such a hydroxyl group is esterified with sulfuric acid. In these cases, the compounds have only one sugar residue at C-3.

Marthasterone (5α-cholesta-9(11),24-dien-3β,6α-diol-23-one) and dihydromarthasterone (5α-cholesta-9(11)-en-3β,6α-diol-23-one) are the sapogenins of saponins from the ice starfish *Marthasterias glacialis*, the cushion starfish *Culcita novaeguineae*, and many others (*Meyenaster gelatinosus, Luidia maculata, Nardoa gomophia*, or the Northern Pacific sea star *Asterias amurensis*) [56, 78].

Thornasterols A and B are present as aglycones in saponins of *Acanthaster planci* [42] and many other sea stars (*Asterias vulgaris, Patiria pectinifera, Luidia clathrata, Pisaster* sp., or the sunflower sea star *Pycnopodia helianthoides*). Pregnane derivatives were also identified as aglycones of saponins in different starfish species.

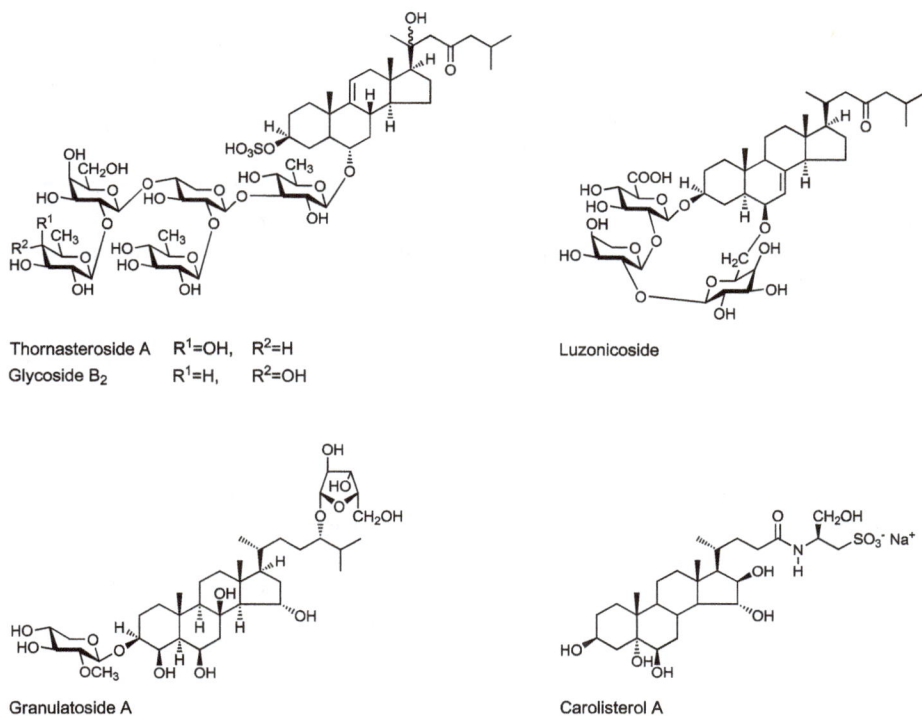

Thornasteroside A R^1=OH, R^2=H
Glycoside B$_2$ R^1=H,   R^2=OH

Luzonicoside

Granulatoside A

Carolisterol A

Fig. 2.25: Saponin-like steroid derivatives of sea stars.

Asterone (pregn-9(11)-en-3β,6α-diol-20-one) was found in *Marthasterias glacialis* and several other species, pregnanediolone was isolated from *Asterias rubens*, and asterogenol was isolated from *Asterias vulgaris* and others [8]. Furthermore, 3β, 20β, 23β-trihydroxy, 16,22-epoxy-6-sulfoxy-stimast-9(11)-en occurs as the aglycone of downeyoside A in *Henricia downeyae* [58]. Δ^7-3β, 6β-Dihydroxysteroid derivatives were found in *Echinaster* species. The oligosaccharide chains in these molecules, which are often composed of three monosaccharide residues, form ring structures between C-3 and C-6. Examples for such structures are sepositoside A from *Echinaster sepositus* and luzonicoside from *Echinaster luzonicus* (Fig. 2.25) [16, 47].

Polyhydroxysteroids are important aglycones of cytotoxic defense molecules in many different starfish species. They carry four to seven hydroxyl groups [59] in addition to hydroxyl groups at C-24 in case of many certonardosterols [73]. Some are conjugated with long-chain fatty acids and inhibit cell proliferation and cell motility in model cell lines [51].

In addition to or instead of saponins, the polar fraction of starfish extracts may also contain free polyhydroxysteroid derivatives with up to nine hydroxyl groups and their sulfates [5]. The content in the animals may reach up to 4.5% of the total body mass [28]. Some of them have unusual side chains. Examples are 24-methylcholestane derivatives (ergostane derivatives) [66], 24-substituted cholestane and stigmastane derivatives [28, 62], 25-desmethyl-24-methyl-cholestane derivatives (amuresterol [43]), 25-didesmethyl-24-methylcholestane derivatives (asterosterol [44]), and 24,26-cyclocholestane derivatives (phrygiasterol 1 [49]). Esterifications with sulfuric acid may be present in steroid sulfates at the carbon atom 3 or at other OH groups within the ring systems or within the side chains. 3,21-Disulfates of Δ^5-trihydroxysteroids, which are similar to those found in brittle stars and have hemolytic activity on mouse erythrocytes, were also found in sea stars [38].

Derivatives of polyhydroxycholanic acid which include amide-like linkages to cysteinolic acid at the C-24 carboxyl group include the carolisterols A to C. Such substances have been isolated from the starfish *Styracaster caroli* [15, 37].

The saponins of the starfish and the polyhydroxysteroids have hemolytic, cytotoxic, antiviral, antibacterial, antimycotic, and ichthyotoxic properties. The structure–activity study has shown that the sulfo-groups in anasteroside A, anasteroside B, as well as in versicoside A seem to be critical for their cell-lytic properties as their removal renders the molecules inefficient as antimycotics [10].

A standard measure of the ability of chemicals to induce cell lysis is the hemolysis index [50], which determines the number of destroyed erythrocytes as a percentage of all cells in the assay. However, also other assay systems are in use, e.g., the sea urchin egg assay. This assay measures inhibition rates of the first cleavage of sea urchin zygotes in the absence or presence of certain amounts of test substances. Using this latter system, the ED_{50} values of cellular inhibition were between 5 and 50 µg/mL when examining various sulfated oligoglycoside saponins from Japanese starfish [26]. The antibacterial activity is particularly high in representatives of such compounds carrying a 7α-OH group. Besides cytotoxicity and antibacterial effects, such substances may also show antitumor activities [5]. Irreversible destruction of the neuromuscular junction has been reported for the asterosaponins A and B from the Northern Pacific sea star *Asterias amurensis* [23]. The starfish saponins are highly toxic for fish and oysters but also for other marine animals [5].

🐾 Poisoning in humans may occur upon contact of bruised skin with the slime of some starfish species. Penetration of saponins into deeper tissue layers may result in local erythema or edema, or, upon absorption, also in systemic effects like vomiting and paralysis. Such symptoms, however, seem to be dependent on interactions of saponins with other ingredients of the toxin mixtures, especially proteotoxins.

🚑 Poisoning by starfish in humans is generally treated symptomatically upon removal of the poisonous mucus.

Brittle stars contain polyhydroxycholestane disulfates as defensive molecules. Examples are 3α,21-disulfates of polyhydroxycholestane derivatives, e.g., (20*R*)-cholesta-5,

24-diene-2β,3α,21-triol-2,21-disulfate found in *Ophiopholis aculeata* (Pacific Ocean), *Astrotoma agassizii*, and other brittle stars [21, 67] which may exert growth inhibitory and cytotoxic effects on eukaryotic cells [3]. Similar substances isolated from *Ophiarachna incrassata* (collected in the area of the Palau Islands) exert inhibitory effects on protein tyrosine kinases in animal and human cells [24]. Antiviral effects have been detected by the application of 3α,21-disulfocholestane derivatives isolated from *Ophioplocus januarii* collected at the Patagonian coast [68].

In contrast to the starfish (Asteroidea) which produces mainly steroid saponins, the sea cucumbers mainly produce triterpene saponins (Fig. 2.22) [72]. The biosynthesis pathways are not fully known, but seem likely that the triterpene precursor of saponins in sea cucumbers (e.g., in *Holothuria floridea* or *Actinopyga agassize*) is parkeol (lanost-9(11)-en-3β-ol) [41].

The approximately 100 known saponins of sea cucumbers are predominantly glycosides of derivatives of the hypothetical triterpene base holostane (20S-hydroxy-5α-lanostan-18-acid (18→20) lactone). The lactone ring may also be missing if the side chain is short as in koreoside A, a triterpene glycoside from *Cucumaria koraiensis* [7]. In addition to triterpenes, the following aglycones were identified in holothurian saponins: nortriterpenes, e.g., derivatives of 23,24,25,26,27-pentanorlanostane [6], and C_{26}, C_{28}, and C_{29} compounds with sterane as a basic structure [19]. In the ring system, the holostane derivatives usually have a double bond between the carbon atoms 9 and 11, rarely between C-7 and C-8. A selection of sapogenins from sea cucumbers is shown in Tab. 2.3.

The sugar chain is almost always attached to the hydroxyl group on C-3. It consists of up to six monosaccharide residues and is, in some cases, branched (holotoxin A from *Holothuria* sp., Fig. 2.22). Sugar residues adjacent to the sapogenin may carry a sulfate group (echinoside A from *Bohadschia* sp. or *Actinopyga* sp.). Monosaccharide components may be D-glucose, D-xylose, 3-*O*-methyl-D-glucose, 3-*O*-methyl-D-xylose, or D-quinovose.

Tab. 2.3: Examples of sapogenins of triterpene saponins of sea cucumbers (data from several authors).

Holothurian saponin	Sapogenin	Occurrence (genus)
Frondoside A	Holost-7(11)-en-3β-ol	*Cucumaria*
Bivvitoside D	Holost-9(11)-en-3β-ol	*Bohadschia*
Bivvitoside C	Holost-9(11)-en-3β,12α-diol	*Bohadschia*
Holothurin A_2	Holost-9(11)-en-3β,12α,17α-triol	*Holothuria, Bohadschia*
Echinoside A, B	Holost-9(11)-en-3β,12α,17α-triol	*Bohadschia, Actinopyga*
Holothurin A_1	Holost-9(11)-en-3β,12α,17α,22-tetraol	*Holothuria*

Tab. 2.3 (continued)

Holothurian saponin	Sapogenin	Occurrence (genus)
Holothurin A, B	Holost-9(11)-en-3β,12α,17α-triol-22,25-epoxide	*Holothuria, Actinopyga*
Holotoxin A, B, A₁, B₁	Holosta-7,25-dien-3β-ol-16-one	*Holothuria*
Cucumarioside A₂	Holosta-9(11),25-dien-3β-ol-16-one	*Cucumaria*
Cucumarioside G₁	16β-Acetoxy-holosta-7,24-dien-3β-ol	*Cucumaria*
Cucumarioside G₂	23,24,25,26,27-Pentanor-ianosta-7(8),20(22)-dien-3β-ol-8(16)-lactone	*Eupentacta*
Stichoposide A–E	23(S)-Acetoxy-holost-7-en-3β-ol	*Stichopus*
Thelenotosides A, B	23(S)-Acetoxy-holost-7,25-dien-3β-ol	*Actinopyga*

Investigations of structure and function relationships have shown that the toxicity of sea cucumber saponins increases with the number of monosaccharide residues bound to the sapogenins [39]. The disruption of neuromuscular signal transmission by holothurian saponins is a result of the lysis of the motor neuron synapse and the loss of ion gradients and of plasma membrane potential [13, 79]. In particular, the sensitive surfaces of fish gills may be harmed by the saponins contained in the toxin mixtures and signal to the fish that this organism is unpalatable [80]. Ichthyotoxicity results from the ability of holothurian saponins to disturb the electrolyte balance of the fish [53] and, in case of intense contact, to lyse epithelial cells in the gills of fish [18].

However, recent research has revealed that holothurian saponins may have also beneficial effects as antioxidants and prevent apoptotic cell death by attenuation of caspase activity [55]. Moreover, antitumor activity was reported for saponin extracts of several holothurians [3, 4], e.g., for *Apostichopus japonicus* [11].

The acute toxicity of holothurian saponins does not seem to be very high, especially in humans. In mice, LD_{50} values (p.o.) of >500 mg/kg BW have been reported for ethanolic extracts of sea cucumber skin [31]. A mixture of different types of sea cucumber, known as trepang (bêche-de-mer), is regularly used as food in East Asian countries upon removal of the saponin-rich internal organs.

2.6.5 Saponin-Like Steroid Derivatives of Cartilaginous Fish (Chondrichthyes) and Bony Fish (Osteichthyes)

In rare cases, poisoning occurs upon consumption of dogfish (*Squalus* sp., Chondrichthyes). Squalamine (Fig. 2.26), a natural cationic steroid, has been isolated from the liver tissue of the spiny dogfish (*Squalus acanthias*) that may be responsible for such

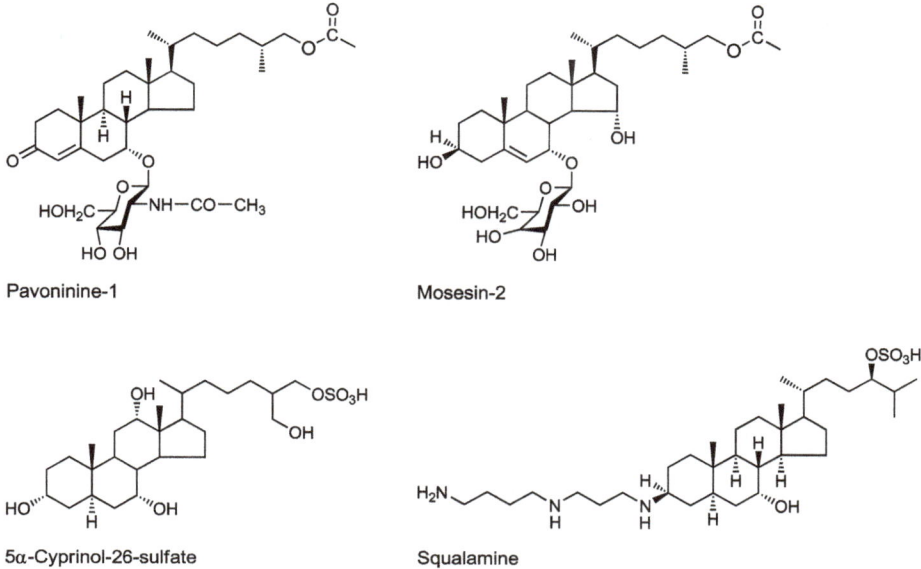

Fig. 2.26: Saponin-like steroid derivatives of fish.

cases because it inhibits the brush-border Na^+/H^+ exchanger (NHE3), a sodium–hydrogen exchanger present on cell surfaces that regulate intracellular pH [1]. In vitro, this amino steroid has bactericidal, antiangiogenic, and cytotoxic activities [65].

Some bony fishes (Osteichthyes; see Section 3.2.11) produce toxic substances in the dorsal skin glands which are released in case of danger by predators. These substances are very efficient in repelling sharks. Examples of such ichthyocrinotoxic fish species are the Moses soles *Pardachirus marmoratus* that occurs in the Red Sea and *P. pavonicus* in the Pacific and Indian Oceans. Skin secretions of these fishes contain the pardaxins P-1, P-2, and P-3, lyobipolar polypeptides composed of 33 aminoacyl residues, and saponin-like steroid glycosides, the pavoninines 1–6 and the mosesins 1–5 (Fig. 2.26) [82]. Pavoninines in the surrounding medium are able to kill killifish (*Oryzias latipes*) at a concentration of 8.5 μg/mL. In vitro, these substances show hemolytic activities [76].

Of toxicological interest for humans is 5α-cholestane-3α,7α,12α,26,27-pentol-26-sulfate (5α-cyprinol-26-sulfate) (Fig. 2.26) [35], a component of the bile of various carp-like freshwater fish species (Cypriniformes). Liver tissue of the Asian grass carp, *Ctenopharyngodon idellus*, is rich in this steroid saponine. Fatal poisoning has been observed in humans who ate grass carp which had not been properly cooked or in which fragments of the gall bladder were still present [40]. Cyprinol and cyprinol sulfate are hemolytic and, in vivo, damage several organs. They are nephrotoxic, cardiotoxic and hepatotoxic.

⚕ Initial symptoms of carp bile poisoning develop 4–6 h after ingestion. Vomiting, abdominal pain, and diarrhea have been described. Within several days, severe functional disorders of the kidneys (acute renal failure, ARF) may occur due to the destruction of proximal kidney tubules. This may be accompanied by liver damage, and, in some cases, also by myocarditis and gastrointestinal problems [83]. Lethality is approximately 20% of all cases of carp bile poisoning.

🚑 Mild cases of carp bile poisoning may be cured by hemodialysis, but severe poisoning may require kidney transplantation.

References

[1]   Akhter S et al. (1999) Am J Physiol 276(1): C136
[2]   Amano T et al. (1992) Biochem Biophys Res Commun 187(1): 274
[3]   Aminin DL et al. (1995) Comp Biochem Physiol C Pharmacol Toxicol Endocrinol 112(2): 201
[4]   Aminin DL et al. (2015) Marine Drugs 13(3): 1202
[5]   Andersson L et al. (1989) Toxicon 27(2): 179
[6]   Avilov SA et al. (1994) J Nat Prod 57(8): 1166
[7]   Avilov SA et al. (1997) J Nat Prod 60(8): 808
[8]   Burnell DJ et al. (1986) Comp Biochem Physiol B 85(2): 389
[9]   Cheng J-F et al. (1988) J Chem Soc, Perkin Trans 1(8): 2403
[10]   Chludil HD et al. (2002) J Nat Prod 65(2): 153
[11]   Dai Y-L et al. (2020) J Food Sci Technol 57(6): 2283
[12]   D'Auria MV et al. (1987) J Org Chem 52(18): 3947
[13]   de Groof RC, Narahashi T (1976) Eur J Pharmacol 36(2): 337
[14]   Demeyer M et al. (2014) Comp Biochem Physiol Part B Biochem Mol Biol 168: 1
[15]   De Riccardis F et al. (1993) Tetrahedron Lett 34(27): 4381
[16]   Dong G et al. (2011) Chem Biodivers 8(5): 740
[17]   Duh CY et al. (1998) J Nat Prod 61(8): 1022
[18]   Eeckhaut I et al. (2015) Biol Bull 228(3): 253
[19]   Elyakov GB et al. (1980) Comp Biochem Physiol B Biochem Mol Biol 65(2): 309
[20]   Ermolenko EV et al. (2020) Marine Drugs 18(12): 613
[21]   Fedorov SN et al. (1994) J Nat Prod 57(12): 1631
[22]   Francis G et al. (2007) Br J Nutr 88(6): 587
[23]   Friess SL (1972) Federation Proc 31(3): 1146
[24]   Fu X et al. (1994) J Nat Prod 57(11): 1591
[25]   Fusetani N et al. (1981) Tetrahedron Lett 22(21): 1985
[26]   Fusetani N et al. (1984) J Nat Prod 47(6): 997
[27]   Garneau F-X et al. (1989) Comp Biochem Physiol B Comp Biochem 92(2): 411
[28]   Goodfellow RM, Goad LJ (1983) Comp Biochem Physiol B Biochem Mol Biol 76(3): 575
[29]   Gunasekera SP et al. (1989) J Nat Prod 52(4): 757
[30]   Gunasekera SP et al. (1994) J Nat Prod 57(12): 1751
[31]   Hanafi PS et al. (2020) Asian J Pharm Clin Res 13(2): 150
[32]   Hoshi M et al. (1994) Int J Dev Biol 38(2): 167
[33]   Hoshi M et al. (2000) Zygote 8(S1): S26

[34] Hostettmann K, Marston A (1995) Chemistry & Pharmacology of Natural Products: Saponins. Cambridge University Press, New York.
[35] Hwang DF et al. (2001) Toxicon 39(2–3): 411
[36] Ikegami S et al. (1979) Tetrahedron Lett 20: 1769
[37] Iorizzi M et al. (1994) J Nat Prod 57(10): 1361
[38] Ivanchina NV et al. (2003) J Nat Prod 66(2): 298
[39] Kalinin VI et al. (1996) Toxicon 34(4): 475
[40] Karatas AD et al. (2010) Int J Clin Pract 64(1): 129
[41] Kerr RG, Chen ZJ (1995) J Nat Prod 58(2): 172
[42] Kitagawa I et al. (1975) Tetrahedron Lett 16(11): 967
[43] Kobayashi M, Mitsuhashi H (1974) Tetrahedron 30(14): 2147
[44] Kobayashi M et al. (1973) Tetrahedron 29(9): 1193
[45] Koehn FE et al. (1991) J Org Chem 56(3): 1322
[46] Kong FM, Andersen RJ (1993) J Org Chem 58(24): 6924
[47] Laires R et al. (2013) Tandem Mass Spectrom – Mol Charact 5: 117
[48] Lee HS et al. (2000) J Nat Prod 63(7): 915
[49] Levina EV et al. (2005) J Nat Prod 68(10): 1541
[50] Lippi G et al. (2014) Clinica Chimica Acta 429: 143
[51] Malyarenko TV et al. (2020) Marine Drugs 18(5): 260
[52] Masuno MN et al. (2004) J Nat Prod 67(4): 731
[53] Mebs D (2002) Venomous and Poisonous Animals. CRC Press, Boca Raton, London, New York, Washington D.C
[54] Minale L et al. (1995) Structural studies on chemical constituents of echinoderms. In: Atta-ur-Rahman (ed.) Studies in Natural Products Chemistry, Vol. 15. Elsevier, Amsterdam, Lausanne, New York, Oxford, Shannon, Tokyo
[55] Moghadam FD et al. (2016) Res Pharm Sci 11(2): 130
[56] Nicholson SH, Turner AB (1976) J Chem Soc, Perkin Trans 1(12): 1357
[57] Ochi M et al. (1991) Chem Lett 20(3): 427
[58] Palagiano E et al. (1996) J Nat Prod 59(4): 348
[59] Peng Y et al. (2010) Chem Pharm Bull 58(6): 856
[60] Plasman V et al. (2000) J Nat Prod 63(5): 646
[61] Podolak I et al. (2010) Phytochem Rev: Proc Phytochem Soc Eur 9(3): 425
[62] Popov RS et al. (2016) Metabolomics 12(2): 21
[63] Qi J et al. (2006) ChemMedChem 1(12): 1351
[64] Qureshi A, Faulkner DJ (2000) J Nat Prod 63(6): 841
[65] Rao MN et al. (2000) J Nat Prod 63(5): 631
[66] Riccio R et al. (1988) J Nat Prod 51(5): 1003
[67] Roccatagliata AJ et al. (1998) J Nat Prod 61(3): 370
[68] Roccatagliata AJ et al. (1996) J Nat Prod 59(9): 887
[69] Seo Y et al. (1995) J Nat Prod 58(8): 1291
[70] Sheu JH et al. (2000) J Nat Prod 63(1): 149
[71] Shin J et al. (2001) J Nat Prod 64(6): 767
[72] Sroyraya M et al. (2018) Microsc Res Tech 81(10): 1182
[73] Stonik VA et al. (2008) Nat Prod Commun 3(10): 1587
[74] Sun X-P et al. (2015) Steroids 94: 7
[75] Sunassee SN et al. (2014) J Nat Prod 77(11): 2475
[76] Tachibana K et al. (1984) Science 226(4675): 703
[77] Takada K et al. (2002) J Nat Prod 65(3): 411
[78] Tang HF et al. (2006) Fitoterapia 77(1): 28

[79]   Thron CD et al. (1964) Toxicolo Appl Pharmacol 6(2): 182
[80]   Van Dyck S et al. (2011) J Exp Biol 214(8): 1347
[81]   Verkade PE (1969) Biochemistry 8(6): 2227
[82]   Williams JR, Gong H (2004) Lipids 39(8): 795
[83]   Xuan BH et al. (2003) Am J Kidney Dis 41(1): 220

2.7 Phenylpropanoids

2.7.1 General

Phenylpropanoids are natural products with the carbon skeletons of 1-phenylpropane (propylbenzene). Biogenesis occurs by the shikimate pathway, starting from erythrose-4-phosphate and phosphoenolpyruvate, two intermediates of the carbohydrate metabolism. Many animals use such defensive substances that they excrete to the environment. Such substances function as repellents of potential attackers. They may just transport information or may induce harm to predators.

Some insects provide chemical weaponry that is truly poisonous or damaging to target organisms by interfering with specific physiological processes (inhibiting enzyme activities or sensory cell functions) or behaviors (impairment of movements of body parts) of targeted animals. On the other hand, repellents function in the distance or upon physical contact by signaling toxicity or non-palatability of the producer to the receiver, e.g., with pungent smell or taste [17]. These categories are not mutually exclusive, as some substances can exert multiple effects. In both cases, the producer has a benefit from using these substances. Thus, they can be considered allomones [15]. Aposematism, i.e., display of obvious warning signals in body shape or pigmentation, is utilized by some but not all of the non-palatable species as a warning to potential predators [12, 18].

Phenylpropanoids are a class of compounds whose members fulfill these criteria especially well.

2.7.2 Phenylpropanoids in Arthropod Defense Secretions

Phenylpropanoids (Fig. 2.27) are widespread in different arthropod species (see Sections 3.2.7–3.2.10) as ingredients of defense secretions [2]. They are often accompanied by aliphatic hydrocarbons or ketones, which are believed to serve as vehicles. Together, many of these substances are volatile and are carried to the target organism by diffusion or by convection in air. Alternatively, these substances may be sprayed onto the attacker.

In harvestmen (Opiliones, Arachnida), 1,4-benzoquinone (Fig. 2.27) and its alkyl derivatives occur in defense secretions. Individuals of *Vonones sayi* produce a secretion in special glands that is rich in 2,3-dimethyl-1,4-benzoquinone and 2,3,5-trimethyl-1,4-

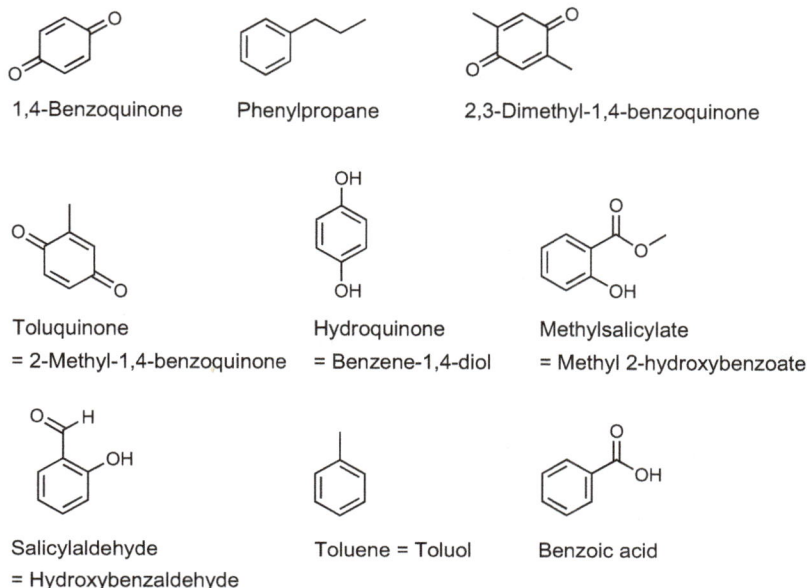

1,4-Benzoquinone   Phenylpropane   2,3-Dimethyl-1,4-benzoquinone

Toluquinone   Hydroquinone   Methylsalicylate
= 2-Methyl-1,4-benzoquinone   = Benzene-1,4-diol   = Methyl 2-hydroxybenzoate

Salicylaldehyde   Toluene = Toluol   Benzoic acid
= Hydroxybenzaldehyde

Fig. 2.27: Phenylpropanoids.

benzoquinone. When attacked, they squeeze drops of liquid out of the paired glands, dilute the secretion with regurgitated gut content, and hurl the drops at the attacker using their front feet [9]. Individuals of *Acanthopachylus aculeatus* use 2,3-dimethyl-1,4-benzoquinone, 2,5-dimethyl-1,4-benzoquinone, and 2,3,5-trimethyl-1,4-benzoquinone as ingredients of their defensive secretions [10]. All of these phenylpropanoids are irritants with a pungent smell that create stinging sensations on moist parts of the skin in target organisms. They seem to also repel spiders, ants, or cockroaches [13].

In millipedes (Diplopoda), paired defense glands in each segment are able to secrete fluids containing 1,4-benzoquinone, toluquinone (2-methyl-1,4-benzoquinone), 3-methyl-1,4-benzoquinone, 2,3-dimethyl-1,4-benzoquinone, 2,5-dimethyl-3-methoxy-1,4-benzoquinone, 2-methoxy-3-methyl-1,4-benzoquinone, or/and 2,3-dimethoxy-1,4-benzoquinone [1, 6]. Capuchin monkeys (*Cebus* sp.) have been observed to successfully protect themselves from mosquito (*Aedes aegypti*) bites by rubbing their fur with millipedes [23].

In insects (Hexapoda), defense against attackers using phenylpropanoids plays a major role in beetles (Coleoptera). 1,4-Benzoquinone, alkyl derivatives of 1,4-benzoquinone, phenol and derivatives, hydroxy derivatives of benzaldehyde, and phenolic carboxylic acids occur frequently in defensive secretions [2].

De novo-synthesized salicylaldehyde and methylsalicylate (Fig. 2.27) are also important ingredients of defensive secretions in carabid beetles. They have been found in the pygidial gland secretions of ground beetles (Carabidae), e.g., in *Idiochroma dorsalis* [20]. Individuals of *Calosoma prominens* can fend off ants or even vertebrates (insectivorous mammals or birds) by directly spraying defense fluids containing salicylaldehyde onto the attacker [11]. The salicylaldehyde in defensive secretions of some members of the leaf beetle family (Chrysomelidae), on the other hand, is generated using the alcoholic β-glucoside salicin as precursor obtained from forage plants (poplars and willows) [4, 16].

The defense secretions of rove beetles (Staphylinidae) as well as those of darkling beetles (Tenebrionidae) may contain 2-methyl-1,4-benzoquinone (toluquinone; Fig. 2.27), 2-ethyl-1,4-benzoquinone, 2-*n*-propyl-1,4-benzoquinone and 2-methyl-3-methoxy-1,4-benzoquinone, or 1,4-benzoquinone. These substances are generally accompanied by aliphatic hydrocarbons and/or ketones. Staphylinid beetles of the genus *Bledius* generate *p*-toluquinone in their defensive glands [21]. The desert stink beetle *Eleodes hispilabris*, a tenebrionid, sprays such pungent substances from an abdominal gland at predators [3].

Diving beetles (Dytiscidae) use benzoic acid (Fig. 2.27), 4-hydroxybenzoic acid, 4-hydroxybenzoic acid methyl ester, 4-hydroxybenzaldehyde, 3,4-dihydroxybenzoic acid or 4-dihydroxy-phenylacetic acid methyl ester, and hydroquinone as 'antifouling' agents to protect the body surface from being overgrown by bacteria, fungi, or algae. The substances are produced in the pygidial glands and distributed over the body surface by movements of the forewings [7].

Hydroquinone and toluhydroquinone are produced and stored together with hydrogen peroxide (H_2O_2) in reservoirs of abdominal pygidial glands by bombardier beetles (Brachynidae), e.g., *Brachinus* sp. or *Aptinus* sp. In addition to the storage compartment, their glands contain a reaction chamber in which portions of stored material may be mixed with catalase and peroxidase which are secreted on demand from the epithelial cells lining the chamber when the animals feel threatened by predators. The enzymes mediate the generation of toxic oxidation products of the quinones (1,4-benzoquinone and toluquinone) and of water and large volumes of oxygen (O_2) from hydrogen peroxide. The hot mixture (up to 100 °C) leaves the reaction chamber explosively as a spray that can be precisely targeted by the beetle toward the attacker [5, 8, 19].

Earwigs (Dermaptera), e.g., the common earwig *Forficula auricularia*, defend themselves with secretions of stink glands which are located on the dorsal rear borders of the third and fourth abdominal body segments. The secretions emitted by the alarmed animals contain 2-methyl- and 2-ethyl-1,4-benzoquinone [22].

Individuals of some termite species (Isoptera) use a rapidly polymerizing mixture of benzoquinone, toluquinone (Fig. 2.27), and proteins secreted from defensive labial glands to immobilize attacking insects [14].

References

[1]   Attygalle AB et al. (1993) J Nat Prod 56(10): 1700
[2]   Blum MS (1981) Chemical Defense of Arthropods. Academic Press, New York
[3]   Blum MS, Crain RD (1961) Ann Entomol Soc Am 54(4): 474
[4]   Brückmann M et al. (2002) Insect Biochem Mol Biol 32(11): 1517

[5]   Dean J et al. (1990) Science 248(4960): 1219

[6]   De Bernardi M et al. (1982) Naturwissenschaften 69(12): 601

[7]   Dettner K (2014) Chemical ecology and biochemistry of Dytiscidae. In: Yee DA (ed.) Ecology, Systematics, and the Natural History of Predaceous Diving Beetles (Coleoptera: Dytiscidae). Springer, Dordrecht, p. 235

[8]   Eisner T, Aneshansley DJ (1999) Proc Natl Acad Sci U S A 96(17): 9705

[9]   Eisner T et al. (1971) Science 173(3997): 650

[10]   Eisner T et al. (2004) J Exp Biol 207(8): 1313

[11]   Eisner T et al. (1963) Ann Entomol Soc Am 56(1): 37

[12]   Huheey JE (1984) Warning coloration and mimicry. In: Bell WJ, Cardé RT (eds.) Chemical Ecology of Insects. Chapman and Hall, London, p. 257

[13]   Machado G et al. (2005) J Chem Ecol 31(11): 2519

[14]   Maschwitz U et al. (1972) J Insect Physiol 18(9): 1715

[15]   Müller-Schwarze D (2006) Chemical Ecology of Vertebrates. Cambridge University Press, Cambridge

[16]   Pasteels JM et al. (1982) Tetrahedron 38(13): 1891

[17]   Pasteels JM et al. (1983) Annu Rev Entomol 28: 263

[18]   Richardson JML, Anholt BR (2010) Defensive morphology. In: Breed MD, Moore J (eds.) Encyclopedia of Animal Behaviour. Academic Press, Oxford., p. 493

[19]   Schildknecht H et al. (1968) Zeitschrift für Naturforschung 23b: 1213

[20]   Schildknecht H et al. (1968) Zeitschrift für Naturforschung 23b: 46

[21]   Steidle JLM, Dettner K (1995) Biochem Syst Ecol 23(7): 757

[22]   Walker KA et al. (1993) J Chem Ecol 19(9): 2029

[23]   Weldon PJ et al. (2003) Naturwissenschaften 90(7): 301

2.8 Amines

2.8.1 Tetramethylammonium

The simplest amine used by animals for defense purposes is tetramethylammonium hydroxide (tetramine, TMAH, TMAOH) (Fig. 2.28). Tetramethylammonium hydroxide is a quaternary ammonium salt that is odorless in its pure form. When it occurs in animals it is usually accompanied by trimethylamine which has a strong fishy smell. Tetramethylammonium and its derivatives are readily absorbed in animal intestines [48]. Transepithelial transport may occur by the same mechanisms that are responsible for choline uptake from the gut. Although small molecules like tetramethylammonium ions are filtered in the kidneys, they may stay in the animal for quite a while because they are effectively reabsorbed in the proximal tubule [44]. While some TMAH is excreted in the urine, most of it leaves the body in the bile [30]. When animals are exposed to high concentrations of tetramethylammonium ions, paralysis and possibly death may occur (LD_{50} is approximately 50 mg/kg BW, rat, p.o. [3]) because, due to TMAH's structural similarity to acetylcholine, it may block the signal transmission at the neuromuscular junction by binding to and interfering with the function of the nicotinic acetylcholine receptor [4]. Initially, TMAH activates these receptors, but desensitizes them in the continued presence of the agonist [9].

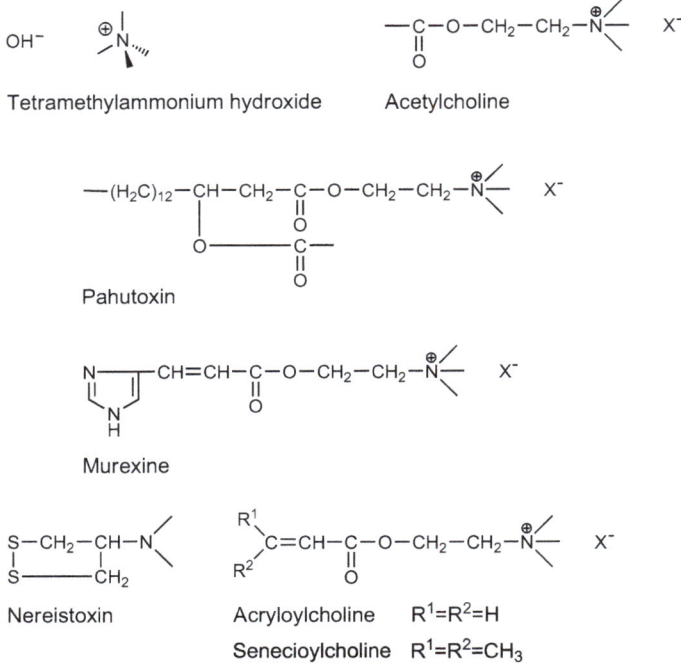

Fig. 2.28: Amines in animal toxins I.

TMAH gets accumulated at high concentrations besides trimethylamine in some marine and estuarine invertebrates because they are used as compatible osmolytes protecting protein conformation from the deleterious actions of other osmolytes or cryoprotectant molecules that are accumulated in these animals under hyperosmotic stress or extreme temperatures [54]. In addition, it serves as a deterrent against potential predators. It is common in the cnidarians (Cnidaria), in mollusks (Mollusca), or in moss animals (Bryozoa). It was also found in some representatives of the Hydrozoa and Scyphozoa. TMAH concentrations of 300 or 900 mg/kg FW, respectively, have been found in tissues of sea anemones or snails of the genus *Neptunea* [1, 3].

☠ Possibilities of poisoning exist above all when consuming commercially available *Neptunea* species (Neptune's horn snails) that are considered edible after proper pretreatment (cooking for long periods). Numerous poisonings after eating *Neptunea arthritica* have been observed in Hokkaido [3]. Symptoms of TMAH intoxication (e.g., visual and respiratory disorders, severe headaches, and paralysis) may occur upon ingestion of Neptune's horn snails which had not been properly cooked.

❢ Two specimens of *Neptunea antiqua*, purchased at a fishmonger in Copenhagen in 1985, were cooked only briefly for 30 min. A 31-year-old female who had consumed these snails experienced short periods of blurred vision and blindness. Other symptoms of poisoning were vomiting and headache. The patient recovered without medical treatment within 1 h.

2.8.2 Acetylcholine

Acetylcholine (ACh; Fig. 2.28), which functions as a neurotransmitter in humans and animals, occurs in venoms of some bees and wasps (Hymenoptera; see Section 3.2.9). It is present in the venom of the hornet (*Vespa crabro*) in relatively high concentrations [8]. Injection of acetylcholine induces pain and muscle cramps (activation of muscarinic as well as nicotinic ACh receptors), but intoxication is only transient due to the presence of acetylcholine esterases in the interstitial space which rapidly degrade ACh into acetate and choline [15], which are not biologically active as neurotransmitters.

2.8.3 Other Choline Esters

Of greater toxicological relevance are certain choline esters in toxin mixtures of animals because they are much less susceptible to acetylcholinesterase and have a higher degree of lipophilicity compared with ACh. Pahutoxin is a representative of this group of toxins which contains a 3-acetoxy-palmitic acid moiety (Fig. 2.28). Homopahutoxins (choline esters containing 3-propyloxy-, 3-butyryloxy-, 3-valeroyloxy-, or 3-caproyloxy-palmitic acid), palmitoylcholine, and choline esters of C_{14}, C_{17}, or C_{18} fatty acids are other examples. Pahutoxin was first isolated from the slimy secretion of the boxfish *Ostracion lentiginosus* which occurs in tropical regions of the Pacific Ocean [10]. Similar compounds with ichthyotoxic properties have been isolated from the skin secretions of other members of the Ostraciidae family of fishes (genera *Aracana*, *Lactophrys*, *Lactoria*, or *Anoplocapros*) [22, 23]. Pahutoxin is poisonous for other fish species. It has no ACh-like biological activities, but instead is hemolytic and cytotoxic. Due to its amphiphilic nature, it is able to interact with plasma membrane lipids and acts as a surfactant like the steroidal saponins of echinoderms. It is interesting that pahutoxin-producing fish species are insensitive to their own toxin, which led to the hypothesis that other fish may have specific receptors for pahutoxin in the plasma membranes of target cells such as gill epithelial cells [34].

A number of other choline esters were found in the secretion of the hypobranchial glands of muricid sea snails [52]. Murexine or urocanylcholine (Fig. 2.28) is a typical ingredient of secretions of these glands. In vertebrates and invertebrates, murexine affected neuromuscular signal transmission and other nicotinic actions but was almost devoid of effects on muscarinic ACh receptors [19]. The lethal dose in the mouse is 8.5 mg/kg BW [18]. Another species of the snail family Muricidae, *Stramonita floridana*, produces a similar substance, senecioylcholine (Fig. 2.28). The common whelk, *Buccinum undatum* (Buccinidae), which inhabits the North Sea and the western Baltic Sea, uses acryloylcholine [52]. It is still unclear whether these substances are used by the producers as deterrents against potential predators or as ingredients of venoms to effectively hunt for prey.

2.8.4 Nereistoxin

A sulfur-containing amine, nereistoxin (Fig. 2.28), has been found in the skin mucus of *Kuwaita heteropoda* (see Section 3.2.5), a jawless predatory polychaete worm (Annelida, Polychaeta) [28]. The muscular proboscis penetrates the body surface of prey organisms and secretes a toxin mixture containing nereistoxin into the wound [12]. Nereistoxin activates nicotinic acetylcholine receptors (nAChRs) in neurons [43]. Large doses of the neurotoxin may result in desensitization of the receptors. As insects use acetylcholine as a transmitter in their central nervous system, synthetic analogues of nereistoxin, e.g., thiocyclam oxalate (*N,N*-dimethyltrithian-5-amine oxalate) or Cartap (*N,N*-dimethyl-1,2-dithiolan-4-amine), have been used as broadband insecticides [43].

☠ Exposure of animals or man to nereistoxin (e.g., by handling worms that are used as bait by fishermen) may cause increased saliva production, miosis (narrowing of the pupilla), headaches, and vomiting. The LD_{50} (rabbit, s.c.) is 1.8 mg/kg BW [28].

2.8.5 Phenylalkylamines

Amines containing aromatic ring systems (phenylalkylamines) occur as ingredients in toxin mixtures of many animal species. These toxins are structurally identical with endogenous substances, the catecholamines (Fig. 2.29). They are used as neurotransmitters or as hormones in virtually all animal species. The catecholamines comprise

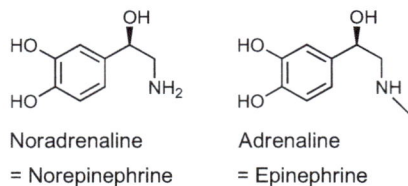

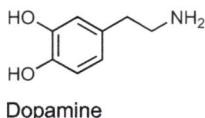

Noradrenaline
= Norepinephrine

Adrenaline
= Epinephrine

Dopamine

Fig. 2.29: Catecholamines.

noradrenaline (= norepinephrine; 1-1-(3,4-dihydroxyphenyl)-2-aminoethanol), adrenaline (= epinephrine), and dopamine (3,4-dihydroxy phenethylamine) [5].

Significant amounts of these compounds are found in venoms of bees and wasps (Hymenoptera). Venom of honeybees (*Apis mellifera*) contains up to 1.7 mg/g noradrenaline and up to 1.4 mg/g dopamine. In the venom of the North American yellowjacket (*Dolichovespula arenaria*), 8.4 mg/g noradrenaline and 2.5 mg/g dopamine were detected. The venom of the German yellowjacket, *Vespula germanica*, however, does not contain any adrenaline but 0.18 mg/g dopamine [39].

High concentrations of catecholamines are present in skin gland secretions of some toad species, e.g., in the South American toad *Rhinella arenarum* (4 mg adrenaline per g of dried secretion) [32]. In other toad species, e.g., in the common toad, *Bufo bufo*, such substances are absent.

Octopamine (norsynephrine; Fig. 2.30) is closely related to norepinephrine. It is biosynthesized in a similar metabolic pathway. Octopamine serves as an important neurotransmitter and as a neurohormone in many invertebrate species. It was first identified as a signaling molecule in the salivary glands of the octopus (Mollusca, Cephalopoda). It is, however, also present in the salivary toxin mixture of the octopus [11, 16] and in the venoms of some species of orb-weaving spiders [35]. In vertebrates, octopamine functions as a sympathomimetic drug by activating adrenergic receptors on neuronal cells which may result in a certain degree of overexcitation. In invertebrates, especially in crabs, lobsters, or crayfish, but also in mollusks, insects, and nematodes, octopamine alters many functions of the central nervous system related to behavior [41].

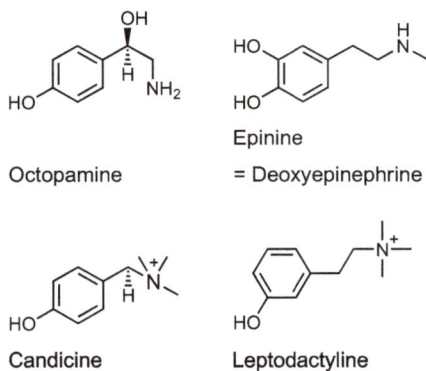

Fig. 2.30: Amines in animal toxins II.

Other phenylalkylamines that are utilized as ingredients of toxin mixtures by certain animal species are epinine and candicine (Fig. 2.30). Epinine (= deoxyepinephrine = *N*-methyldopamine) is present in skin secretions of toads [37]. Candicine (2-(4-hydroxyphenyl)-*N,N,N*-trimethylethan-1-aminium) is a quaternary ammonium salt. It is present in skin gland secretions of the smoky jungle frog (*Leptodactylus pentadactylus*) at a concentration of 45 µg/g skin. Its occurrence, however, is limited to only a few species, while its positional isomer, leptodactyline, is much more widespread among different species of amphibians [20, 45].

☠ In case of bee or wasp stings, the phenylalkylamines affect the cardiovascular system and cooperate with other venom ingredients for triggering states of shock in the target organism. In addition, they enhance pain sensations in the target organisms that are induced by other amines in the venoms, serotonin, and histamine (Fig. 2.31) [21, 47].

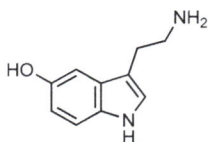

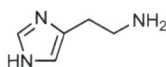

Serotonin
= 5-Hydroxytryptamine

Histamine

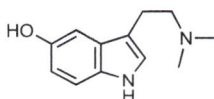

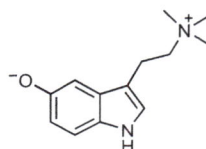

Bufotenin

Bufotenidine

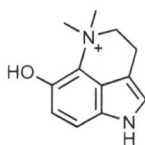

Dehydrobufotenidine

Fig. 2.31: Indole or imidazole alkylamines.

2.8.6 Indole and Imidazole Alkylamines

Biogenic amines like serotonin (5-hydroxytryptamine) and histamine (Fig. 2.31) belong to the substance class of indole or imidazole alkylamines, respectively. They are important signaling molecules in animals and humans and may be utilized as neurotransmitters or as hormones to relay information between different cells within the organism [13]. Both agents mediate many different changes in cell function depending on the type of target tissue. A universal feature, however, is the induction of pain sensations if they locally occur at high concentrations in animal or human tissues [6, 46].

Moreover, serotonin mediates the contraction of smooth muscle cells surrounding blood vessels which restrict the blood flow. Histamine increases the permeability of the vessel walls for water and small solutes which may create local edema [7]. This may provide an explanation as to why many toxic animals use serotonin as well as histamine in their venoms. It may be beneficial for an animal to use high concentrations of these mediators to manipulate monoaminergic signaling systems in target organisms, for both offensive and defensive purposes [50].

Serotonin is an ingredient in the venoms of cnidarians (Cnidaria) [33], mollusks (Mollusca) [27], and many arthropods (Arthropoda), such as spiders [51], scorpions [2], and insects (Hexapoda), e.g., in venoms of wasps and bees [8, 31, 40]. It was also detected in relatively large amounts in the skin secretions of amphibians [17, 45].

Histamine has been found widespread as serotonin in the toxin mixtures of various animal species. It was found in sponges (Porifera), e.g., 100 mg/kg in the siliceous sponge *Geodia cydonium*) [1], in cnidarians (Cnidaria) [33], in mollusks (Mollusca) [42], and in arthropods (Arthropoda). Histamine has been detected in the defense gland secretions of the centipede *Scolopendra subspinipes* [24], in the nettle hairs of butterfly caterpillars, e.g., in the genus *Dirpha* [14, 49], as well as in the venoms of bees and wasps [8, 31]. Histamine has also been detected in skin secretions of vertebrates, e.g., in frogs (genus *Leptodactylus*) [45] and fishes (*Trachinus draco*) [25].

The indole alkylamines of frog's skin gland secretions [26, 45] have been particularly well studied. These secretions may contain high concentrations of serotonin, e.g., up to 10% of the dry weight of secretions produced by the European fire-bellied toad, *Bombina bombina*. Besides serotonin, toad's skin secretions may contain bufotenin, dehydrobufotenidine, or bufotenidine (Fig. 2.31) [37, 38, 53]. Different compositions of these substances were found in the skin and parotid gland secretions of Brazilian bufonids. Differences occurred also between different individuals of the same species [36] depending on the living and feeding conditions of the animals.

> ♟ Symptoms of poisoning with indole alkylamines from the toad skin in humans are hallucinations, rapid heart rate, extreme cyanosis, auricular fibrillation, or cardiac arrest. Deliberate ingestion of toad skin secretions is practiced by drug abusers for achieving hallucinogenic effects. Accidental poisoning of domestic animals or humans upon consumption of toads may be fatal [29].

References

[1]   Ackermann D, List PH (1957) Naturwissenschaften 44(6): 184
[2]   Adam KR, Weiss C (1956) Nature 178(4530): 421
[3]   Anthoni U et al. (1989) Toxicon 27(7): 717
[4]   Anthoni U et al. (1989) Toxicon 27(7): 707
[5]   Axelsson J (1971) Annu Rev Physiol 33(1): 1
[6]   Bannister K, Dickenson AH (2016) Curr Opin Support Palliat Care 10(2): 143
[7]   Bernstein JA et al. (2017) Int J Emerg Med 10(1): 15

[8]   Bhoola KD et al. (1961) J Physiol 159: 167
[9]   Blanchet C, Dulon D (2001) Brain Research 915(1): 11
[10]   Boylan DB, Scheuer PJ (1967) Science 155(3758): 52
[11]   Cooke IR et al. (2017) Toxicity in cephalopods. In: Malhotra A, Gopalakrishnakone P (eds.) Evolution of Venomous Animals and Their Toxins. Springer, Dordrecht, Netherlands, p. 125
[12]   Cuevas N et al. (2018) Sci Rep 8(1): 7635
[13]   D'Aniello E et al. (2020) Front Ecol Evol 8(322): 587036
[14]   Dinehart SM et al. (1987) J Investig Dermatol 88(6): 691
[15]   Dvir H et al. (2010) Chem Biol Interact 187(1–3): 10
[16]   Erspamer V (1952) Nature 169(4296): 375
[17]   Erspamer V (1961) Fortschritte Arzneimittelforschung 3: 151
[18]   Erspamer V, Benati O (1953) Biochemische Zeitschrift 324(1): 66
[19]   Erspamer V, Glässer A (1957) Br J Pharmacol 12(2): 176
[20]   Erspamer V et al. (1963) Life Sci 11: 825
[21]   Fitzgerald KT, Flood AA (2006) Clin Tech Small Anim Pract 21(4): 194
[22]   Fusetani N, Hashimoto K (1987) Toxicon 25(4): 459
[23]   Goldberg AS et al. (1988) Toxicon 26(7): 651
[24]   Gomes A et al. (1982) Indian J Med Res 76: 888
[25]   Haavaldsen R, Fonnum F (1963) Nature 199(489): 286
[26]   Habermehl G (1969) Naturwissenschaften 56(12): 615
[27]   Hartman WJ et al. (1960) Ann NY Acad Sci 90(3): 637
[28]   Hashimoto Y, Okaichi T (1960) Ann NY Acad Sci 90: 667
[29]   Hitt M, Ettinger DD (1986) N Engl J Med 314(23): 1517
[30]   Hughes RD et al. (1973) Biochem J 136(4): 967
[31]   Jaques R, Schachter M (1954) Br J Pharmacol 9(1): 53
[32]   Jensen H (1935) J Am Chem Soc 57(10): 1765
[33]   Jouiaei M et al. (2015) Toxins 7(6): 2251
[34]   Kalmanzon E (2003) Toxicon 42(1): 63
[35]   Langenegger N et al. (2019) Toxins 11(10): 611
[36]   Maciel NM et al. (2003) Comp Biochem Physiol B Biochem Mol Biol 134(4): 641
[37]   Märki F et al. (1962) Biochimia Biophysica Acta 58: 367
[38]   Märki F et al. (1961) J Am Chem Soc 83(15): 3341
[39]   Owen MD, Bridges AR (1982) Toxicon 20(6): 1075
[40]   Owen MD, Sloley BD (1988) Toxicon 26(6): 577
[41]   Pflüger H-J, Stevenson PA (2005) Arthropod Struct Dev 34(3): 379
[42]   Ponte G, Modica MV (2017) Front Physiol 8: 580
[43]   Raymond Delpech V et al. (2003) Invert Neurosci 5(1): 29
[44]   Rennick BR (1981) Am J Physiol Renal Physiol 240(2): F83
[45]   Roseghini M et al. (1986) Comp Biochem Physiol C: Comp Pharmacol 85(1): 139
[46]   Shim W-S, Oh U (2008) Mol Pain 4: 29
[47]   Sommer C (2010) Handbook of Behavioral Neuroscience 21: 457
[48]   Tsubaki H, Komai T (1986) J Pharmacobio Dyn 9(9): 747
[49]   Valle JR et al. (1954) Archives Internationales de Pharmacodynamie et de Thérapie 98(3): 324
[50]   Weisel-Eichler A, Libersat F (2004) J Comp Physiol A 190(9): 683
[51]   Welsh JH, Batty CS (1963) Toxicon 1(4): 165
[52]   Whittaker VP (1960) Ann NY Acad Sci 90: 695
[53]   Wieland H et al. (1934) Justus Liebigs Annalen der Chemie 513(1): 1
[54]   Yancey PH (2005) J Exp Biol 208(15): 2819

2.9 Alkaloids

2.9.1 General

Alkaloids are a class of naturally occurring heterocyclic molecules that contain at least one nitrogen atom. Most of the alkaloids have alkaline properties when dissolved in aqueous solutions and taste bitter. Due to the high degree of diversity in N-containing molecules, it is not easy to discriminate between alkaloids in sensu strictu and other nitrogen-containing natural compounds. Natural compounds containing nitrogen in the exocyclic position (serotonin, dopamine, etc.) are usually classified as amines (see Section 2.8) rather than as alkaloids. Protoalkaloids are naturally occurring nitrogenous compounds which, in contrast to alkaloids in sensu strictu and pseudoalkaloids, are characterized by the absence of a nitrogenous heterocycle. They are secondary metabolites in plants and animals and are formed from amino acids by decarboxylation or N-methylation. Polyamines as putrescine, spermidine, and spermine are examples of protoalkaloids used by animals as toxic secondary metabolites [23, 24]. As they are also used as intermediate metabolites for the biosynthesis of complex alkaloids, the protoalkaloids with relevance for the toxicity in animals will be discussed in this chapter. The description of animal toxins with pseudoalkaloid character, whose basic structures are not directly derived from amino acids as in true alkaloids, is integrated into this chapter as well.

The alkaloids are a structurally highly diverse group of substances (Fig. 2.32). In addition to carbon, hydrogen, and nitrogen, alkaloids may also contain other elements such as oxygen, chlorine, bromine, or sulfur [56, 216]. Alkaloids in sensu strictu are biosynthesized from amino acid precursors, e.g., ornithine, lysine, phenylalanine, tyrosine, tryptophan, histidine, nicotinic acid, or anthranilic acid [57, 95, 161]. Indole alkaloids are synthesized from tryptophan, quinoline alkaloids from tyrosine or phenylalanine, and quinazoline alkaloids from phenylalanine or ornithine. Imidazole alkaloids are synthesized from histidine, and pyridine alkaloids from nicotinic acid which is a derivative of aspartic acid. Typical initial reactions in the biosynthesis of classical alkaloids are the synthesis of a Schiff base (from an aliphatic or aromatic amine and a carbonyl compound) or the Mannich reaction [26].

Alkaloids are produced as secondary metabolites by organisms of different systematic groups, especially in plants, bacteria, or fungi, but also in animals [46, 135, 136, 142, 150, 198]. Alkaloids in animals may be de novo-synthesized or derived from precursors ingested with food of plant origin (in herbivores) or animal origin (in carnivores) [202]. Alkaloids have been identified as defensive substances in sponges (Porifera) [68, 144], sea anemones and corals (Cnidaria) [18, 22, 77], moss animals (Bryozoa) [209], gastropods (Mollusca) [184], insects (Hexapoda) [21, 42], tunicates (Tunicata) [133, 151], as well as in newts, toads, and frogs (Amphibia) [51, 87]. Although the biological significance of such compounds as defensive molecules is not clear in any case, alkaloids are mainly used in passively poisonous animals to protect them from attacks by predators or overgrowth by

Imidazole

Pyrrole

Pyrrolizidine

Acridine

Piperidine

Quinazoline

Pyridine

Pyrimidine

Quinoline

Quinolizidine

Isoquinoline

Indole

Purine

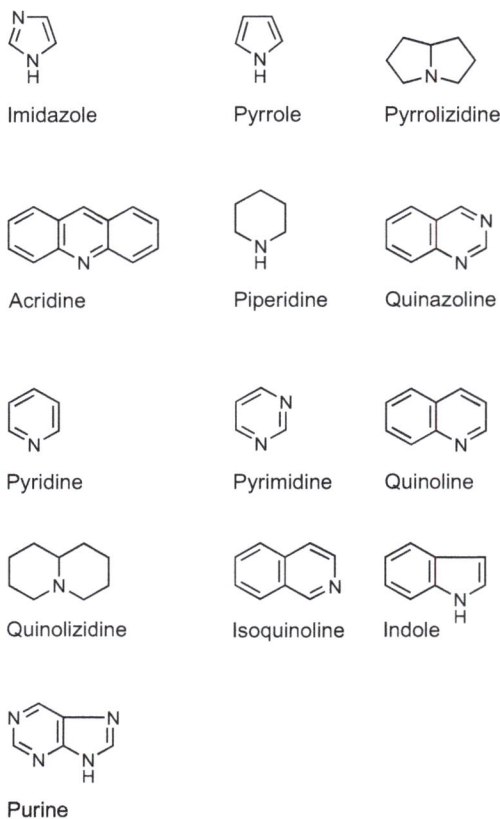

Fig. 2.32: Basic structures of alkaloids.

microorganisms. Alkaloids are only rarely used as ingredients of animal venoms. Venoms of some species of fire ants [73] or scorpions [15], however, contain alkaloids.

Most animals use alkaloids as ingredients of defensive secretions or they use them to impregnate their body surfaces or tissues. The bitter taste of most of the alkaloids serves as a warning sign for potential predators. An interesting example is the ornate moth, *Utetheisa ornatrix*. The larva and even the adults of this species are rejected by many of their potential enemies because they contain the pyrrolizidine alkaloid monocrotaline (see Section 2.9.4). Monocrotaline is originally a plant product and is accumulated in leaves and seeds of the smooth crotalaria, *Crotalaria pallida*, to fend off herbivores. *Utetheisa ornatrix* larva, however, have a preference for this host plant because they are insensitive to the toxin. They express a strong oxidase activity in their cells and sequester most of the toxin in the form of the moderately less toxic *N*-oxide (Fig. 2.33) [121]. Predators avoid *Utetheisa ornatrix* larva or adults that have accumulated this toxin. Spiders that caught these alkaloid-loaded moths in their webs have been observed to cut them out of their nets and releasing them entirely un-harmed [81]. The monocrotaline concentration in the body fluids of mated female moths is especially high because they receive extra portions of this alkaloid from the males as a nuptial gift during copulation [64]. Monocrotaline (Fig. 2.33) is per se not a strong toxin in animals but is converted to dehydromonocrotaline by the cytochrome P450 monooxygenase system in target organisms [208]. This substance functions as a highly effective cytotoxin in animal and human cells, especially in liver cells as their monooxygenase activities are rather high. Moreover, dehydromonocrotaline acts as a mutagen and a carcinogen in different types of animal and human cells.

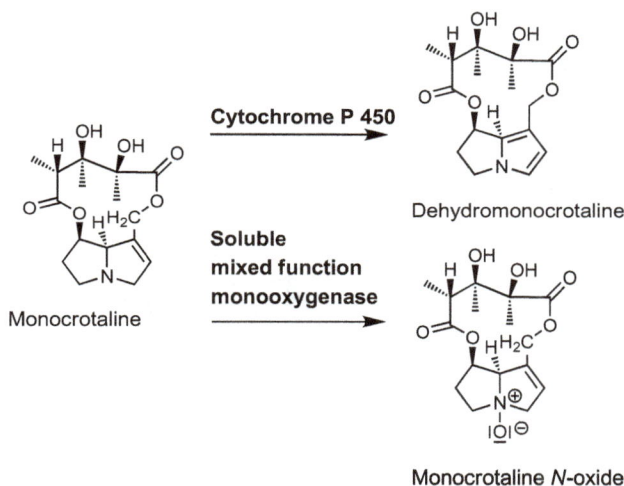

Fig. 2.33: Monocrotaline and its metabolism to toxic (dehydromonocrotaline) or moderately less toxic (monocrotaline N-oxide) derivatives.

Excretion of alkaloid-containing body fluids or impregnation of the integument with alkaloids may help some animals to prevent the growth of fungi or other microorgan-isms on their body surfaces. Such antifouling functions have also been described for indole and purine alkaloids of corals, e.g., *Paramuricea clavata* [152].

Alkaloids are only rarely used in animal venoms. Exceptions are the solenopsins (Fig. 2.34) which are relatively simple piperidine alkaloids produced by fire ants of the genus *Solenopsis*. They are used to repel or to attack vertebrates and inverte-brates. For example, isosolenopsin A (*cis*-2-methyl-6-undecylpiperidine) has been demonstrated to have strong insecticidal effects [75]. In addition to their general toxic-ity, solenopsins have a number of specific biological activities in target organisms. They are antiangiogenic agents, possibly by inhibiting the phosphoinositide 3-kinase signaling pathway [10], and they inhibit neuronal nitric oxide synthase [210]. Due to their anti-inflammatory properties, analogs of solenopsins are being tested for their potential in the treatment of psoriasis [11]. Another alkaloid, (Z)-N-(2-(1H-imidazol-4-yl)ethyl)-3-(4-hydroxy-3-methoxyphenyl)-2-methoxyacrylamide (Fig. 2.34), was recently discovered as an ingredient of the venom of the Mexican scorpion, *Megacormus gertschi* [15]. Its specific function is still unknown.

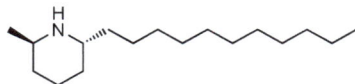

Solenopsin A in the venom
of the fire ant *Solenopsis* sp.

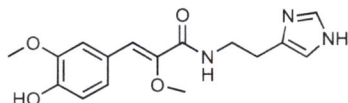

(Z)-N-(2-(1H-imidazol-4-yl)ethyl)-3-(4-hydroxy-3-methoxyphenyl)-2-methoxyacrylamide
in the venom of the scorpion *Megacormus gertschi*

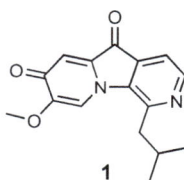

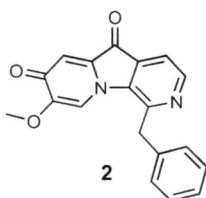

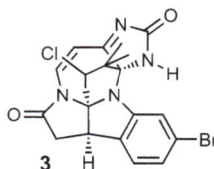

Bryozoan alkaloids: Pterocellin A (**1**) and B (**2**) of *Pterocella vesiculosa*
and securamine C (**3**) of *Securiflustra securifrons*

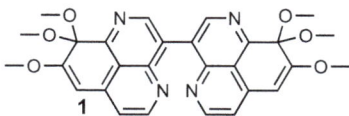

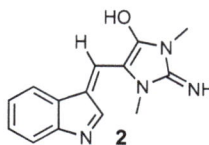

Sponge alkaloids: Suberitine C of *Aaptos suberitoides* (**1**)
and aplysinopsin of *Aplysina* sp. (**2**)

Fig. 2.34: Examples of toxic alkaloids in animals.

Although the exact physiological functions of alkaloids in other animal taxa remain unclear, it is likely that many of them have been developed as defense weapons against predation. Such an option is of great importance for sessile or slow moving organisms as they are usually not able to flee from predators or to actively defend themselves.

These defensive alkaloids belong to different classes (Fig. 2.32), e.g., acridine, dimeric aaptamine, imidazole, indole, piperidine, pyrimidine, pyridine, pyrrole, quinoline, and quinolizidine alkaloids [68]. To mention just two examples, the bryozoan *Pterocella vesiculosa* accumulates pterocellins A and B with antimicrobial properties [209], while *Securiflustra securifrons* contains securamines, halogenated hexacyclic indole–imidazole alkaloids that display various degrees of cytotoxicity against human cancer cell lines [90] (Fig. 2.34). The marine sponge *Aaptos suberitoides* accumulates the dimeric aaptamine alkaloids suberitine A, B, C, and D [123]. The Okinawan marine sponge *Aplysina* sp. contains high amounts of aplysinopsin, a tryptophan-derived alkaloid [113]. Many of these sponge alkaloids have been characterized as potent cytotoxins in animal and human cells.

Spongivorous gastropods as the Mediterranean opisthobranch *Tylodina perversa* feed on sponges rich in alkaloids. They are not only able to tolerate the alkaloids but even utilize these substances to become themselves unpalatable for potential predators. Snails feeding on the sponge *Aplysina aerophoba* accumulate brominated isoxazoline alkaloids, while those feeding on *Aplysina cavernicola* accumulate aerothionin [71, 184].

🦑 The intestinal absorption of alkaloids upon oral administration occurs efficiently and rapidly (except those containing quaternary nitrogen atoms or many free phenolic OH groups). Some alkaloids are also absorbed through moist parts of the skin, but at much lower rates. Due to their lipophilic nature, alkaloids are able to pass the blood–brain barrier. Some alkaloids have structural similarities with neurotransmitters which enable them to interact with ionotropic or metabotropic receptors in the peripheral or central nervous system or with neurotransmitter-processing enzymes. Inhibition of neurotransmitter-degrading enzymes results in accumulation of transmitters in the synaptic cleft and erratic signaling at the postsynaptic cell membrane. Depending on what kind of receptor is targeted and which alkaloid is present, interactions may trigger depolarization of the plasma membrane potential and overexcitation in neuronal cells (e.g., batrachotoxins and voltage-sensitive sodium channels) or block signal transduction through these receptors (e.g., tetrodotoxin and voltage-sensitive sodium channels). These properties of alkaloids are essential for their function as fast-acting paralyzing toxins in animal poisons and venoms because they are able to promptly change the behavior of the target organism.

2.9.2 Imidazole Alkaloids

Imidazole alkaloids contain an imidazole nucleus and are derived from the amino acid histidine. These alkaloids occur as components of defensive substance mixtures in sponges (Porifera), crust anemones (Cnidaria), and snails or slugs (Mollusca) which feed on such organisms. Spongy tissues are especially rich in imidazole alkaloids. Keramadine (Fig. 2.35) was isolated from *Agelas* sp. [141]. Hymenine as well as hymenidine are defensive alkaloids in *Hymeniacidon* sp. [111, 112]. Kermadine and hymenidine are efficient blockers of serotonin receptors in animal and human cells, while hymenine blocks alpha-adrenergic receptors. The naamidines A–G have been found in the Indonesian sponge *Leucetta chagosensis* [93]. They are cytotoxic to animal and human cells and are contact poisons for marine invertebrates. Tissues of the same sponge species hold a variety of other imidazole alkaloids. Structurally, some of them possess an unusual skeleton featuring imidazole and oxazolone rings linked via a nitrogen atom (e.g., leuchagodine A), whereas others bear an intriguing guanylurea-substituted imidazole ring (e.g., leuchagodine B). They are able to suppress production and secretion of the inflammatory mediator interleukin-6 from cells of the human monocytic leukemia cell line THP-1. Moreover, they displayed cytotoxicity against human cells with IC_{50} values of 5–10 μmol/ L [183]. Murexine (see also Section 2.8.3, Fig. 2.28) is broadly distributed as a defense substance in the hypobranchial body of marine Gastropoda, e.g., in *Bolinus brandaris* [164, 165]. Egg-laying animals are especially rich in this toxin which is also used to impregnate the egg masses which may deter predators and provide antifouling (inhibition of organismal growth on surfaces) for the eggs [19]. Murexine inhibits muscular nicotinic acetylcholine receptors in humans and animals [70].

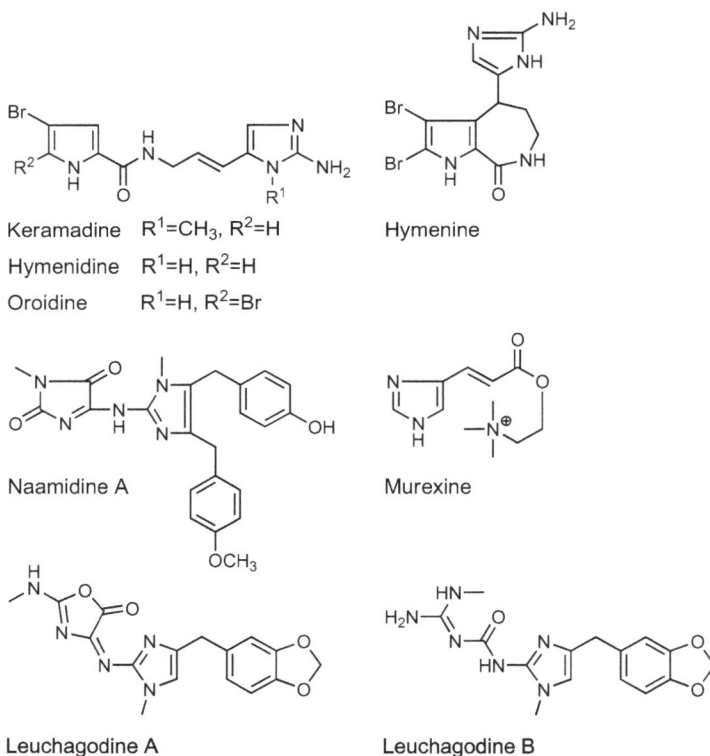

Keramadine R^1=CH$_3$, R^2=H
Hymenidine R^1=H, R^2=H
Oroidine R^1=H, R^2=Br

Hymenine

Naamidine A

Murexine

Leuchagodine A

Leuchagodine B

Fig. 2.35: Imidazole alkaloids.

2.9.3 Zoanthamine Alkaloids

Zoanthamine alkaloids or zoanthamines are natural products isolated from the polyps of marine zoanthids (crust anemones found in coral reefs). Gorgonian corals, stony corals, as well as soft corals (Cnidaria) are sessile animals (see Section 3.2.2). Their tissues contain different types of alkaloids that have cytotoxic properties with respect to animal and human cells but may have antimicrobial effects as well [77]. This indicates that the biological significance of these substances lies in the protection of the animals against predators or overgrowth. Zoanthamine (Fig. 2.36) is a typical example of alkaloids accumulated in the tissues of marine zoanthids [18, 158]. Variants of these densely functionalized heptacyclic molecules have been chemically synthesized for the selection of candidates for drug screening [212]. Zoanthamine alkaloids have been shown to exert neuroinflammatory effects mediated by microglial cells and suppress the generation of reactive oxygen species (ROS) and nitric oxide (NO) [85].

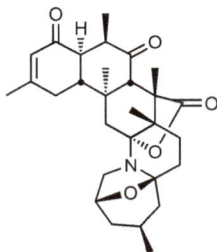

Fig. 2.36: Zoanthamine.

2.9.4 Pyrrolizidine Alkaloids

Pyrrolizidine alkaloids are a group of naturally occurring alkaloids based on the structure of pyrrolizidine. They are synthesized as secondary metabolites in more than 6,000 plant species (e.g., Boraginaceae, Asteraceae, Orchidaceae, and Fabaceae) as defense substances against insect herbivores [174, 179]. Most of these alkaloids belong to the family of necines which are derived from L-arginine or L-ornithine [162] with putrescine, spermidine, and retronecine as intermediate products. Retronecine (Fig. 2.37) is the basic structure from which most of the plant pyrrolizidine alkaloids are derived by esterification of the two hydroxyl groups. These alkaloids are stored in the plants in the form of respective *N*-oxides and generally function as feeding deterrents against herbivorous animals, especially against insects.

Some insect species (butterflies and beetles; see Section 3.2.9), however, have developed evolutionary adaptations to tolerate ingestion of plant materials containing pyrrolizidine alkaloids. Moreover, some species, especially butterflies (Lepidoptera), rear their offspring on food plants that contain high concentrations of these alkaloids. The caterpillars even accumulate the plant alkaloids in their own tissues rendering them unpalatable

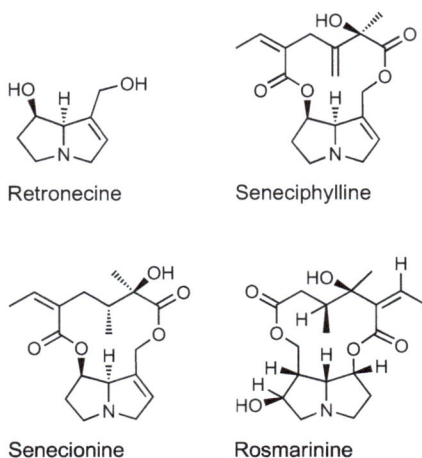

Retronecine   Seneciphylline

Senecionine   Rosmarinine   Fig. 2.37: Pyrrolizidine alkaloids of insects.

for potential predators [162]. The attractiveness of plants containing pyrrolizidine alkaloids for female butterflies is illustrated by the fact that males secrete metabolites of such alkaloids on their hair-pencils as pheromones [61]. Moths of the genus *Rhodogastria* produce defensive secretions containing pyrrolizidine alkaloids.

The caterpillars of the butterflies *Danaus chrysippus* and *D. plexippus* obtain pyrrolizidine alkaloids (see Section 2.9.4), seneciphylline, senecionine, and rosmarinine (Fig. 2.37) [180] from their host plants and store these substances within their bodies (up to 250 µg per individual [102]), making them toxic to and unpalatable for predators [62].

Larvae and adults of the glasswing butterfly (*Greta oto*) use pyrrolizidine alkaloids from nightshades (Solanaceae) to impregnate their tissues for defense against potential predators. The bitter taste of these alkaloids protects them reliably from being attacked by bullet ants (*Paraponera clavata*) [200]. The same applies to larvae of the cinnabar moth (*Tyria jacobaeae*) which obtain pyrrolizidine alkaloids from ragwort plants (genus *Senecio*). The animals are well protected against attacks from the European red wood ant (*Formica polyctena*) [125].

Ladybirds (e.g., *Coccinella septempunctata*) accumulate pyrrolizidine alkaloids (up to 5 mg/g FW) that they take up from their prey, the greenfly *Aphis jacobaeae*, which feeds on sap from ragwort (*Senecio jacobaea*) [204].

Tunicates are animals with limited mobility which use chemical weapons for defense purposes (see Section 3.2.6). The Australian ascidian *Didemnum chartaceum* accumulates different types of alkaloids. Among them are pyrrolizidine alkaloids. Examples for this substance class are the lamellarins [53], which contain also sulfated derivatives (Fig. 2.38).

🐝 Pyrrolizidine alkaloids are metabolized to hepatotoxic substances [162] which may harm target organisms by disruption of signal transduction in liver cells [195] or by forming DNA adducts [139] (see Section 2.9.1). While intoxication occurs from time to time when herbs are erroneously ingested, consumption of insects containing these alkaloids is not of any relevance for humans.

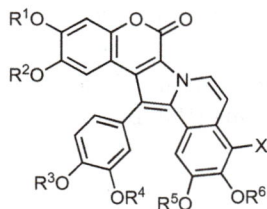

Lamellarins of the ascidian
Didemnum chartaceum

lamellarin	R^1	R^2	R^3	R^4	R^5	R^6	X
B	H	CH_3	H	CH_3	CH_3	CH_3	OCH_3
B 20-sulfate	SO_3^-	CH_3	H	CH_3	CH_3	CH_3	OCH_3

Fig. 2.38: Pyrrolizidine alkaloids of ascidians.

2.9.5 Pyridoacridine Alkaloids

Pyridoacridines are a class of colored, polycyclic aromatic natural products from marine organisms. Tetracyclic members of the pyridoacridine family of alkaloids, the cystodytines (Fig. 2.39), were initially discovered in the tunicate *Cystodytes dellechiajei* [110]. Biological testing of these substances in human and animal cells indicated that they exert anti-infectious, anticancer, neurological, psychotropic, cardiovascular, or immune effects. Analogues of such alkaloids are therefore being tested for potential use in medicine [40].

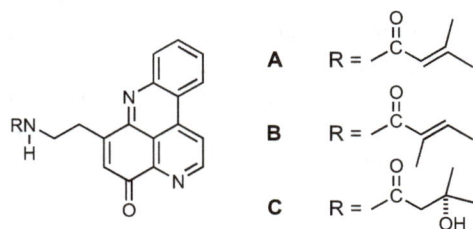

Cystodytines A-C
of the tunicate *Cystodytes dellechiajei*

Fig. 2.39: Pyridoacridine alkaloids.

2.9.6 Piperidine Alkaloids

Alkaloids with piperidine as the central building block are termed piperidine alkaloids. Insects usually acquire their defensive alkaloids from food plants and sequester them directly or as derivatives in defensive secretions or in their tissues [92, 121, 149, 202]. However, alkaloids may also be de novo generated by the animals themselves, which may be the case when alkaloids are used as venom ingredients (as observed in fire ants) [30]. Piperidine alkaloids (solenopsins; Fig. 2.34) are produced in abdominal venom glands of these ants (genus *Solenopsis*) and injected into the body of an enemy or a prey organism through a sting [114]. Due to the antimicrobial potency of solenopsins, they may also be used for impregnation of the body surface of the ant or of the eggs which may reduce bacterial adhesion and biofilm formation on these surfaces [55].

🐜 Fire ants attack in groups so that the victim generally receives a few or even many stings. In most people, stings produce only minor skin irritation. Pain or itching and skin blisters usually appear and fade within an hour. Most stings heal without any treatment. However, some human victims, especially those who have received many stings, report hallucinations and similar symptoms indicating that the venom affects the central nervous system.

✱ People with severe allergies to fire ant venom typically develop symptoms within a few minutes after being stung. Anaphylaxis occurs only rarely but can be life-threatening if left untreated.

2.9.7 Quinazoline Alkaloids

Quinazoline alkaloids are natural products which are chemically derived from quinazoline. Quinazoline alkaloids are produced as deterrents of herbivores in plants, but some microorganisms living in symbioses with animals generate them as well [127, 155, 171] and provide the respective animals with efficient chemical protection against predators. Some terrestrial animals like amphibians, however, may be able to generate such toxins on their own [88]. Especially important quinazoline alkaloids are tetrodotoxin (TTX) and its analogues like chiriquitoxin (Fig. 2.40). They occur in amphibians like the Panamanian golden frog *Atelopus zeteki* [214] or in newts (e.g., *Notophthalmus viridescens*) [213] but also in Cephalopoda and puffer fish [14]. Snakes feeding on tetrodotoxin-containing newts may become poisonous as well [199].

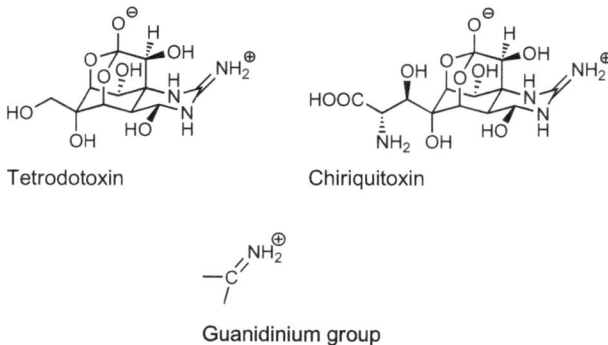

Tetrodotoxin

Chiriquitoxin

Guanidinium group

Fig. 2.40: Alkaloid toxins with guanidinium groups.

Tetrodotoxin (TTX) and its relatives carry a characteristic guanidinium group that is positively charged at physiological pH. Due to the positive charge and its exposed position, this functional group is able to enter the vestibule of voltage-activated sodium channels in target cells. The box-shaped rest of the molecule sticks to the surface of the vestibule of the ion channel and blocks it efficiently ($IC_{50} = 11.7 \pm 1.4$ nmol/L) [196].

🦑 Human intoxications and fatalities occur when people are bitten by the Australian blue-ringed octopus, *Hapalochlaena lunulata*, or ingest TTX-contaminated puffer fish (genus *Takifugu*). Thin slices of *Takifugu* muscle tissue ('*fugu*') are a culinary delicacy, especially in Japan. Due to the low TTX content in the muscle tissue, consumers feel just a tingling sensation on lips and tongue. However, if not properly prepared, TTX may be transferred from the skin or the internal organs (which are usually rich in TTX) causing death through respiratory arrest. There is no known antidote to TTX. Puffer fish raised under sterile conditions are free of TTX, indicating that the origin of the toxin in the fish is in symbiontic bacteria (e.g., *Vibrio* strains) [127].

Symptoms of TTX poisoning appear about 30 min after ingestion. First, a tingling in the mouth region is noticed. This is followed by numbness in the mouth area, general weakness with drowsiness, paleness, dizziness, and paresthesia. Later, there is a drop in blood pressure, cyanosis, coordination disorders, and progressive muscle paralysis. Death occurs after several hours from paralysis of ventilatory muscles.

Diverse substitutions in the P-loop regions of skeletal muscle and neuronal voltage-gated sodium channels have been detected in animals that accumulate TTX. Such mutations render the channels resistant to the toxin (e.g., in puffer fish, octopuses, garter snakes, and soft-shell clams) [89, 176].

2.9.8 Pyridine Alkaloids

The basic structure of pyridine alkaloids is the pyridine ring (Fig. 2.32). The biosynthesis of pyridine alkaloids starts from nicotinic acid. Nicotinic acid is itself synthesized from aspartic acid and dihydroxyacetone phosphate with quinolinic acid as an intermediate [47]. Pyridine alkaloids are present in plants and animals and generally utilized as repellents against herbivores or carnivorous predators.

Anabaseine (3,4,5,6-tetrahydro-2,3′-bipyridine) and the reduced form anabasine are alkaloid venom components produced by ribbon worms (Nemertea; see Section 3.2.5) [104, 105] and ants [197]. Anabaseine consists of a nonaromatic tetrahydropyridine ring connected to a pyridyl ring (Fig. 2.41). It is structurally similar to nicotine but possesses an imine double bond in the piperidine ring. It acts as an agonist on nicotinic acetylcholine receptors (nAChRs) in the peripheral nervous system (PNS) and in the central nervous system (CNS) of target animals [103]. While anabaseine has the ability to activate different nAChR subtypes, it has a preference for those skeletal muscle and brain subtypes that also display high affinities for the snake toxin α-bungarotoxin (BTX). In vertebrates, anabaseine may cause peripheral neuromuscular blockage and respiratory arrest. In invertebrates (e.g., potential prey organisms of nemerteans like annelids or crustaceans), anabaseine affects nAChRs in the CNS causing behavioral changes or even paralysis [105]. Besides anabaseine, the two most abundant alkaloids with pyridinyl rings occurring in the hoplonemertine *Amphiporus angulatus* are nemertilline and 2,3'-bipyridyl [104]. The latter is the main paralytic constituent which is most effective in activating nAChRs in invertebrates. Most nemerteans are characterized by an eversible proboscis carrying a stylet that is used to hunt for prey or for defense. The proboscis may actively transfer the toxins to prey or to an attacker while the thick mucus layer covering the body surface provides the animals with passive toxicity to deter potential predators [78]. The mucus may also contain tetrodotoxin [193]. Ants of the genus *Aphaenogaster* and those of the genus *Messor* are able to spray anabasine- and anabaseine-containing fluids on attackers or on prey [118, 197]. Analogues of anabaseine which target human nAChRs in the brain are currently tested as potential psychotropic drugs [105]. Cytotoxic pyridine alkaloids (Fig. 2.41) occur in sessile marine invertebrates, e.g., in sponges (Porifera), cnidarians (Cnidaria), and tunicates (Tunicata). Sponges of the genera *Xestospongia* and *Amphimedon* contain 3-alkylated pyridine alkaloids, the hachijodines [187]. Corydendramines are present in the hydrozoan *Corydendrium parasiticum* (Cnidaria) [122]. The scleractinian coral *Montipora* sp. contains montipyridine [6].

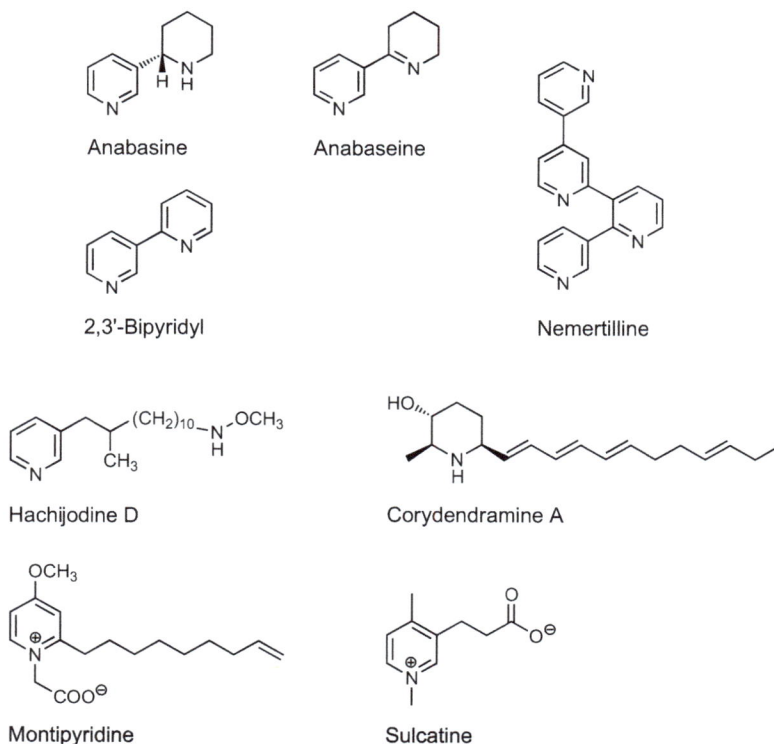

Fig. 2.41: Pyridine alkaloids.

Sulcatine is a typical defense substance accumulated in tunicates, e.g., in *Microcosmus vulgaris*. Sulcatine inhibits cell proliferation and is toxic to animal and human cells [2].

Some amphibian species, especially the poison dart frogs (Dendrobatidae; see Section 3.2.12), impregnate their tissues, body fluids, and skin secretions with alkaloids that they obtain from their food, especially from ants and mites [157]. Alkaloid-loaded frogs become entirely unpalatable for potential predators [168]. Moreover, due to the antimicrobial potencies of their alkaloids, the frogs are well protected against skin infections [126]. Besides piperidines and decahydroquinolines, the alkaloid mixtures of these frogs include batrachotoxins, steroidal alkaloids which are potent and selective activators of voltage-sensitive sodium channels [106], the histrionicotoxins which are activators of the nicotinic acetylcholine receptor (AChR) and accelerate desensitization of the receptor [33], the pumiliotoxins which activate sodium channels and enhance myotonic and cardiotonic activities [50], and epibatidine which is a chlorinated alkaloid that is secreted by skin glands of the Ecuadoran frog *Epipedobates anthonyi* when it feels threatened (Fig. 2.42). Epibatidine has a strong antinociceptive effect in vertebrates [177]. It is more than 200-fold more potent than morphine in suppressing pain sensations [51]. Molecular targets of epibatidine, however, are nicotinic acetylcholine receptors (nAChRs) that modulate ascending nociceptive information processing in the central nervous system [186]. Epibatidine preferentially activates α4β2 nAChRs in GABAergic and glycinergic neurons which mediate tonic inhibitory effects on ascending nociceptive signals [181, 190]. Synthetic analogues of epibatidine have been tested for their use in medicinal applications, but human clinical trials were stopped due to the unacceptable side effects of the drugs [190]. The frog species that secrete these alkaloids have evolved ion channel proteins with certain mutations that render them insensitive to the toxins [215].

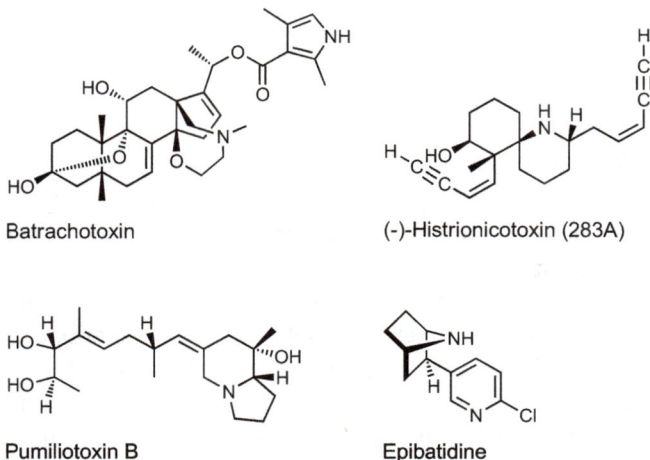

Batrachotoxin

(-)-Histrionicotoxin (283A)

Pumiliotoxin B

Epibatidine

Fig. 2.42: Pyridine alkaloids in dendrobatid frogs.

2.9.9 Pyrimidine Alkaloids

The pyrimidine (1,3-diazine) ring carries two nitrogen atoms in the six-membered ring at positions 1 and 3 (Fig. 2.32). The biosynthesis of pyrimidines starts from L-aspartic acid and carbamoyl phosphate with orotic acid as an intermediate [25]. Three of the four types of nucleobases are pyrimidine derivatives: cytosine (C), thymine (T), and uracil (U). The latter is most likely precursor of all pyrimidine alkaloids. Examples of these alkaloids used as predator repellents by sponges (Fig. 2.43) are the fluorouracil derivatives of *Phakellia fusca* [207] or spongouridine and spongothymidine of *Tectitethya crypta* [79]. Analogues of the latter two substances are used as anticancer agents [145].

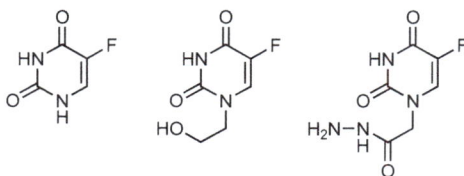

Fluorouracil derivates of the marine sponge *Phakellia fusca*

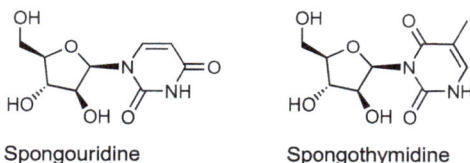

Spongouridine   Spongothymidine

Fig. 2.43: Pyrimidine alkaloids.

2.9.10 Quinoline Alkaloids

Quinoline alkaloids have the basic structure 1-benzopyridine in common (Fig. 2.32). Additional ring systems may be annealed. The precursors for the biosynthesis of quinoline alkaloids in plants are in most cases anthranilic acid or the aromatic amino acid tryptophan. Animals may synthesize quinoline alkaloids by themselves or acquire these substances by feeding on plant materials. Quinoline alkaloids have been found in defensive secretions of stick insects (Phasmatodea) and beetles (Coleoptera), e.g., fireflies (Lampyridae) and ladybirds (Coccinellidae). The venom of *Scolopendra* sp. (Chilopoda) also contains quinoline alkaloids. Sponge (Porifera) and sea squirt (Tunicata) tissues may also contain these alkaloids to deter potential predators [134, 136].

Quinoline (Fig. 2.44) is used as a component of secretions from prothoracal defensive glands in stick insects (Phasmatodea), e.g., in *Oreophoetes peruana*. The compound proved irritant or repellent in assays with ants, spiders, cockroaches, and frogs [65].

Jineol is a quinoline alkaloid found in the venom of the Chinese red-headed centipede, *Scolopendra subspinipes mutilans* (Chilopoda). It has the ability to trap free radicals and has strong antioxidant effects on low-density lipoprotein complexes. It may be suitable as an agent to prevent atherosclerosis [211]. Pyrroloiminoquinoline alkaloids with anti-tumor properties like the discorhabdines were isolated from the deep-sea sponge *Batzella* sp. [67]. Uranidine from the Australian sponge *Oceanapia* sp. was discovered to be an inhibitor of the mycobacterial mycothiol *S*-conjugate amidase [143]. The tunicates *Aplidium tabascum*, *Clavelina lepadiformis*, and *Clavelina picta* protect themselves against predators by accumulating strongly cytotoxic decahydroquinoline alkaloids, the lepadines [54]. It has been recently discovered that lepadine A mediates immunogenic cancer cell death [146]. Tissues of the Australian ascidian *Botrylloides perspicuus* contain the cytotoxin perspicamide A [131]. The firefly *Photuris versicolor* renders itself and its eggs unplatable for predators by tissue accumulation or impregnation of egg clutches, respectively, with mixtures of toxins including the quinoline alkaloid *N*-methylquinolinium 2-carboxylate [80]. The tricyclic alkaloid gephyrotoxin has been

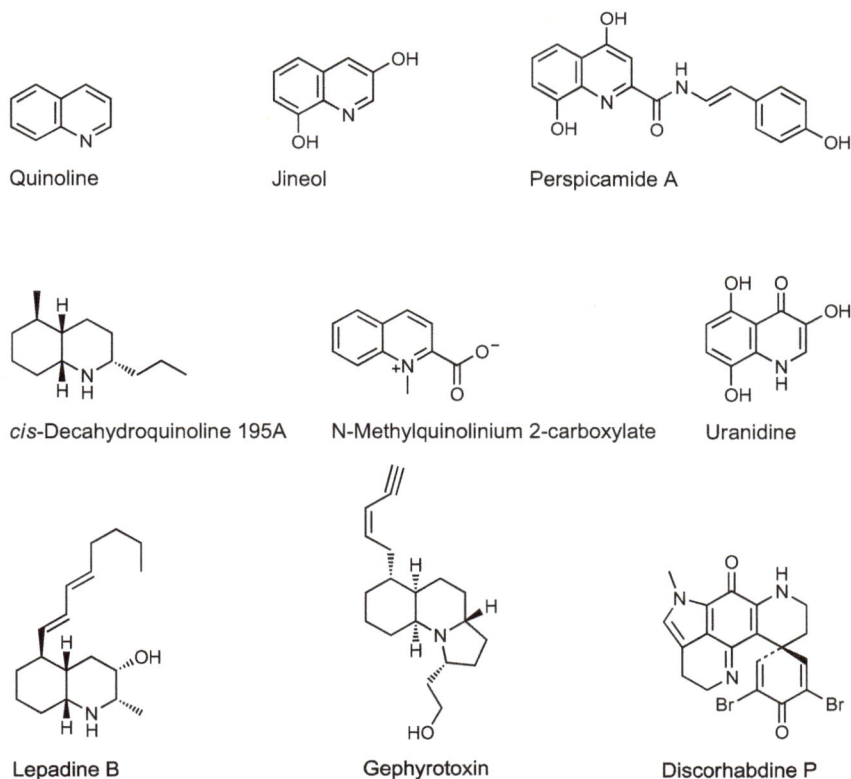

Quinoline   Jineol   Perspicamide A

cis-Decahydroquinoline 195A   N-Methylquinolinium 2-carboxylate   Uranidine

Lepadine B   Gephyrotoxin   Discorhabdine P

Fig. 2.44: Quinoline alkaloids.

isolated from the skin of the Colombian harlequin poison frog, *Oophaga histrionica*. Gephyrotoxin is an antagonist at muscarinic acetylcholine receptors [52].

Several decahydroquinoline alkaloids have been detected as components of skin gland secretions in amphibians [51]. These alkaloids are common in neotropical dendrobatid frogs. Furthermore, the *cis*-decahydroquinoline 195A (Fig. 2.44) occurs also in the skin of mantellid frogs (Mantellidae) and toads (Bufonidae). These alkaloids are obtained from myrmicine ants which are the principal food source for these anurans, sequestered in tissues, and secreted in glandular liquids. They may have antimicrobial and antifungal functions [126].

2.9.11 Quinolizidine Alkaloids

The basic molecular structure occurring in quinolizidine alkaloids (Fig. 2.45) is the quinolizidine ring system (Fig. 2.32). Biosynthesis of these alkaloids is probably confined to plant cells and starts from L-lysine [201]. Quinolizidine alkaloids that are

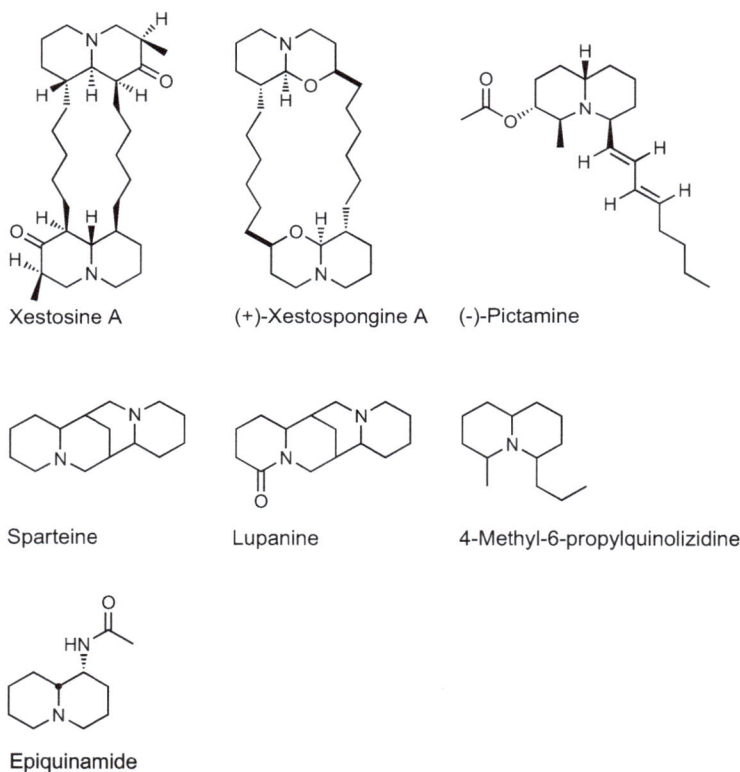

Xestosine A   (+)-Xestospongine A   (-)-Pictamine

Sparteine   Lupanine   4-Methyl-6-propylquinolizidine

Epiquinamide

Fig. 2.45: Quinolizidine alkaloids.

present in animals, e.g., in sponges (Porifera), insects (Hexapoda), tunicates (Tunicata), and amphibians (Amphibia), are likely to originate from plants and are supposedly taken up via the food web. They serve as defensive substances against potential predators.

Marine sponges of the genus *Xestospongia* contain, among others, the bisquinolizidine alkaloids xestosine A and xestospongine A [97, 138]. Other sponge species accumulate petrosines. These substances have been characterized as ichthyotoxins (fish poisons). They function as vasodilators in mammals and inhibit the viral reverse transcriptase [29, 83]. Several aphid species have been described, which are specialized to feed on alkaloid-rich lupins or broom plants. These animals accumulate quinolizidine alkaloids (up to 0.5 mg/g FW) which they obtain from the phloem sap of their host plants [203]. Among the alkaloids in the aphids are 17-oxosparteine, sparteine, and 12,13-dehydrosparteine (Fig. 2.45).

Besides other alkaloids, a quinolizidine alkaloid has been identified in the tunicate *Clavelina picta*. (–)-Pictamine blocks neuronal nicotinergic acetylcholine receptors containing alpha-4 or alpha-7 subunits in animal and human cells [188].

Numerous different types of quinolizidine alkaloids occur in defensive skin secretions of dendrobatid frogs. These substances are venom components of ants (e.g., fire ants of the genus *Solenopsis*) which are a major food source of these frogs [178]. Examples are 4-methyl-6-propylquinolizidine [99] and the nicotinic agonist epiquinamide (Fig. 2.45) from an Ecuadoran poison dart frog, *Epipedobates tricolor* [72].

2.9.12 Isoquinoline Alkaloids

Isoquinoline alkaloids form one of the largest groups of alkaloids. They are generally derived from the amino acid tyrosine and have either an isoquinoline or an 1,2,3,4-tetrahydroisoquinoline basic structure to which further ring systems may be attached (Fig. 2.46). These alkaloids have been detected in moss animals (Bryozoa), tunicates (Tunicata), amphibians (Anura), and humans. Examples of isoquinoline alkaloids in moss animals are caulibugulone A in *Caulibugula inermis* [137] or the perfragilins A and B in *Membranipora perfragilis* [44]. All these compounds have cytotoxic effects on human or animal cells. Very complex alkaloids comprising three isoquinoline rings and sulfur atoms have been detected in tunicates (Tunicata) from Thailand, e.g., ecteinascidin (Fig. 2.46) from *Ecteinascidia turbinata* [117]. These molecules are products of symbiontic bacteria and accumulate in the host animals. They have cytotoxic properties in human and animal cells and are being tested for their suitability as potential anticancer drugs [13]. Ecteinascidin (trabectedin) is an approved drug for the treatment of adult patients with advanced soft tissue cancer or ovarian cancer [82].

A surprising finding was that morphine (Fig. 2.46) occurs in skin glands of toads (*Rhinella marina*) [148]. Traces of de novo-synthesized morphine were also detected in human cells [156], but their role remains unclear.

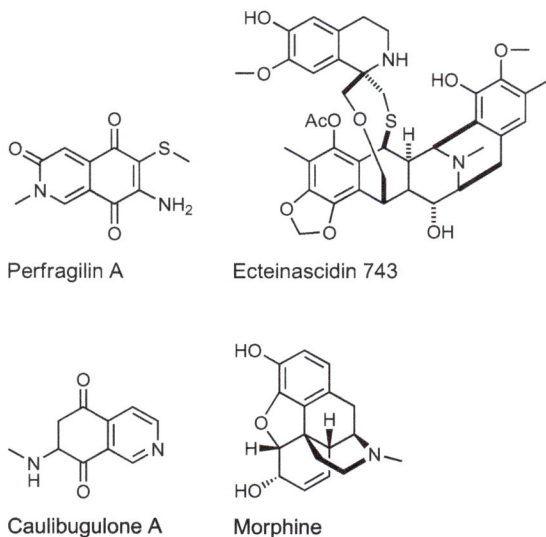

Perfragilin A   Ecteinascidin 743

Caulibugulone A   Morphine

Fig. 2.46: Isoquinoline alkaloids.

2.9.13 Indole Alkaloids

Another group of very diverse alkaloids occurring in various organisms are the indole alkaloids which may contain one or several indole ring systems. They are synthesized from L-tryptophan. Some members of this molecule family are synthesized in animals. An example is pseudophrynamine A (Fig. 2.47) in the skin of the red-backed toadlet *Pseudophryne coriacea* which is native to eastern Australia [49]. Some marine gastropod species, e.g., *Stylocheilus striatus*, feed on stinging seaweed (*Lyngbya majuscula*), a filamentous species of cyanobacteria, which contains indole alkaloids, e.g., lyngbyatoxins. These sea hares accumulate lyngbyatoxin A in their tissues and release this toxin with their defensive ink when threatened [35].

☠ Intensive skin contact with sea hares can cause the same symptoms as direct contact with *Lyngbya*, namely itching, erythema, skin blisters (seaweed dermatitis), and systemic reactions such as headaches and drowsiness [98].

Many indole alkaloids have been detected in sponges (Porifera; see Section 3.2.1). They seem to be of defensive relevance for these animals and are mainly directed against fish. They may also have antifungal functions. The indole alkaloids generally contain multiple indole ring systems (Fig. 2.47). Examples are the hamacanthins A and B isolated from *Hamacantha* sp. [86] and topsentin A from *Haliclona lacazei* [37]. Relatively simple indole alkaloids, 6-bromoindoles, have been isolated from *Pseudosuberites hyalinus* [160]. Some

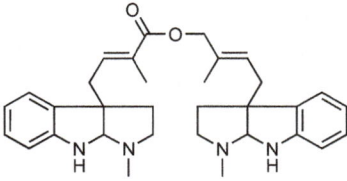

Pseudophrynamine A

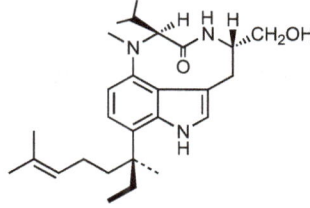

Lyngbyatoxin A
(Teleocidin A-1)

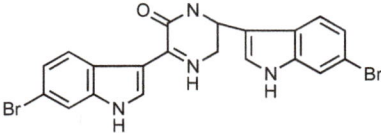

Hamacanthin A

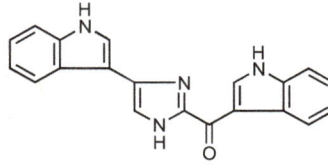

Topsentin A

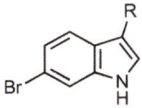

6-Bromoindoles
1 R=CH₂CN
2 R=CH₂CONH₂
3 R=CH₂COOCH₃
4 R=COOH
5 R=CHO

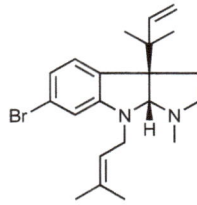

Flustramine A

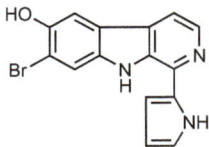

Eudistomin A

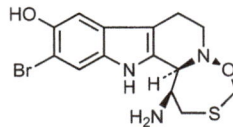

Eudistomin C

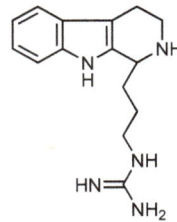

Trypargine

Fig. 2.47: Indole alkaloids.

of these indole alkaloids of sponges have been described as antiproliferative or cytostatic agents in animal cells [37]. Moss animals (Bryozoa) produce many different indole alkaloids as defensive substances [45], among them the flustramines in *Flustra foliacea* which function as antimicrobials [36, 116] by interfering with the communication between bacterial cells through the *N*-acyl-homoserine lactone-dependent quorum-sensing system [153]. Some gastropods, e.g., *Peringia ulvae* or *Gibulla ardens*, as well as the sea star *Asterias rubens* which feed on moss animals are able to accumulate these indole alkaloids in their own tissues [154]. Indole alkaloids are directly synthesized in several species of the Muricidae family of predatory marine snails. An example is 6-bromo-2-methylthio-indoxyle-3-sulfate (tyrindoleninone) which is a precursor of the dye 6,6′-dibromoindigotin (Tyrian purple) that is isolated from the hypobranchial glands of these snails [48]. It is used by egg-laying females as an antimicrobial agent to impregnate the egg masses [20]. This substance has been found to induce apoptosis in human cancer cells [63].

Several different forms of highly cytotoxic indole alkaloids have also been isolated from tunicates (Tunicata; see Section 3.2.6). It is assumed that they accumulate these toxins to repel fishes. Eudistomins and trypargine (Fig. 2.47) have been detected in *Eudistoma olivaceum* and *Ritterella sigillinoides*. These substances have antimicrobial, antifungal, as well as antiviral properties, which may act as cytotoxins or interfere with calcium signal transduction in animal and human cells [1, 58, 159, 191].

2.9.14 Purine Alkaloids

The central structure of purine alkaloids is purine (7H-imidazo [4,5-d]pyrimidine), a compound with a six-membered pyrimidine ring and a five-membered imidazole ring. Purine alkaloids are generated by plants or microorganisms and accumulate in animals. Examples are saxitoxins and gonyautoxins (Fig. 2.48).

Since purine alkaloids are not derived from amino acids but from adenosine or guanosine monophosphates via xanthosine [12], they are considered to be 'pseudoalkaloids' by some authors although parts of the bases are derived from amino acids as well. Besides plants and algae, cyanobacteria and dinoflagellates (Dinophyceae) are able to synthesize purine alkaloids [9, 39, 41, 192]. Dinoflagellates are able to increase the synthesis rate of these toxins in the presence of predators or grazers, e.g., copepods [4, 169]. Grazers or filter feeders ingesting these dinoflagellates may accumulate high amounts of these toxins due to evolutionary development of resistance against these alkaloids. Mutations in the amino acid sequences of voltage-gated sodium channels, the main targets of purine alkaloids [38], have rendered these channels insensitive to the toxins [31]. Saxitoxins and similar compounds block the channel pore of these ion channels, which results in the inability of excitable human or animal cells (nerve or muscle cells) to generate action potentials. The consequence is paralysis which is one of the symptoms of 'paralytic shellfish poisoning' (PSP) caused by the ingestion of animals (crabs, mussles, etc.) containing these toxins. Even in plankton-eating fish like mackerels

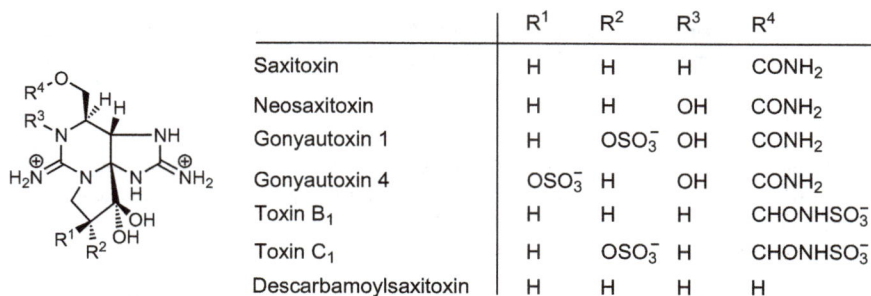

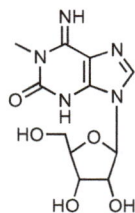

	R^1	R^2	R^3	R^4
Saxitoxin	H	H	H	CONH$_2$
Neosaxitoxin	H	H	OH	CONH$_2$
Gonyautoxin 1	H	OSO$_3^-$	OH	CONH$_2$
Gonyautoxin 4	OSO$_3^-$	H	OH	CONH$_2$
Toxin B$_1$	H	H	H	CHONHSO$_3^-$
Toxin C$_1$	H	OSO$_3^-$	H	CHONHSO$_3^-$
Descarbamoylsaxitoxin	H	H	H	H

Doridosine

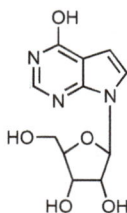

7-Deazainosine

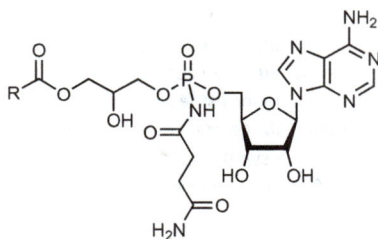

Dinogunellin

R = stearic or palmitic acid

Fig. 2.48: Purine alkaloids.

(*Scomber scombrosus*), saxitoxin accumulates to high concentrations which may occasionally cause poisoning and death of humpback whales (*Megaptera novaeangliae*) [170]. Thus, the purine alkaloids are usually subsumed under the term 'paralytic shellfish toxins' [175]. The LD$_{50}$ of saxitoxin (p.o.) in guinea pigs is 135 µg/kg BW. A dose of 1 mg of saxitoxin is deadly for humans [130].

⚹ Symptoms of poisoning from consumption of saxi- or gonyautoxin-containing marine animals appear rapidly (within 30 min). They begin with tingling and numbness in the lips, then in the tongue, throat, and extremities. Numbness in the arms, legs, and neck area develops gradually. Finally, dizziness and general weakness sets in. These symptoms may be accompanied by increased heart rate, intense thirst, muscle pain, and visual disturbances including temporary blindness. Death occurs from paralysis of ventilatory muscles. If the victim survives the first 12 h, the symptoms subside after 3–4 days.

📧 The first measures to be taken after the occurrence of paresthesia upon eating marine animals containing saxi- or gonyautoxin are induction of vomiting, gastric lavage, and activated charcoal instillation. Intubation is required if swallowing and breathing difficulties occur. There are no specific antidotes to the toxins available.

Besides the paralytic shellfish poisons, invertebrates may contain purine alkaloids with different target mechanisms in animals and humans. Some sponge species, e.g., *Tedania anhelans*, as well as nudibranch gastropods, e.g., *Montereina nobilis*, contain doridosine (N^1-methylisoguanosine) (Fig. 2.48). Doridosine is a strong agonist at adenosine receptors and relaxes vascular smooth muscle cells. It has an antihypertensive effect in animals and humans [76].

Tunicates contain purine alkaloids as deterrents against predators. An example is 7-deazainosine in *Aplidium pantherinum* [108]. Dinogunellin, a lysoglycerophosphatide bound to an adenosine or amidoasparaginyl residue, is present in the roe of some marine or freshwater fish species, e.g., in the prickleback, *Stichaeus grigorjewi*, as well as in the sculpin, *Scorpaenichthys marmoratus*, and in eel blood (*Anguilla anguilla*) [129].

🐟 Ichthyotoxins like dinogunellin may cause inflammation when they come into contact with mucous membranes of other animals or humans. If swallowed they cause diarrhea, nausea, and abdominal pain. They may also cause fast and irregular pulse, cyanosis, fever, and dizziness. Countermeasures after consumption of ichthyotoxins include immediate gastric emptying and administration of activated charcoal. There is no specific antidote, so the symptoms of poisoning mentioned earlier are treated symptomatically. Victims usually recover from ichthyotoxic poisoning after a few days.

2.9.15 Terpene Alkaloids

Terpene alkaloids are pseudoalkaloids as they are not derived from amino acids. Their basic structures are mono-, sesqui-, or diterpenes. The nitrogen sources are methylamine, ethylamine, or β-aminoethanol [43]. Most natural terpene alkaloids occur in plants and only some are found in animals (Fig. 2.49). Examples are the sesquiterpene alkaloids oceanapamine isolated from the marine sponge *Oceanapia* sp. that has antibiotic properties [28] and the sollasins of *Characella pachastrelloides* which may have antifungal properties and inhibit the growth of cancer cells [107]. Diterpene alkaloids have been identified in tunicates of the genus *Lissoclinum*. The haterumaimides as well as chloro- and dichlorolissoclimides may have antitumor effects in animals and man [189].

Oceanapamine   Sollasin b

Haterumaimide F   Dichlorolissoclimide

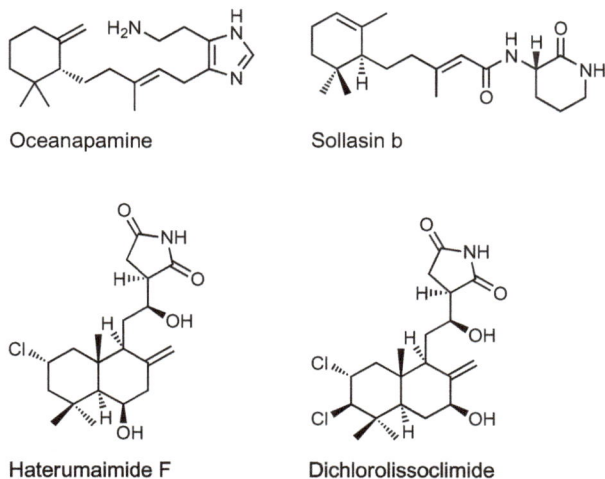

Fig. 2.49: Terpene alkaloids.

2.9.16 Steroid Alkaloids

Common to all steroid alkaloids is the sterane ring system (Figs. 2.16 and 2.50) usually carrying a side chain at carbon 17. In animals, the triterpene squalene undergoes cyclization which results in the formation of cholesterol via lanosterol (Fig. 2.15) [206]. Plants, however, generally use cycloartenol as an intermediate of cholesterol biosynthesis [167]. Cholesterol-derived steroid hormones (such as progesterone, aldosterone, cortisol, and testosterone) are common in animals and humans, but cholesterol has additional functions in these organisms. The ability to synthesize cholesterol has been clearly shown for vertebrates, tunicates, annelids, nemertines, echinoderms, and some mollusks. However, also animals of other invertebrate taxa, e.g., sponges and cnidarians, typically contain several types of sterols. Not less than 74 different sterols have been identified in the marine sponge *Axinella cannabina* [96]. Whether these are newly synthesized within the animal tissue or are generated by symbiontic microbes is not clear.

Addition of *N*-heterocyclic compounds to the D ring of cholesterol (Fig. 2.15) or the extension of the A ring with nitrogen may occur during the formation of some steroid alkaloids which are, due to this mode of biosynthesis, pseudoalkaloids. While most types of steroidal alkaloids present in animals are synthesized by eukaryotic microorganisms or plants and taken up via the food chain, others may be autonomously synthesized by the animals [84, 185].

Some species of marine sponges contain 'corticium alkaloids' which are 3-amino -23-imino steroids. Examples are plakinamine A and lokysterolamine A (Fig. 2.51) that occur in sponges of the genera *Corticium* or *Plakina*, respectively [100, 115, 119]. Some

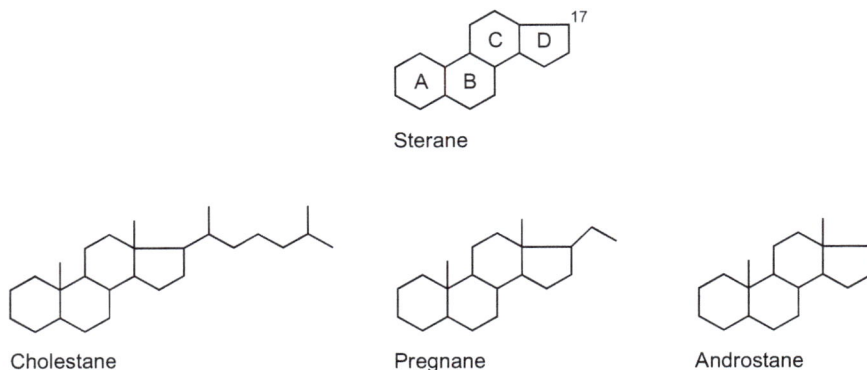

Sterane

Cholestane   Pregnane   Androstane

Fig. 2.50: Basic structures of steroid alkaloids.

of these compounds showed antiproliferative activities in human colon cancer cell lines with IC_{50} values between 1.4 and 11.5 µmol/L [182].

Cephalostatins and ritterazines have been found in acorn worms (Hemichordata) of the Cephalodiscidae family, *Cephalodiscus gilchristi* and *Ritterella tokioka* [140]. These substances are toxic to animal and human cells. Synthetic analogues are being tested for potential use as cytostatic drugs in cancer therapy [74].

Batrachotoxins (Fig. 2.51) are steroidal alkaloids which occur in skin secretions of South American poisonous frogs (Dendrobatidae) like *Phyllobates terribilis* and other species of this genus. Up to 20 µg of batrachotoxin can be extracted from the skin of a single frog [205]. Batrachotoxins have also been isolated from the skin and feathers of certain New Guinean bird species (*Pitohui dichrous* and *Ifrita kowaldi*). The poison protects the birds from being attacked by predators [16, 59, 60]. Batrachotoxins are potent and selective activators of voltage-sensitive sodium channels [106]. They induce rapid overexcitation in the neuromuscular junction and spastic paralysis of the skeletal muscles in target organisms [7, 38].

☠ Batrachotoxin poisoning in humans is a very rare event. Oral ingestion or contact with mucous tissues result in irritation and inflammation at the exposure site. However, fatal outcomes of poisoning may occur if batrachotoxin gets into skin wounds. The LD_{50} in mice has been determined to be 0.5–2 µg/kg BW. Cause of death is the spastic paralysis of the respiratory muscles.

The inner organs, the skin, and skin gland secretions of salamanders (Caudata/Urodela; see Section 3.2.12) contain substances that protect them from being overgrown by microorganisms [124, 194] or being eaten by predators [32, 132]. Among different other ingredients of these secretions, steroid alkaloids are major components (Fig. 2.51): samandarine, samandarone, samandaridine, samandenone, and samandinine. A nitrogen atom (probably from glutamine) is inserted into the A ring of these steroid alkaloids. In many of these compounds (e.g., in samandarine), there is an oxygen atom forming a

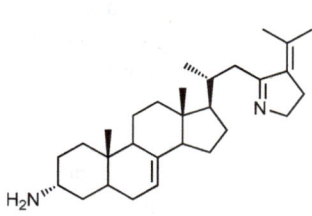

Plakinamine A

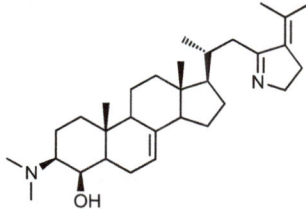

Lokysterolamine A

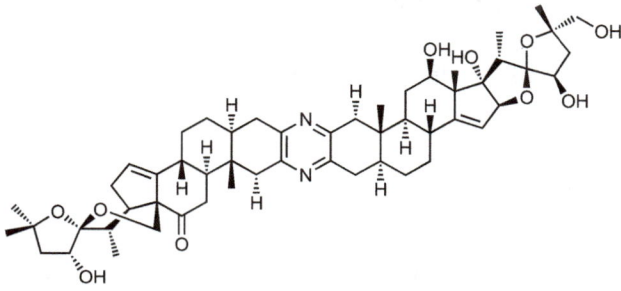

Cephalostatin 1

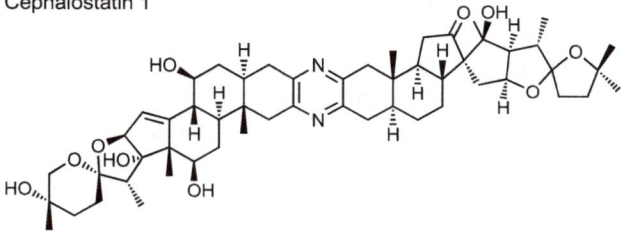

Ritterazine A

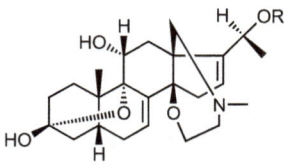

Batrachotoxin   R =

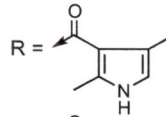

Homobatrachotoxin   R =

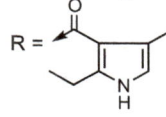

Batrachotoxinin A   R = H

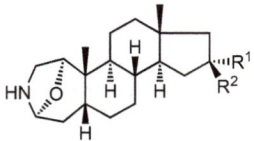

Samandarine   R¹ = H, R² = OH   Samandaridine
Samandarone   R¹ + R² = O

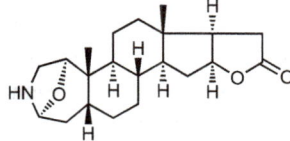

Fig. 2.51: Steroid alkaloids.

bridge across the A ring. The total amount of steroid alkaloids that can be extracted from a single fire salamander (*Salamandra salamandra*) may sum up to 35 mg [17]. The LD_{50} values (mouse, s.c.) of each of these compounds are 1.2–1.5 mg/kg BW [17, 87, 124].

☠ The salamander alkaloids are strong analeptics which cause severe neuronal overexcitation and spastic paralysis in skeletal muscles. In humans, samandarine causes high blood pressure, cardiac arrhythmias, hemolysis, and respiratory arrest due to paralysis of ventilatory muscles. It is a powerful local anesthetic. Attempts to utilize analogues as potential drugs in humans failed due to the extreme small therapeutic concentration range, so they are not used in medicine.

🐈 Deaths of cats and dogs after just chewing or preying on salamanders have been described [69].

2.9.17 Polyamine Alkaloids

Polyamines and polyamine alkaloids are characterized by the presence of two or more amino groups as part of an otherwise aliphatic hydrocarbon chain. The several amino groups are usually separated by three or four methylene units. They have been isolated from all kinds of living organisms such as bacteria, protozoa, fungi, plants, sponges, corals, insects, arthropods, fishes, mammals, and also from humans. Polyamine toxins are known as ingredients of poisons of soft corals [24] and of venoms in spiders, wasps (philanthotoxins), and snakes [3, 66, 172]. Open-chained N^5,N^{10},N^{10}-trimethylated acylspermidine derivatives (Fig. 2.52) were discovered in the soft coral *Sinularia* sp. These sinularins may inhibit certain transport ATPases and possess potent cytotoxic activities. They induce apoptosis in animal and human cells [147].

2.9.18 Azaphenalene Alkaloids

Azaphenalene alkaloids are tricyclic aromatic heterocycles. Other than in the true alkaloids in which the carbon skeleton is derived from amino acids, the carbon skeletons of the pseudoalkaloids occurring in the defensive secretions of ladybirds (Coccinellidae; see Section 3.2.9) are derived from fatty acids. Stearic acid is the precursor of coccinellin (in the seven-spot ladybird *Coccinella septempunctata*) and harmonine (in the Asian ladybird *Harmonia axyridis*), while myristic acid is the precursor of adaline (in the two-spot ladybird, *Adalia bipunctata*) [94]. The nitrogen is secondarily introduced from glutamine by an amidotransferase. The azaphenalene alkaloids may form complex ring structures [8], while others are aliphatic [5]. A biosynthetic pathway for the circular azaphenalene alkaloids involving oxidation steps of the fatty acid precursors has been proposed [8, 94] (Fig. 2.53).

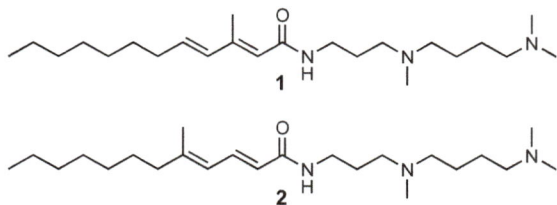

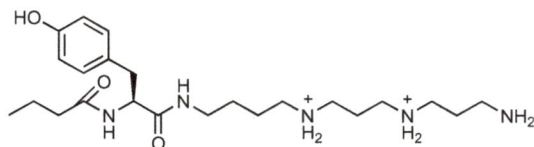

Sinularins isolated from the soft coral *Sinularia* sp.

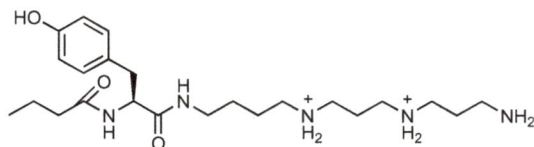

Philanthotoxin 433 of the digger wasp *Philanthus triangulum*

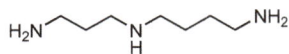

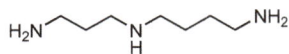 Cadaverine

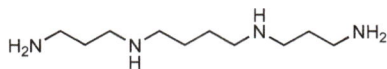 Putrescine

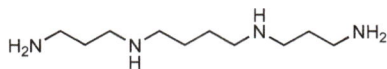 Spermidine

Spermine

Fig. 2.52: Polyamines which may also be considered as protoalkaloids.

Various azaphenalene alkaloids are found in the body fluids, tissues, and the defensive gland secretions of beetles. Especially ladybirds (Coccinellidae) are rich in these substances [109]. When the animals feel threatened, they release oily, alkaloid-enriched secretions from defense glands ('reflex bleeding' [91]) to deter potential attackers. Defense secretions are also used by female beetles to impregnate their eggs to protect them from predators. This ability is the reason why ladybirds or their eggs are generally not considered prey by other insects, reptiles, or birds [27, 128, 166] with only a few exceptions [173].

The main ingredients of defense substance mixtures in ladybirds are the perhydro-9b-azaphenalene derivatives precoccinellin and its *N*-oxide coccinellin (Fig. 2.54). They occur, for example, in *Coccinella undecimpunctata*, *C. septempunctata*, and *C. quinquepunctata*. The stereoisomers of precoccinellin, myrrhine, and hippodamine, as well as additional similar compounds occur in related species from the genera *Anisosticta*, *Hippodamia*, *Myrrha*, and *Propylea*. The bicyclic homotropane derivative adaline is produced by species of the genus *Adalia*.

A

B

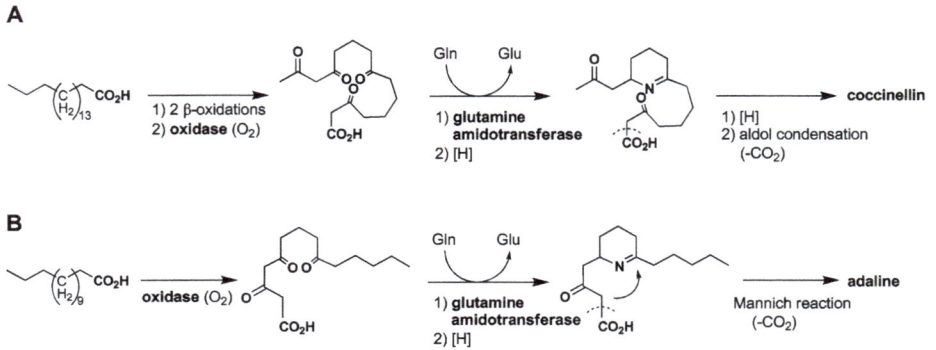

Fig. 2.53: Suggested synthesis pathway of defensive alkaloids in ladybirds: (A) Biosynthesis of coccinellin in *Coccinella septempunctata;* (B) Biosynthesis of adaline in *Adalia bipunctata*. Based on investigations using ^{14}C-labeled sodium acetate, it was initially assumed that biosynthesis of coccinellin started from a β-polyketo-acid generated from seven acetate units. Subsequent studies using ^{3}H-labeled stearic acid have resulted in an improved model which points to fatty acids as precursors for azaphenalene alkaloids [8, 94] (image adapted from [94]).

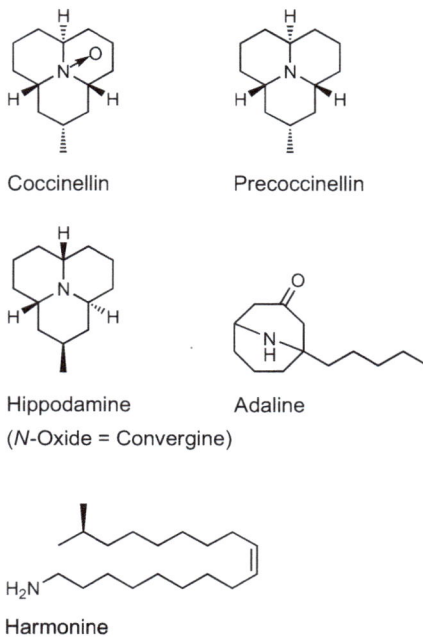

Coccinellin   Precoccinellin

Hippodamine   Adaline
(*N*-Oxide = Convergine)

Harmonine

Fig. 2.54: Azaphenalene alkaloids in ladybirds (Coccinellidae).

While the defense substances of ladybirds have repellent effects on ants, spiders, amphibians, reptiles, and birds, they are harmless for humans apart from their unpleasant odor due to the methoxypyrazine ingredients [34] and bitter taste due to the presence of azaphenalene derivatives. Precoccinellin, hippodamine, or coccinellin are able to block nicotinic acetylcholine receptors (nAChRs) when applied to isolated cells [120]. This is the reason why such alkaloids are being tested as potential anesthetics for use in medicine [190]. Harmonine has been shown to have a broad antimicrobial activity against pro- and eukaryotes [5, 163]. It is toxic to parasitic trematodes of the genus *Schistosoma* and is able to inhibit acetylcholine esterase [101].

🐎 Ingestion of azaphenalene alkaloids has been shown to induce liver damage in birds [128]. For humans, this is usually not a problem because only minor amounts may be taken up accidentally.

References

[1]   Adesanya SA et al. (1992) J Nat Prod 55(4): 525
[2]   Aiello A et al. (2000) J Nat Prod 63(4): 517
[3]   Aird SD et al. (2016) Toxins 8(10): 279
[4]   Akbar MA et al. (2020) Marine Drugs 18(2): 103
[5]   Alam N et al. (2002) Bull Korean Chem Soc 23(3): 497
[6]   Alam N et al. (2001) J Nat Prod 64(7): 956
[7]   Albuquerque EX et al. (1971) Science 172(3987): 995
[8]   Alujas-Burgos S et al. (2018) Org Biomol Chem 16(37): 8218
[9]   Anderson DM et al. (1989) Toxicon 27(6): 665
[10]   Arbiser JL et al. (2007) Blood 109(2): 560
[11]   Arbiser JL et al. (2017) Sci Rep 7(1): 11198
[12]   Ashihara H et al. (2013) Adv Bot Res 68: 111
[13]   Aune GJ et al. (2002) Anticancer Drugs 13(6): 545
[14]   Bane V et al. (2014) Toxins 6(2): 693
[15]   Banerjee S et al. (2018) J Nat Prod 81(8): 1899
[16]   Bartram S, Boland W (2001) Chembiochem 2(11): 809
[17]   Becker H (1986) Pharmazie in unserer Zeit 15(4): 97
[18]   Behenna DC et al. (2008) Angew Chem – Int Ed 47(13): 2365
[19]   Benkendorff K et al. (2000) J Chem Ecol 26(4): 1037
[20]   Benkendorff K et al. (2001) Molecules 6(2): 70
[21]   Bernays EA et al. (2003) J Chem Ecol 29(7): 1709
[22]   Bialonska D, Zjawiony JK (2009) Marine Drugs 7(2): 166
[23]   Bienz S et al. (2002) Alkaloids Chem Biol 58: 83
[24]   Bienz S et al. (2005) Nat Prod Rep 22(5): 647
[25]   Blanco A, Blanco G (2017) Purine and pyrimidine metabolism. In: Blanco A, Blanco G (eds.) Medical
       Biochemistry. Academic Press, London, San Diego, Cambridge, Oxford, Chapt. 18, p. 413
[26]   Blicke FF (2011) Org React 1(10): 303
[27]   Blum MS, Hilker M (2002) Chemical protection of insect eggs. In: Hilker M, Meiners T (eds.)
       Chemoecology of Insect Eggs and Egg Deposition. Blackwell, Berlin, Germany, p. 61
[28]   Boyd KG et al. (1995) J Nat Prod 58(2): 302
[29]   Braekman JC et al. (1982) Tetrahedron Lett 23(41): 4277
[30]   Brand JM et al. (1972) Toxicon 10(3): 259
[31]   Bricelj VM et al. (2005) Nature 434(7034): 763

[32]   Brodie ED et al. (1991) Biotropica 23(1): 58
[33]   Burgermeister W et al. (1977) Proc Natl Acad Sci U S A 74(12): 5754
[34]   Cai L et al. (2007) J Chromatogr A 1147(1): 66
[35]   Capper A et al. (2005) J Chem Ecol 31(7): 1595
[36]   Carle JS, Christophersen C (1979) J Am Chem Soc 101(14): 4012
[37]   Casapullo A et al. (2000) J Nat Prod 63(4): 447
[38]   Catterall WA (1980) Annu Rev Pharmacol Toxicol 20: 15
[39]   Cembella AD et al. (1987) Biochem Syst Ecol 15(2): 171
[40]   Chabowska G et al. (2021) Molecules 26(14): 4324
[41]   Chang FH et al. (1997) Toxicon 35(3): 393
[42]   Chen L et al. (2010) J Agric Food Chem 58(22): 11534
[43]   Cherney EC, Baran PS (2011) Isr J Chem 51(3–4): 391
[44]   Choi YH et al. (1993) J Nat Prod 56(8): 1431
[45]   Christophersen C (1985) Acta Chem Scand B – Org Chem Biochem 39(7): 517
[46]   Clark AM, Hufford CD (1992) Antifungal alkaloids. In: Cordell GA (ed.) The Alkaloids: Chemistry and
       Pharmacology, Vol. 42. Academic Press, New York, USA, p. 117
[47]   Cleaves HJ, Miller SL (2001) J Mol Evol 52(1): 73
[48]   Cooksey CJ (2001) Molecules 6(9): 736
[49]   Daly JW et al. (1990) J Nat Prod 53(2): 407
[50]   Daly JW et al. (1985) J Med Chem 28(4): 482
[51]   Daly JW et al. (2005) J Nat Prod 68(10): 1556
[52]   Daly JW et al. (1977) Helvetica Chimica Acta 60(3): 1128
[53]   Davis RA et al. (1999) J Nat Prod 62(3): 419
[54]   Davis RA et al. (2002) J Nat Prod 65(4): 454
[55]   De Carvalho DB et al. (2019) Toxins 11(7): 420
[56]   Dembitsky VM (2002) Russ J Bioorg Chem 28(3): 170
[57]   Dewick PM (2001) Alkaloids. In: Dewick PM (ed.) Medicinal Natural Products. Wiley, Hoboken, New
       Jersey, USA, Chapt. 6, p. 291
[58]   Diaz-Sylvester PL et al. (2014) Mol Pharmacol 85(4): 564
[59]   Dumbacher JP et al. (1992) Science 258(5083): 799
[60]   Dumbacher JP et al. (2000) Proc Natl Acad Sci U S A 97(24): 12970
[61]   Edgar JA (1982) J Zool 196: 385
[62]   Edgar JA et al. (1976) Experientia 32(12): 1535
[63]   Edwards V et al. (2012) Marine Drugs 10(1): 64
[64]   Eisner T, Meinwald J (1995) Proc Natl Acad Sci U S A 92(1): 50
[65]   Eisner T et al. (1997) J Exp Biol 200(19): 2493
[66]   Eldefrawi AT et al. (1988) Proc Natl Acad Sci U S A 85(13): 4910
[67]   El-Demerdash A et al. (2018) Nutrients 10(1): 33
[68]   Elissawy AM et al. (2021) Biomolecules 11(2): 258
[69]   Erjavec V et al. (2017) Medycyna Weterynaryjna – Veterinary Medicine – Science and Practice 73(3): 186
[70]   Erspamer V, Glässer A (1957) Br J Pharmacol 12(2): 176
[71]   Fattorusso E et al. (1970) J Chem Soc D: Chem Commun 1970(12): 752
[72]   Fitch RW et al. (2003) J Nat Prod 66(10): 1345
[73]   Fitzgerald KT (2013) Insects - Hymenoptera. In: Peterson ME, Talcott PA (eds.) Small Animal
       Toxicology, 3rd ed. W. B. Saunders, Saint Louis, USA, p. 573
[74]   Flessner T et al. (2004) Cephalostatin analogues - Synthesis and biological activity. In: Herz W, Falk H,
       Kirby GW (eds.) Progress in the Chemistry of Organic Natural Products, Vol. 87. Springer, Vienna, p. 1
[75]   Fox EGP et al. (2019) Toxicon 158: 77
[76]   Fuhrman FA et al. (1980) Science 207(4427): 193

[77]   Gao C et al. (2013) Chem Biodivers 10(8): 1435
[78]   Göransson U et al. (2019) Toxins 11(2): 120
[79]   Gomes NGM et al. (2016) Marine Drugs 14(5): 98
[80]   Gonzalez A et al. (1999) J Nat Prod 62(2): 378
[81]   Gonzalez A et al. (1999) Proc Natl Acad Sci U S A 96(10): 5570
[82]   Gordon EM et al. (2016) Advances in Therapy 33(7): 1055
[83]   Goud TV et al. (2003) Biol Pharm Bull 26(10): 1498
[84]   Gu Y et al. (2021) Bioresour Bioprocess 8(1): 110
[85]   Guillen PO et al. (2018) Marine Drugs 16: 242
[86]   Gunasekera SP et al. (1994) J Nat Prod 57(10): 1437
[87]   Habermehl G (1969) Naturwissenschaften 56(12): 615
[88]   Hanifin CT (2010) Marine Drugs 8(3): 577
[89]   Hanifin CT, Gilly WF (2015) Evolution 69(1): 232
[90]   Hansen Kø et al. (2017) J Nat Prod 80(12): 3276
[91]   Happ GM, Eisner T (1961) Science 134(347): 329
[92]   Hartmann T et al. (2005) Insect Biochem Mol Biol 35(10): 1083
[93]   Hassan W et al. (2004) J Nat Prod 67(5): 817
[94]   Haulotte E et al. (2012) Eur J Org Chem 2012(10): 1907
[95]   Herbert RB (1971) Biosynthesis. In: Saxton JE (ed.) The Alkaloids, Vol. I. The Chemical Society, Burlington House, London, UK, p. 1
[96]   Itoh T et al. (1983) J Chem Soc Perkin Trans 1983(0): 147
[97]   Iwagawa T et al. (2000) J Nat Prod 63(9): 1310
[98]   Izumi AK, Moore RE (1987) Clin Dermatol 5(3): 92
[99]   Jones TH et al. (1999) J Chem Ecol 25(5): 1179
[100] Jurek J et al. (1994) J Nat Prod 57(7): 1004
[101] Kellershohn J et al. (2019) PLoS Negl Trop Dis 13(3): e0007240
[102] Kelley RB et al. (1987) Experientia 43(8): 943
[103] Kem WR et al. (1997) J Pharmacol Exp Ther 283(3): 979
[104] Kem WR et al. (1976) Experientia 32(6): 684
[105] Kem W et al. (2006) Marine Drugs 4(3): 255
[106] Khodorov BI (1985) Prog Biophys Mol Biol 45(2): 57
[107] Killday KB et al. (1993) J Nat Prod 56(4): 500
[108] Kim JW et al. (1993) J Nat Prod 56(10): 1813
[109] King AG, Meinwald J (1996) Chem Rev 96(3): 1105
[110] Kobayashi J et al. (1988) J Org Chem 53(8): 1800
[111] Kobayashi J et al. (1986) Experientia 42(10): 1176
[112] Kobayashi J et al. (1986) Experientia 42(9): 1064
[113] Kondo K et al. (1994) J Nat Prod 57(7): 1008
[114] Kugler C (1978) Studia Entomologica (N S) 20: 413
[115] Laird DW et al. (2007) J Nat Prod 70(5): 741
[116] Laycock MV et al. (1986) Can J Chem 64(7): 1262
[117] Le VH et al. (2015) Nat Prod Rep 32(2): 328
[118] Leclercq S et al. (2001) J Chem Ecol 27(5): 945
[119] Lee HS et al. (2001) J Nat Prod 64(11): 1474
[120] Leong RL et al. (2015) Neurochem Res 40(10): 2078
[121] Lindigkeit R et al. (1997) Eur J Biochem 245(3): 626
[122] Lindquist N et al. (2000) J Nat Prod 63(9): 1290
[123] Liu C et al. (2012) Org Lett 14(8): 1994
[124] Lüddecke T et al. (2018) Naturwissenschaften/Sci Nat 105(9): 56

[125] Macel M (2011) Phytochem Rev: Proc Phytochem Soc Eur 10(1): 75

[126] Macfoy C et al. (2005) Zeitschrift für Naturforschung C 60(11–12): 932

[127] Magarlamov TY et al. (2017) Toxins 9(5): 166

[128] Marples NM et al. (1989) Ecol Entomol 14(1): 79

[129] Matsunaga S et al. (2009) Pure Appl Chem 81(6): 1001

[130] McFarren EF et al. (1961) Adv Food Res 10: 135

[131] McKay MJ et al. (2005) J Nat Prod 68(12): 1776

[132] Mebs D, Pogoda W (2005) Toxicon 45(5): 603

[133] Menna M et al. (2011) Molecules 16(10): 8694

[134] Michael JP (2002) Nat Prod Rep 19(6): 742

[135] Michael JP (2003) Nat Prod Rep 20(5): 458

[136] Michael JP (2008) Nat Prod Rep 25(1): 166

[137] Milanowski DJ et al. (2004) J Nat Prod 67(1): 70

[138] Moon SS et al. (2002) J Nat Prod 65(3): 249

[139] Moreira R et al. (2018) Int J Mol Sci 19(6): 1668

[140] Moser BR (2008) J Nat Prod 71(3): 487

[141] Nakamura H et al. (1984) Tetrahedron Lett 25(23): 2475

[142] Neelabh S (2019) Alkaloid. In: Vonk T, Shackelford T (eds.) Encyclopedia of Animal Cognition and Behaviour. Springer International Publishing, Cham, Switzerland, p. 1

[143] Nicholas GM et al. (2001) Org Lett 3(10): 1543

[144] Novanna M et al. (2019) Mini Rev Med Chem 19(3): 194

[145] Nuijen B et al. (2000) Anticancer Drugs 11(10): 793

[146] Nuzzo G et al. (2022) Biomolecules 12(2): 246

[147] Ojika M et al. (2003) Biosci Biotechnol Biochem 67(6): 1410

[148] Oka K et al. (1985) Proc Natl Acad Sci U S A 82(6): 1852

[149] Opitz SEW, Müller C (2009) Chemoecology 19(3): 117

[150] Palkar MB et al. (2015) Anticancer Agents Med Chem 15(8): 947

[151] Palanisamy SK et al. (2017) Nat Prod Bioprospecting 7(1): 1

[152] Pénez N et al. (2011) J Nat Prod 74(10): 2304

[153] Peters L et al. (2003) Appl Environ Microbiol 69(6): 3469

[154] Peters L et al. (2004) J Chem Ecol 30(6): 1165

[155] Piel J (2004) Nat Prod Rep 21(4): 519

[156] Poeaknapo C (2005) Med Sci Monit 11(5): MS6

[157] Pröhl H et al. (2010) Amphibia-Reptilia 31(2): 217

[158] Rao CB et al. (1985) J Org Chem 50(20): 3757

[159] Rashid MA et al. (2001) J Nat Prod 64(11): 1454

[160] Rasmussen T et al. (1993) J Nat Prod 56(9): 1553

[161] Roberts MF, Wink M (1998) Alkaloids: Biochemistry, Ecology, and Medicinal Applications. Springer, Boston, MA, USA

[162] Robins DJ (1984) Nat Prod Rep 1(3): 235

[163] Röhrich CR et al. (2012) Biology Letters 8(2): 308

[164] Roseghini M et al. (1996) Toxicon 34(1): 33

[165] Rudd D et al. (2015) Sci Rep 5(1): 13408

[166] Santi F, Maini S (2006) Bull Insectology 59(1): 53

[167] Schaller H (2003) Prog Lipid Res 42(3): 163

[168] Schulte LM et al. (2017) Biotropica 49(1): 23

[169] Selander E et al. (2006) Proc R Soc Lond Biol Sci 273(1594): 1673

[170] Shimizu Y et al. (1989) Pure Appl Chem 61(3): 513

[171] Simidu U et al. (1987) Appl Environ Microbiol 53(7): 1714

[172] Skinner WS et al. (1990) Toxicon 28(5): 541
[173] Sloggett JJ, Davis AJ (2010) J Exp Biol 213: 237
[174] Smith LW, Culvenor CCJ (1981) J Nat Prod 44(2): 129
[175] Sobel J, Painter J (2005) Clin Infect Pract 41(9): 1290
[176] Soong TW, Venkatesh B (2006) Trends Genet 22(11): 621
[177] Spande TF et al. (1992) J Am Chem Soc 114(9): 3475
[178] Spande TF et al. (1999) J Nat Prod 62(1): 5
[179] Stegelmeier BL et al. (1999) J Nat Toxins 8(1): 95
[180] Stirling IR et al. (1997) J Chem Soc – Perkin Trans 1(5): 677
[181] Sullivan JP, Bannon AW (1996) CNS Drug Rev 2(1): 21
[182] Sunassee SN et al. (2014) J Nat Prod 77(11): 2475
[183] Tang W-Z et al. (2018) J Nat Prod 81(4): 894
[184] Thoms C et al. (2003) Zeitschrift für Naturforschung C 58(5–6): 426
[185] Torres JPSchmidt EW (2019) J Biol Chem 294(46): 17684
[186] Traynor JR (1998) Br J Anaesth 81(1): 69
[187] Tsukamoto S et al. (2000) J Nat Prod 63(5): 682
[188] Tsuneki H et al. (2005) Biol Pharm Bull 28(4): 611
[189] Uddin MJ et al. (2001) J Nat Prod 64(9): 1169
[190] Umana IC et al. (2013) Biochem Pharmacol 86(8): 1208
[191] Van Wagoner RM et al. (1999) J Nat Prod 62(5): 794
[192] Verma A et al. (2019) Microorganisms 7(8): 222
[193] Vlasenko AE, Magarlamov TY (2020) Toxins 12(12): 745
[194] von Byern J et al. (2017) Toxicon 135: 24
[195] Wang W et al. (2020) J Appl Toxicol 40(11): 1534
[196] Weiser T (2004) J Neurosci Methods 137(1): 79
[197] Wheeler JW et al. (1981) Science 211(4486): 1051
[198] Wibowo JT et al. (2021) Marine Drugs 20(1): 3
[199] Williams BL et al. (2004) J Chem Ecol 30(10): 1901
[200] Willmott KR, Freitas AVL (2006) Cladistics 22(4): 297
[201] Wink M (1987) Planta Medica 53(6): 509
[202] Wink M (1998) Chemical ecology of alkaloids. In: Roberts MF, Wink M (eds.) Alkaloids: Biochemistry, Ecology and Medicinal Applications. Plenum Press, New York, USA, Chapt. 11, p. 265
[203] Wink M et al. (1982) Zeitschrift für Naturforschung C 37(11–12): 1081
[204] Witte L et al. (1990) Naturwissenschaften 77(11): 540
[205] Witkop B (1971) Experientia 27(10): 1121
[206] Woodward RB et al. (1957) J Chem Soc 1957(0): 1131
[207] Xu XH et al. (2003) J Nat Prod 66(2): 285
[208] Yan CC, Huxtable RJ (1995) Toxicolo Appl Pharmacol 130(1): 1
[209] Yao B et al. (2003) J Nat Prod 66(8): 1074
[210] Yi GB et al. (2003) Int J Toxicol 22(2): 81
[211] Yoon MA et al. (2006) Biol Pharm Bull 29(4): 735
[212] Yoshimura F et al. (2012) Acc Chem Res 45(5): 746
[213] Yotsu-Yamashita M et al. (2012) Toxicon 59(2): 257
[214] Yotsu-Yamashita M et al. (2004) Proc Natl Acad Sci U S A 101(13): 4346
[215] Yuan ML, Wang IJ (2018) PLoS ONE 13(3): e0194265
[216] Zhang X et al. (2021) Org Lett 23(8): 2858

2.10 Cyanogenic Compounds

Cyanogenic glycosides are amino acid-derived nitrogen-containing secondary metabolites of plants that have the ability to produce hydrogen cyanide upon enzymatic cleavage [15]. The typical cyanogenic glycoside consists of an α-hydroxynitrile as the aglycone moiety and a mono- or disaccharide residue that is β-glycosidically bound to the α-hydroxy group of the aglycone (Fig. 2.55). The sugar residues are usually glucose molecules, but other monosaccharides occur as well. The sugar moieties may be derivatized by addition of malonic, benzoic, or sulfuric acid. The cyanogenic glycosides and the degrading enzymes (β-glucosidase and hydroxynitrile lyase) are usually sequestered in different subcellular compartments. Only when the plant tissue is injured (e.g., by herbivores), enzymes and substrates are getting in contact and hydrogen cyanide is produced (Fig. 2.56).

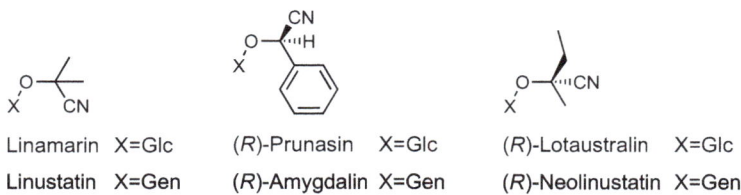

Linamarin X=Glc   (R)-Prunasin X=Glc   (R)-Lotaustralin X=Glc

Linustatin X=Gen   (R)-Amygdalin X=Gen   (R)-Neolinustatin X=Gen

Fig. 2.55: Cyanogenic glycosides. Linamarin and linustatin are derived from valine, prunasin, and amygdalin from phenylalanine, lotaustralin, and neolinustatin from isoleucine. Glc, glucose; Gen, gentiobiose.

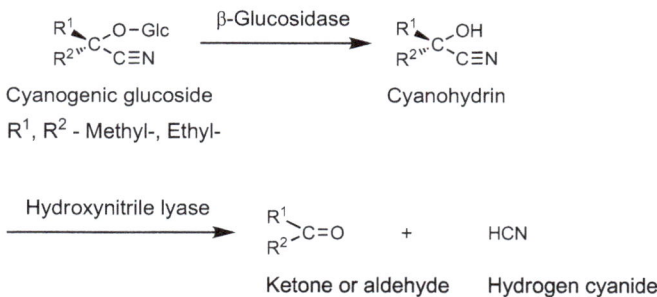

Cyanogenic glucoside   Cyanohydrin

R^1, R^2 - Methyl-, Ethyl-

Ketone or aldehyde   Hydrogen cyanide

Fig. 2.56: Production of hydrogen cyanide from cyanogenic glycosides.

Cyanide inhibits the action of the mitochondrial cytochrome C oxidase, the terminal enzyme of the respiratory chain. The complex of cytochrome oxidase and cyanide maintains the heme iron within the oxidase in its trivalent form, impeding electron transport and blocking the respiratory chain. Inhibition of the respiratory chain results in cell death [7]. Due to this mode of action in animals, hydrogen cyanide produced in injured plant tissue is usually an efficient deterrent of herbivores.

Some herbivorous animals, however, have evolved the ability to detoxify cyanide by forming nontoxic adducts. Others are able to consume material from cyanogen-containing plants and utilize these compounds to render themselves unpalatable for potential predators. Hydrogen cyanide is used as a defense poison in some species of leaf beetles (Chrysomelidae). *Paropsis atomaria,* a species occurring in eastern Australia and feeding on Australian eucalyptus, uses autonomously synthesized mandelonitrile as a defense poison. Mandelonitrile is the aglycone part of the cyanogenic glycosides prunasin and amygdalin (Fig. 2.55) [8].

The larvae of the five-spot burnet, *Zygaena trifolii,* a European moth species (Lepidoptera), are feeding on legumes (Fabaceae) producing cyanogenic glycosides (linamarin and lotaustralin; Fig. 2.55) and sequester these substances within their own tissues. Interestingly, the larvae are also able to directly synthesize these glucosides from valine and isoleucine [10]. Thus, both sequestration and biosynthesis of the same compounds can occur in animals. In addition to linamarin and lotaustralin, all cyanogenic butterflies also contain the neurotoxic β-cyanoalanine, a fusion product of hydrogen cyanide and cysteine, which, however, also occurs in many non-cyanogenic species [14]. The caterpillars have defense glands, from which secretions enriched in cyanogenic glycosides and β-cyanoalanine are released when the animals are attacked by predators [9].

🐎 Cases of miscarriages in horses (MRLS, mare reproductive loss syndrome) have been reported in the United States that may be associated with the uptake of cyanogenic glycosides. The horses consumed large numbers of caterpillars of the American ring moth, *Malcosoma americanum,* sitting on the leaves of their forage plant, the highly cyanogenic black cherry tree (*Prunus serotina*) [5, 13].

Many centipedes (Chilopoda) and millipedes (Diplopoda) carry segmental defense glands whose secretions contain hydrogen cyanide in addition to benzaldehyde, mandelonitrile, mandelonitrile benzoate, or benzoylnitrile (Fig. 2.57) [2, 4]. An individual of *Pachymerium ferrugineum,* a European ground runner (Geophilidae), may produce and release more than 2 µg hydrogen cyanide when threatened [12]. Not only in insects but also in chilopods and diplopods, the cyanogenic compounds are autonomously synthesized, since all the required enzymes have been found in these species [17]. The catabolic pathway is also present in these animal species which is exemplified by the observation that hydroxynitrile lyase is expressed in an Asian millipede, *Chamberlinius hualienensis* [3].

🧍 Accidental hydrogen cyanide (HCN) poisoning in humans from handling or ingesting toxic animals is a rare event. Generally, early symptoms of cyanide poisoning are headache, dizziness, fast heart rate, shortness of breath, and vomiting. This phase may be followed by seizures, slow heart rate, low blood pressure, loss of consciousness, and cardiac arrest. Onset of symptoms usually occurs within a few minutes [6, 16]. The lethal dose upon oral uptake of humans is approximately 1.5 mg/kg BW [11].

Mandelonitrile   Benzoylnitrile

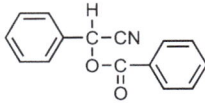

Mandelonitrile benzoate   Fig. 2.57: Cyanogens.

🚑 Treatment of acute cyanide poisoning includes immediate gastric lavage with 5% sodium thiosulfate solution. Di-cobalt EDTA, 600 mg in 40 mL saline over 1 min, may be applied intravenously. Inhalation of pure oxygen may improve tissue oxygenation. Acidosis should be treated with sodium bicarbonate infusion [1].

References

[1]   Aronson JK (2014) Plant poisons and traditional medicines. In: Farrar J, Hotez PJ, Junghanss T et al. (eds.) Manson's Tropical Infectious Diseases, 23. Ed.W. B. Sauders, London, UK, p. 1128
[2]   Blum MS, Woodring JP (1962) Science 138(3539): 512
[3]   Dadashipour M et al. (2015) Proc Natl Acad Sci U S A 112(34): 10605
[4]   Duffey SS et al. (1977) J Chem Ecol 3(1): 101
[5]   Fitzgerald TD et al. (2002) J Chem Ecol 28(2): 257
[6]   Hamel J (2011) Crit Care Nurs 31(1): 72
[7]   Leavesley HB et al. (2008) Toxicol Sci 101(1): 101
[8]   Nahrstedt A (1987) Recent developments in chemistry, distribution and biology of the cyanogenic glycosides. In: Biologically Active Natural Products Hostettmann K, Lea PJ (ed.) Annual Proceedings of the Phytochemical Society of Europe. Clarendon Press, Oxford, UK, Chapt. 27, p. 213
[9]   Nahrstedt A (1988) Biologie in unserer Zeit 18(4): 105
[10] Nahrstedt A, Davis RH (1986) Phytochemistry 25(10): 2299
[11] Newhouse K, Chiu N (2010) Toxicological Review of Hydrogen Cyanide and Cyanide Salts.. U.S. Environmental Protection Agency, Washington, DC, USA, EPA/635/R-08/016F
[12] Schildknecht H et al. (1968) Naturwissenschaften 55(5): 230
[13] Webb BA et al. (2004) J Insect Physiol 50(2–3): 185
[14] Witthohn K, Naumann CM (1987) J Chem Ecol 13(8): 1789
[15] Yamane H et al. (2010) Chemical defence and toxins of plants. In: Liu H-W, Mander L (eds.) Comprehensive Natural Products II. Elsevier, Oxford, UK, 339
[16] Yen D et al. (1995) Am J Emerg Med 13(5): 524
[17] Zagrobelny M et al. (2018) Insects 9(2): 51

2.11 Glucosinolates

Glucosinolates are C-substituted S-(β-D-glucopyranosyl)-methanethiohydroximic acid-O-sulfates. They are derived from glucose and an amino acid [8]. More than 130 different members of this family of natural substances are known with the largest number occurring in plants [6]. Glucosinolates are plant secondary metabolites which are also called 'mustard oil glucosides' as they are especially abundant in plant species belonging to the order of Brassicales. The pungent smell and taste that can be sensed when such plant material (e.g., mustard, *Sinapis* sp.) is chewed or cut is due to liberation of mustard oils that are enzymatically produced from glucosinolates (Fig. 2.58). These substances and their breakdown products, e.g., isothiocyanates, thiocyanates, and nitriles, are aimed at herbivores and serve as antifeedants [19] or protect plants from being attacked by microorganisms or fungi [5, 16]. The enzymes that degrade glucosinolates are thioglucoside glucohydrolases (E.C. 3.2.1.147) of which many isoforms exist in different plant species. One of the widely distributed enzymes is myrosinase [25]. In the plant, this thioglucosidase is present in a different compartment as its substrates the glucosinolates. In *Arabidopsis thaliana*, the glucosinolates are located in S cells adjacent to the phloem, while the myrosinase is stored in special myrosin cells that are scattered in the adjacent cortical tissue. However, when plant tissue is injured, e.g., by insect damage, the glucosinolates come into contact with myrosinase and are instantaneously broken down [9]. Due to the fact that the toxic compounds are promptly produced at the time of demand, this two-component defense system has been named 'mustard oil bomb'.

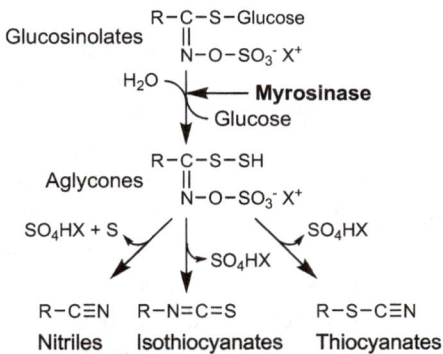

Glucosinolates
$$R-C-S-Glucose$$
$$||$$
$$N-O-SO_3^- \ X^+$$

H_2O ⟶ **Myrosinase**
Glucose

Aglycones
$$R-C-S-SH$$
$$||$$
$$N-O-SO_3^- \ X^+$$

$SO_4HX + S$ SO_4HX
SO_4HX

$R-C{\equiv}N$ $R-N{=}C{=}S$ $R-S-C{\equiv}N$
Nitriles   Isothiocyanates   Thiocyanates

Fig. 2.58: Glucosinolates and their cleavage products. The figure shows the basic structure of glucosinolates and their myrosinase-mediated degradation to the potential products, nitriles, isothiocyanates, or thiocyanates. Generally, the nitriles are less toxic and often produced as detoxification products of glucosinolates by special enzymes in insect larvae feeding on glucosinolate-containing plant material. These detoxification enzymes, the nitrile-specifier proteins (NSPs), were originally described in caterpillars of pierid butterflies feeding on cabbage [23]. Thiocyanates and isothiocyanates, however, are highly toxic metabolites of glucosinolates used by plants of the order Brassicales as antifeedants.

Upon ingestion by animals or humans, the enzymes in the plant material, including the plant-borne myrosinase, undergo denaturation in the acidic stomach compartment. If the plant material contains residual glucosinolates, bacterial β-thioglucosidases in the colon cleave this material. The resulting metabolites may have roles in colon cancer prevention [1, 15, 22]. Specifically, they seem to induce the expression of detoxifying enzymes, e.g., epoxide hydrolases, glutathion-S-transferases, and glucuronosyl-transferases, which process different xenobiotics by adding hydrophilic groups. More hydrophilicity of the xenobiotics favors their excretion from the body via the kidneys contrary to the xenobiotics in their original, lipophilic form [11].

Coevolutionary adaptations of insect herbivores (see Section 3.2.9) and their host plants have led to the situation that each insect species uses only one or some specific host plant species containing very specific compositions of glucosinolates. Females of such insects actually search for such food plants to deposit their eggs on the leaves. These insect species have overcome the deterrent function of the glucosinolate/myrosinase system or may even exploit it for establishing chemical defense for themselves or their offspring [25]. A well-established example is the green-veined white (*Pieris napi*), a butterfly of the family Pieridae. Larvae overcome the toxicity of glucosinolate hydrolysis products by expressing nitrile-specifier proteins (NSP) and major allergen (MA) proteins in their gut cells. The exact mechanism of action of these proteins is not yet known, but it seems possible that they change the substrate preference of the myrosinases. Another possibility is that they induce a shift of the glucosinolate degradation pathway away from highly toxic products (isothiocyanates and thiocyanates) to less toxic ones (nitriles) (Fig. 2.58). The biological significance of these substances is indicated by the fact that *Pieris* larvae regulate NSP and MA gene expression depending on the glucosinolate profiles of their host plants [12].

In generalist herbivores, the presence of glucosinolates in potential food plants usually keeps them from using them as nutritional resource [7]. However, some herbivores have evolved means of circumventing the antifeeding strategy of the plants. An example is the phloem-feeding silverleaf whitefly, *Bemisia tabaci*, which expresses glucosinolate sulfatase genes in gut cells and secretes these enzymes into the intestines. The sulfatase converts glucosinolates into inactive desulfo-glucosinolates that cannot any longer be converted to the biologically active degradation products, isothiocyanates, thiocyanates, or nitriles by myrosinases as these enzymes accept only sulfur-containing molecules as substrates [10]. A similar detoxification system is utilized by other plant-feeding insects, e.g., by moth larvae of the genus *Spodoptera* [14].

Researchers have been puzzled for a long time that some secondary metabolites, e.g., terpenes, cyanogenic glycosides, benzoic acid derivatives, benzoquinones, and naphthoquinones, occur in plants as well as in certain insect species. Moreover, enzyme systems that activate two-component defense systems, such as glucosinolates, occur in species of both groups of organisms despite their far phylogenetic distance. However, the biosynthetic pathways for the production of such metabolites may be very different in insects and plants [2].

Many herbivorous insects make use of plant-borne compounds by ingesting and storing them in their own tissues or in the integument. Sequestration of toxic metabolites may deter potential predators or may protect eggs from being preyed upon or degraded by microorganisms. More than 250 insect species have been shown to utilize plant metabolites for defensive purposes [13].

Some herbivorous insects selectively accumulate glucosinolates in their own tissues. Unfortunately, little is known about the processes that enable insects to absorb and store such compounds in the body. The horseradish flea beetle (*Phyllotreta armoraciae*) accumulates glucosinolate obtained from host plants (Brassicaceae) in the hemolymph. To achieve this, these beetles have evolved glucosinolate-specific transporters which are expressed in the Malpighian tubules, the excretory system of insects. Glucosinolates are excreted from the animal's body at much faster rates when the expression of these transporters is suppressed indicating that these transport proteins usually reabsorb glucosinolates from the Malpighian tubule lumen to prevent their loss through the excretory system [26].

The cabbage aphid, *Brevicoryne brassicae*, sequesters glucosinolates from its host plants [3] and expresses its own myrosinase that can be activated when the animal has to defend itself against predators. Surprisingly, the amounts of toxic isothiocyanates that were produced during simulated attacks on the aphids were quantitatively outweighed by the generation of less toxic nitriles. Nevertheless, the defensive secretions of the aphids impaired the development of the predatory larvae of the lacewing, *Chrysoperla carnea* [18].

👤🐏 The use of crops containing glucosinolates as primary food source for animals may have negative effects depending on the affected species. Closely related animal species may differ substantially in their tolerance toward glucosinolates [17]. Some glucosinolates have been shown to function as goitrogens and antithyroid agents in both humans and animals when applied over longer periods at high concentrations [20, 21]. A potential explanation for the glucosinolate-mediated enlargement of the thyroid gland is that thiocyanates compete with iodine and suppress the generation of thyroid hormone which prompts the thyroid gland to grow in order to compensate for this deficiency [4, 24].

References

[1]   Barba FJ et al. (2016) Front Nutrit 3: 24
[2]   Beran F et al. (2019) New Phytol 223(1): 52
[3]   Chaplin-Kramer R et al. (2011) J Appl Ecol 48(4): 880
[4]   Di Dalmazi G, Giuliani C (2021) Food Chem Toxicol 152: 112158
[5]   Dubey S et al. (2021) PhytoFrontiers 1(1): 40
[6]   Fahey JW et al. (2001) Phytochemistry 56(1): 5
[7]   Hopkins RJ et al. (2009) Annu Rev Entomol 54(1): 57
[8]   Lockhart J (2019) Plant Cell 31(7): 1429
[9]   Lv Q et al. (2022) Int J Mol Sci 23(3): 1577

[10] Manivannan A et al. (2021) Front Plant Sci 12: 671286
[11] Nho CW, Jeffery E (2001) Toxicolo Appl Pharmacol 174(2): 146
[12] Okamura Y et al. (2019) Sci Rep 9(1): 7256
[13] Opitz SEW, Müller C (2009) Chemoecology 19(3): 117
[14] Ratzka A et al. (2002) Proc Natl Acad Sci U S A 99(17): 11223
[15] Rouzaud G et al. (2004) Cancer Epidemiology, Biomarkers and Prevention 13(1): 125
[16] Saladino F et al. (2017) Antimicrobial activity of the glucosinolates. In: Mérillon J-M, Ramawat KG (eds.) Glucosinolates. Springer International Publishing, Cham, Switzerland, p. 249
[17] Samuni-Blank M et al. (2013) Evol Ecol 27(6): 1069
[18] Sun R et al. (2021) J Pest Sci 94(4): 1147
[19] Textor S, Gershenzon J (2009) Phytochem Rev 8(1): 149
[20] Tripathi MK, Mishra AS (2007) Anim Feed Sci Technol 132(1–2): 1
[21] Vanderpas J (2003) Goitrogens and antithyroid compounds. In: Caballero B (ed.) Encyclopedia of Food Sciences and Nutrition, 2nd ed. Academic Press, Oxford, UK, p. 2949
[22] Verhoeven DT et al. (1996) Cancer Epidemiol Biomarkers Prev 9: 733
[23] Wheat CW et al. (2007) Proc Natl Acad Sci U S A 104(51): 20427
[24] Willemin M-E, Lumen A (2019) Toxicolo Appl Pharmacol 365: 84
[25] Winde I, Wittstock U (2011) Phytochemistry 72(13): 1566
[26] Yang Z-L et al. (2021) Nat Commun 12(1): 2658

2.12 Proteins and Peptides

2.12.1 General

Virtually every cell in a living organism continuously synthesizes peptides and proteins. The synthesis and degradation rates of individual proteins are regulated according to the momentary needs of the cell [142]. Protein synthesis occurs at ribosomes, large molecular aggregates comprising ribosomal RNA and proteins [152]. Since ribosomes translate the information encoded in consecutive nucleotide triplets of the messenger RNA (mRNA) into the respective amino acid sequences, the process of protein biosynthesis is also termed 'translation'. In eukaryotic cells, translation generally occurs in the cytoplasm of the cell. The ribosomes are either floating in the cytosol or they are attached to the endoplasmic reticulum ('rough ER'). The protein synthesis usually starts with a methionine residue that forms the N-terminus in the final protein. According to the mRNA sequence encoding the respective protein, further amino acids are recruited from a pool of 20 (or 21 if selenocysteine is considered independently of cysteine) different proteogenic amino acids and chemically fixed to the growing protein strand (Fig. 2.59) by formation of covalent peptide bonds (Fig. 1.1). The specific sequence of the amino acid stretch of the final protein is called the 'primary sequence' of a protein.

Some proteins carry short stretches of amino acids, so-called signal peptides, at their N-terminal ends (Fig. 2.59), which indicate the cellular protein transport system that the native protein has to be directly imported into the lumen of the ER [99]. Proteins with signal peptides are generally synthesized by ribosomes that are fixed on the cytosolic side of the ER membrane. The signal peptide is recognized by an import

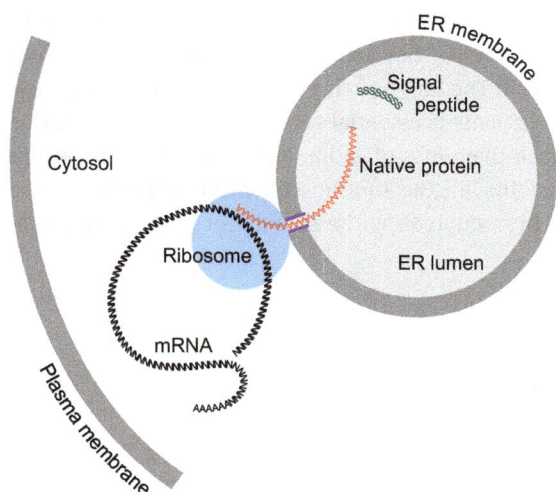

Fig. 2.59: Protein synthesis at an endoplasmic reticulum (ER)-associated ribosome, uptake of the native protein into the ER lumen, and cleavage of the signal peptide.

machinery in the ribosomal membrane, and the N-terminus of the protein is already transported into the ER lumen while the rest of the protein molecule is still in the process of being synthesized. When the signal peptide of the native proteins has been imported into the ER lumen, it is proteolytically cleaved off so that the respective protein gets a new N-terminus (Fig. 2.59). Once the ribosome has completed a protein strand and this has been fully imported into the ER lumen, chaperones and enzymes in the ER lumen assist the protein to obtain its proper folding, i.e., to assume its 3D structure that is essential for the function of the protein [82].

The chemical nature of portions of the protein sequence allows a protein to spontaneously form 'secondary structures' like alpha-helices or beta-sheets. These are structures stabilized by hydrogen bonds forming between the amino acid side chains within the polypeptide. Several of these secondary structures together with unfolded regions in between them form higher level structures, the 'tertiary structure' of the protein. This process is facilitated by chaperones and folding-support proteins in the ER lumen. Secondary and tertiary structural features render the proteins more stable in comparison with the unfolded state and allow specific molecular interactions with binding partners that would not be possible for the unfolded protein [123].

How important the 3D structure of a protein is may be illustrated by an example of structural convergence in animal toxins. In many nonhomologous arachnid and mollusk toxins, a characteristic superstructure has been identified that is characterized by three disulfide bridges between the side chains of cysteine residues, the 'inhibitor cystine knot' [161] that is also called 'knottin' (Fig. 2.60). The primary sequences of the protein toxins vary but the patterns of the cysteines are very similar in these proteins. The presence of this motif renders the respective proteins resistant against thermal and chemical denaturation and against proteolysis. They are stable over long periods also in the foreign milieu of the target organism. The high stability of proteins containing knottin motifs or comparable structural elements is essential for the effectiveness of toxins as they must be stable enough to reach their sites of action within the target organism [93]. These properties make protein motifs like the knottin sequence interesting also for biotechnologists who are interested in manipulating the stability of recombinant proteins [127].

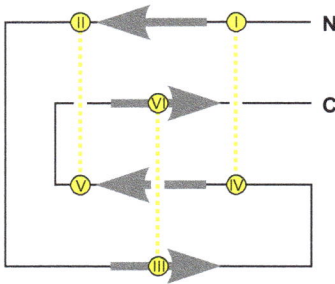

Fig. 2.60: Molecular organization of the inhibitor cystine knot. The cysteine residues I to VI in a given protein strand form disulfide bridges (dotted yellow lines) in a regular pattern (I–IV, II–V, and III–VI). This in conjunction with secondary structures (β-sheets indicated as gray arrows) forms the 'knottin' domain that occurs in protein toxins of many animal species.

While still in the lumen of the ER, virtually every newly synthesized protein undergoes 'posttranslational modification' [128]. There are more than 200 types of chemical alterations that affect 3D structure, further handling of the proteins by the cellular protein trafficking machinery, the ability of the protein to interact with binding partners, or allowing their enzymatic activities. Posttranslational modifications may increase the protein diversity in a cell by the factor of 2–5. There are five basic types of posttranslational modifications:
- Proteolytic cleavage to form two or more biologically active proteins from a single proprotein that may not have any biological activity itself
- Covalent addition of chemical groups (e.g., hydroxylation, sulfatation, methylation, acetylation, or phosphorylation) to the side chains of individual amino acids
- Other chemical modifications of amino acids within the protein (e.g., C-terminal amidation)
- Covalent addition of complex molecules (e.g., lipidation and N-linked or O-linked glycosylation) to amino acid side chains
- Formation of intramolecular covalent bonds (e.g., disulfide bridges and self-ligation to form circular peptides)

Major types of protein modifications are glycosylations which occur frequently in plasma membrane proteins. These proteins carry branched carbohydrate chains which usually stick out into the extracellular space. These protein modifications provide recognition and attachment sites for other molecules. Proteomic investigations have revealed that approximately 50% of all proteins in eukaryotic cells are glycosylated [78].

Certain modifications in proteins like formation of intramolecular disulfide bridges [12], N-terminal acetylation [66], or C-terminal amino acid amidation [103] render these proteins resistant against rapid degradation by proteases. This enhances the protein stability and elongates the lifetime of the respective protein. This is especially important for toxins which are stored for extended periods in gland reservoirs of toxic animals or have to stay active upon transfer to the target organism until the pharmacological effect of the toxin has been achieved.

In primary sequences of peptides or proteins that are designated to be integrated into cellular membranes, there are generally clusters of lipophilic amino acid residues (approx. 20–28) which are interrupted by stretches of hydrophilic amino acid residues. Such a pattern gives the molecule a lyobipolar character. Stretches of lipophilic amino acids form α-helical cylinders (α-helices) whose outer surface interacts with the lipid bilayers of biological membranes and forms transmembrane spanning domains while the hydrophilic intermediates form unstructured loops on alternate sides of the membrane, in the cytosol or the extracellular space, respectively [145]. Membrane proteins may act as receptors, ion channels, transporters, docking proteins, or stationary enzymes in the cells. Some animal toxins, e.g., the pore-forming toxins, have the same structural features and are able to integrate themselves into biological membranes in the target organism. They may form membrane pores, increase membrane permeability to ions and small molecules, and ultimately destroy the respective cell. Examples for such toxins are the main ingredient of bee venom, the lyobipolar protein melittin, or eiseniapore, a pore-forming toxin in earthworms that has a function in predator deterrence.

When toxic peptides or proteins are ingested by animals or humans, they are usually not absorbed as such but digested by proteases and peptidases in the intestines. The resulting di- or tripeptides or individual amino acids are not toxic any more. Thus, protein toxins are usually ineffective if orally taken in. Some peptides and proteins, however, resist breakdown by digestive enzymes, e.g., some cyclic peptides or lectin-like compounds. They usually attach to oligosaccharides or lipids in cell membranes and are taken up into the cells by endocytosis. Such proteotoxins like the toxic lectins of plants are usually made up of a haptomer, an attachment group (usually a lectin), and the toxomer. The haptomer is used to transport the toxomer into the cell. The toxomer exerts the toxic effects inside the cell. This mechanism is also observed in many bacterial AB toxins [126].

Venomous animals make use of peptide and proteotoxins for defense and prey acquisition by actively injecting the toxins into their victims using a venom apparatus and applicators like stingers, spines, or fangs, which bring these substances through the integument into the extracellular space of the target organism. Aquatic animals, however, may be affected by the presence of water-soluble toxic peptides or proteins in the environment as their gill epithelial cells or other delicate epithelia are directly exposed to these toxins. Such toxins may be used as predator repellents by toxic animals when they feel threatened.

The mechanisms of action of peptide and proteotoxins are highly diverse. Since they can only in exceptional cases (see above) penetrate into the cytosol of intact cells, their points of attack are the extracellular domains of molecules within the plasma membrane (e.g., ion channels or membrane receptors) or the lipid bilayer of the plasma membrane itself. The latter may be affected by the integration of lyobipolar peptides or proteins that alter the membrane structure or its permeability. Peptide and proteotoxins trigger reactions in a cell or induce cell damage by:

–   increasing the permeability of the cell membrane or forming pores and inducing nonspecific flow of ions or small molecules in and out of the cell;
–   blocking or erratically activating endogenous ion channels which alter ion gradients across the plasma membrane, and alter the membrane potential and electrical activity in excitable cells (muscle cells and neurons);
–   activation or inhibition of signal transduction through endogenous hormone or transmitter receptors; and
–   enzymatic degradation of endogenous extracellular or cell surface proteins in the target organism.

2.12.2 Enzymes

Animal venoms may contain proteins that have enzymatic activity on substrate molecules in the host organism. These substances may have different functions, among them the facilitation of the distribution of other components of the toxin mixture within the target organism. They may also mediate the uptake of other toxins into the target cells or cooperate in a synergistic manner with other toxins. Especially enzyme constituents in snake or spider venoms are helpful in predigesting tissues of prey animals before the intestinal enzymes take over.

Special cases are the parasitoids which inject venoms into other animals to preserve the victims for later digestion by their own progeny. The components of their venoms regulate the physiology of the victim by transforming the host into a suitable environment for the development of the newborn larvae of the parasitoids. The parasitoid wasp *Torymus sinensis* injects proteases, phosphatases, esterases, nucleases, as well as protease inhibitors alongside with an egg into the galls around the larvae of the chestnut gall wasp, *Dryocosmus kuriphilus* [144]. The *Torymus* venom preserves the larvae of *Dryocosmus* so that the developing *Torymus* larva can use the *Dryocosmus* larva as a food resource during its own development.

There are different classes of enzymes that play roles in degrading proteins, lipids, extracellular matrix materials, or small regulatory molecules in the target organism. Virtually, all enzymes in animal venoms are hydrolases. Most widely distributed in animal toxins are the first three of these enzyme classes:
- Proteases
- Phospholipases
- Hyaluronidases
- Deoxyribonucleases, ribonucleases, and nucleosidases
- Esterases and apyrases
- Amino acid oxidases

2.12.2.1 Proteases

Proteases, also called proteinases or peptidases, are enzymes that mediate proteolysis. They cleave proteins and peptides by a hydrolytic mechanism at the peptide bond (Fig. 1.1) producing smaller peptide fragments of a protein (by endopeptidases) or single amino acids (by exopeptidases). There are many different subtypes of proteases with respect to their evolutionary origins, reaction mechanisms, occurrences, and biological significances [129, 130]. Aspartic, glutamic, and metalloproteases use an activated water molecule as a nucleophile for the attack on the peptide bond. Serine, threonine, and cysteine proteases use an amino acid residue within a catalytic triad as a nucleophile. This residue performs a nucleophilic attack on the peptide bond and links the protease covalently to the substrate protein which results in the release of the first cleavage product of the substrate protein. The complex of enzyme and second cleavage product of the substrate protein is then hydrolyzed by activated water to regenerate the free enzyme.

Most animal venoms contain proteases. They have to be injected into the target organism to become effective. They act locally at the bite or sting site and may also, at least in some instances, be transported to other places in the body with interstitial fluid or blood plasma. Locally around the bite or sting site, they mediate loosening of connective tissue (proteinaceous cell–cell as well as cell–matrix interactions) and degrade proteinaceous components of the extracellular matrix. Especially in animals that use extraintestinal digestion for preparing their prey for consumption (spiders, and partially also snakes), proteases injected into the prey start immediately dissolving proteinaceous material in the body of the prey animal. This is part of or at least preparatory for the digestion processes that is required to provide the predator with dipeptides and amino acids for absorption in the gut. Toxic animals that bite or sting potential predators for defensive purposes may have proteases in their venoms that either kill the attacker or cleave endogenous biologically inactive precursor proteins (proproteins) in the target animal to generate biologically active peptides or proteins that mediate pain sensations and other irritating effects.

A few examples may illustrate the uses of proteases in toxin mixtures of animals. Snake venoms of Viperidae and Elapidae which are injected into the target organism using hollow fangs contain proteases that start digesting proteinaceous tissue components. When humans are affected, large areas of bruises and blood blistering occurs because the proteases diffuse away from the site of injection through the extracellular space or are carried through the entire body if venom is injected into the blood stream. The proteases may destroy the endothelium and blood leaks into the interstitial space. In a prey animal, the proteases start already with the digestion of proteinaceous material before the snake ingests it [17]. A prominent family of proteases in snake venom are the metalloproteases (MPs) that play key roles in subduing prey in many vipers. These proteases are only active in the presence of divalent cations. Genomic analyses of rattlesnakes (*Crotalus* sp.) showed that all MP variants occurring in venoms of extant species are derived from a single, highly conserved disintegrin and metalloprotease (ADAM) gene [56]. Batroxobin (reptilase), a serin protease isolated from the venom of *Bothrops atrox*, has similar activity as thrombin and is clinically used for the improvement of blood fluidity [105].

The skin secretions of two Brazilian hylid frogs (*Corythomantis greeningi* and *Aparasphenodon brunoi*) are more toxic (LD$_{50}$ (mice, i.v.) is approximately 0.2 mg/kg BW) than the venoms of deadly venomous Brazilian pit vipers of the genus *Bothrops*. The skin secretions of these frogs show, among other effects, proteolytic activity. These frogs have well-developed delivery mechanisms, bony spines on the skull that decorate the upper lip. When approached by a potential predator, the frog jumps at it and pierces its skin. The toxin mixture from skin glands is rammed into the attacker's skin which makes these frogs venomous animals. *C. greeningi* has larger head spines and large skin glands producing a greater volume of secretion, while *A. brunoi* has more lethal venom (LD$_{50}$ in mice of 0.16 mg/kg BW (i.p.) or 1.6 mg/kg BW (s.c.)) [74].

Soldier-caste nymphs of the social aphid *Tuberaphis styraci* produce a protease, cathepsin B, in their intestinal cells. When aphid predators threaten the reproductive aphids, the soldier nymphs attack the intruder using their piercing mouthparts and inject the protease. This results in paralysis and death of the attacker [88, 89].

The venom of the Gila monster (*Heloderma* sp.) contains a protease that aims at stunning the attacker by inducing shock and intense pain. The glycoprotein gilatoxin is a variant of kallikrein, an endogenous protease that usually produces bradykinin from circulating plasma kininogen when blood vessels are injured [162]. The presence of gilatoxin in tissues of animals bitten by Gila monsters induces intense pain sensations and a sudden fall in blood pressure (shock). The LD$_{50}$ (mouse, s.c.) is approximately 1 mg/kg BW. A very similar toxin is produced by salivary gland cells of a very small mammal, the American shrew, *Blarina brevicauda*. Blarinatoxin (BLTX) is comprised of 253 amino acids (molecular mass: 28 kDa) and induces pain, swelling, hypotonicity, as well as paralysis in target animals. The LD$_{50}$ (mice, i.p.) is 0.1 mg/kg BW [81].

2.12.2.2 Phospholipases

Phospholipases (PL) are enzymes that cleave fatty acids from membrane-bound phospholipids. There are four major classes (A, B, C, and D), which are distinguished by the type of reaction which they mediate (Fig. 2.61) [136]. Phospholipases A cleave the acyl chain at the SN-1 position of phospholipids (phospholipase A_1, PLA_1) or at the SN-2 position (phospholipase A_2, PLA_2). As the fatty acid in position SN-2 is arachidonic acid (AA) in most membrane lipids, PLA_2 is the relevant enzyme generating free AA when activated by external stimuli and cellular signaling [114]. Phospholipase B cleaves ester bonds of acyl chains at SN-1 or SN-2 acyl chains, resulting in the formation of lysolipids (membrane lipids with one missing acyl chain). Phospholipase C cleaves off polar headgroups in membrane lipids between the lipid backbone and the phosphate bridge to the polar headgroup at SN-3. Phospholipase D leaves the phosphate untouched and cleaves between the phosphate and the polar headgroup. The resulting lipid is termed 'phosphatidic acid'.

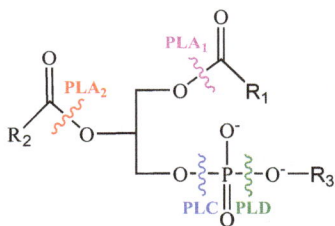

Fig. 2.61: Cleavage sites of different types of phospholipases in a membrane lipid (glycerophospholipid) of animal cells. Enzymes that have activities of both PLA_1 and PLA_2 are termed phospholipase B.

Animal venoms may contain phospholipases A_2, B, or D. Such phospholipases support the degradation of membrane lipids and tissue destruction in preparation for digesting prey organisms. However, there are also important immediate functions of phospholipases in animals that utilize their venoms for defensive purposes. Honeybee (*Apis mellifera*) venom contains a phospholipase A_2 that mobilizes arachidonic acid from membrane lipids of cells in the target organism. Arachidonic acid is then metabolized by endogenous cyclooxygenases [183], resulting in the production of tissue mediators like prostaglandins that have functions in blood pressure regulation, in immune modulation, or in sensitizing pain receptors in tissues of target animals [76].

Phospholipase A_2 activities (E.C. 3.1.1.4) are found in venoms of many different toxic animals [178]. It is present in virtually any snake venom and amplifies the venom toxicity by overactivating endogenous signaling pathways in cells of the target animal [15]. Many snake venom PLA_2s act through enzymatic as well as nonenzymatic mechanisms [77, 79]. Bee venom contains PLA_2 activity as well. The secretory phospholipase A_2 makes up 10–12% of dry bee venom [151]. Moreover, this protein is one of the major allergens in bee venom that may stimulate IgE-mediated overreactions of the immune system [121, 125]. On the other hand, bee venom PLA_2 may be used in medical applications as a tool for suppressing growth of prostate cancer cells [6] and for treating other ailments [95]. Phospholipase A_2 activities have been detected in the toxic saliva of Cephalopoda [27] and Komodo dragons (*Varanus komodoensis*) [51]. Scorpion venoms [86, 163] as well as wasp venoms [28] also contain phospholipases A_2.

Phospholipase B activities have been identified in snake and bee venoms [36]. A phospholipase D has been found in the venoms of American brown spiders (*Loxosceles* sp.) which induce severe dermonecrosis [60]. This enzyme hydrolyzes glycerophospholipids as well as sphingolipids which may explain its high tissue destructive potency. True sphingomyelinases which selectively hydrolyze the sphingolipid sphingomyelin occur in spider and in tick venoms and may have cytotoxic functions in the target animals [50]. The degradation of sphingomyelin which is enriched in the outer leaflet of the plasma membrane [23] may destabilize the plasma membranes of target cells.

2.12.2.3 Hyaluronidases

Hyaluronan (hyaluronic acid, HA) is a linear, non-sulfated polysaccharide composed of repeating disaccharide units comprised of D-glucuronic acid and *N*-acetyl-D-glucosamine [166]. It is an important constituent of the extracellular matrix in human and animal tissues. It maintains tissue elasticity by hydrostatically binding large amounts of non-diffusible water. When venomous animals inject toxins locally into the skin of a target organism, the venom ingredients would more or less stay at the injection site and would not be diffusively distributed over large distances within the body of the target animal. Hyaluronidases in the venom, however, induce the destruction of hyaluronan and liberate the local water molecules. This allows for rapid diffusion of the venom ingredients to more distant tissue areas around the site of injection or their distribution over the entire body.

Hyaluronidases [42] represent a group of glycosidases which mainly degrade hyaluronan. All hyaluronidases in animal toxins seem to belong to the same subclass of hyaluronidases (E.C. 3.2.1.35) [16]. They occur in venoms of arthropods (scorpions, spiders, bees, and wasps), leeches, lizards, and snakes. Hyaluronidases are highly abundant in venoms of scorpions, bees, and wasps [157]. The enzyme is also present in the venom of brown recluse spiders (*Loxosceles reclusa*) [174]. Leech hyaluronidases are β-endoglucuronidases and degrade hyaluronan [100] by producing tetrasaccharides with gluconic acid at the reducing end. They show no activities toward chondroitin and chondroitin-4- or chondroitin-6-sulfates [101]. The toxic saliva of the American desert lizard *Heloderma* sp. is very rich in hyaluronidase. This enzyme is specific for hyaluronan cleavage and has only weak activity toward chondroitin-4-sulfate. It does not cleave chondroitin-6-sulfate [160]. Snake hyaluronidases degrade hyaluronan, chondroitin, chondroitin-4- and chondroitin-6-sulfates, generating various oligosaccharides, mainly tetrasaccharides [39, 57].

2.12.2.4 Deoxyribonucleases, Ribonucleases, and Nucleosidases

Deoxyribonucleases are enzymes that degrade DNA, and ribonucleases (RNase) cleave RNA. Nucleosidases hydrolyze nucleosides generating free purine or pyrimidine bases. These enzymes are present in many animal venoms, especially in those that contribute to initial steps of predigestion of prey animals.

An exonuclease (phosphodiesterase I) from the venom of a rattlesnake, *Crotalus adamanteus*, successively hydrolyzes 5′-mononucleotides from 3′-OH-terminated ribo- and deoxyribo-oligonucleotides [171].

From the venom reservoirs of the spines of the crown-of-thorns starfish *Acanthaster planci* two so-called lethal factors, plancitoxins I and II, have been isolated. Amino acid sequence analysis of plancitoxin I revealed that this protein carries 40–42% of all amino acids at positions where they also occur in mammalian deoxyribonucleases II (DNases II) indicating homology of these proteins [147]. Functional analysis revealed that plancitoxin I has indeed DNA degrading activity. In mice that were i.v.-injected with plancitoxin I liver enzymes increased substantially, indicating that this protein is hepatotoxic. The LD_{50} (i.v., mice) is 140 μg/kg BW.

Virtually, all kinds of snake venom contain ribonucleases, e.g., those of the Vietnam cobra, *Naja atra* [117], which are, however, less well studied than other snake venom components in terms of structure and function.

Nucleosidases are constituents of many snake venoms. It was recently revealed that venoms of rattlesnakes, e.g., of *Crotalus atrox* and *C. oreganus helleri*, contain extracellular vesicles whose content is enriched in 5′-nucleosidase [170].

2.12.2.5 Esterases and Apyrases

Esterases cleave ester bonds or phosphodiester bonds, e.g., in phospholipids or oligonucleotides. They are present in many animal venoms, especially enriched in snake and spider venoms [119].

Acetylcholine esterases (E.C. 3.1.1.7) are present in the venoms of elapid snakes [1, 49]. These enzymes cleave the neuromuscular transmitter acetylcholine in the synaptic cleft and inhibit signal transmission from motor neurons to skeletal muscle cells. This results in rapid paralysis of the target organism.

The roles of nucleotidases (5′-nucleotidase, ATPase, and ADPase) which are widely distributed among animal venoms are not well studied [34]. It is assumed that their major function is in the generation of purines, mainly adenosine, which may be instrumental in rapid prey immobilization by relaxing skeletal muscle cells. This is especially important for many snake species as they usually do not hold on to their prey until the injected venom becomes effective. Thus, applying the bite and prey paralysis should rapidly follow each other so that the prey does not get too far away to be easily retrieved once it is immobilized.

Apyrase (E.C. 3.6.1.5), also called ATP-diphosphatase, adenosine diphosphatase, ADPase, or ATP-diphosphohydrolase, is a calcium-dependent enzyme that hydrolyzes ATP to yield AMP and inorganic phosphate. Apyrases act also on other nucleotides. The salivary apyrases of blood-feeding arthropods like bed bugs (*Cimex lectularius*) are nucleotide-hydrolyzing enzymes that may inhibit host platelet aggregation by degrading ADP [150].

2.12.2.6 Amino Acid Oxidases

L-Amino acid oxidases (LAAOs; E.C. 1.4.3.2) are major components of many snake venoms (up to 30% of the venom dry mass in some snake species). These homodimeric flavoenzymes catalyze the stereospecific oxidative deamination of L-amino acids. LAAOs in snake venom prefer hydrophobic L-amino acids as substrates. Products are alpha-keto acids, ammonia, and hydrogen peroxide [46]. Pure preparations of LAAOs are cytotoxic and induce apoptosis in human and animal cells. They affect platelet aggregation and seem to have bactericidal and antiparasitic activities due to the production of hydrogen peroxide and the induction of oxidative stress in target tissues [186].

2.12.3 Cytotoxins

The cytotoxins among the polypeptides induce structural damage in animal or human cells and tissues. They may directly mediate malfunctions in receptors, ion channels, or other cell surface proteins like enzymes. Alternatively, they may indirectly affect normal cell functions by different mechanisms, such as blocking protein synthesis, attenuating the turnover in energy metabolism, or interfering with the assembly and disassembly of cell junctions or the cytoskeleton. Alternatively, they may induce intense inflammatory responses which induce damage to cells or tissues.

An evolutionarily ancient mechanism of inducing damage to cells of competing organisms, attacking predators, or to cells of potential prey is the generation and secretion of pore-forming toxins (PFTs) [72]. PFTs of very similar molecular structures have been identified in several groups from bacteria [124] like hemolysin A in *Staphylococcus aureus* [58, 69] to animals like sea anemones [110, 140], earthworms (lysenin and eiseniapore) [33, 91, 148], or vertebrates (membrane attack complex of the complement system) [37, 111].

Common to all of these toxins is that PFTs are synthesized and secreted by their producers as water-soluble protein monomers. These molecules recognize their target cells by binding to more or less specific receptors, which may be sugar or lipid derivatizations of plasma membrane proteins, or to cell surface proteins. Binding of many monomers results in a substantial increase in the local concentration of PFTs which facilitates their oligomerization process. Multimeric complexes of PFTs (usually heptamers) undergo a conformational change that forms transmembrane pores in the plasma membranes of target cells. These pores may be permeable to ions, water, and organic molecules of low molecular masses which compromise the cell function or induce cell death.

While PFTs attack microorganisms or many different cell types in a multicellular eukaryotic target organism (epithelial cells, blood cells, etc.), there are protein toxins, especially in animal venoms, which target specific cell types in other animals or humans. Neurotoxins specifically target neurons or other cell types in the nervous system of animals, myotoxins affect smooth or skeletal muscle cells, and cardiotoxins interfere with normal heart function. Hemotoxins interfere with platelet aggregation or blood clotting in affected animals or humans, or they induce blood cell lysis. Many animal toxins contain proteins that, besides having other functions, induce immune responses in the target organism. Recurrent exposure to such agents (allergens) may induce allergic reactions or even anaphylaxis which may be life-threatening. Other protein components of animal toxins specifically recognize certain combinations of sugar residues in glycosylated surface proteins of target cells. These so-called lectins have roles in recognition at the cellular and molecular levels (e.g., self- or non-self-recognition). Upon binding to their target molecules, lectins may cause agglutination of particular cells (e.g., platelets) or precipitation of glycoconjugates and polysaccharides which alters cell function or may even induce damage to the target tissues.

2.12.4 Neurotoxins

Animals rely on their nervous systems to detect potential predators in their habitats (sensory system) and on their skeletal muscle system (motor neurons and skeletal muscle cells) to escape or fight. Thus, toxic predators often rely on neurotoxins and/or myotoxins to hamper proper neural and muscle function of their victims. Inducing severe pain sensations, stunning the target animal, and inducing shock are mechanisms utilized by various venomous animal species like scorpions [118], box jellyfish [30], centipedes [25], and lizards [35]. The transmitters used by motor neurons to activate skeletal muscle cells are taxon-specific. Glutamate is the excitatory transmitter released from motor neurons in arthropods, while acetylcholine is the transmitter that activates skeletal muscle cells in vertebrates, mollusks, and annelids [146]. Toxins may interfere with transmitter release on the presynaptic nerve ending or may affect the postsynaptic side, i.e., the receptor mechanisms of the skeletal muscle cell. Toxins directed against postsynaptic transmitter receptors must, therefore, be different to achieve rapid immobilization of arthropod prey or deterrence of vertebrate predators.

Protein toxins that modulate transmitter release at the presynaptic membrane occur in snake venoms. Mamba (*Dendroaspis* sp.) venoms contain dendrotoxin that structurally resembles pancreatic protease inhibitors but does not display any inhibitory potencies against trypsin or chymotrypsin. Instead, they inhibit voltage-activated potassium channels in the nerve endings (Shaker Kv1.1, Kv1.2, and Kv1.6 KC channels) followed by prolongation of each of the incoming action potentials. This results in ever increasing amplitudes of the action potentials in the nerve ending, copious transmitter release, and a rapid exhaustion of the postsynaptic muscle cell with ultimate flaccid paralysis of the skeletal muscle [143, 149].

Other predators rely on toxins that bind to endogenous docking molecules in plasma membranes of neuronal cells of their victims and destroy such cells. Examples are the neurotoxic phospholipases A_2 in snake venoms [164]. These toxins are thought to inhibit proper presynaptic plasma membrane trafficking associated with exocytotic neurotransmitter release and endocytotic reuptake of membrane material, which rapidly blocks acetylcholine release into the synaptic cleft. These toxins, e.g., notexin of the Australian tiger snake (*Notechis scutatus scutatus*), ultimately destroy the synapses of motor neurons by enzymatic and nonenzymatic mechanisms [137, 177]. There is a remarkable structural diversity among the neurotoxic phospholipases A_2 in venoms of different snake species. Notexin is built from a single polypeptide strand that is stabilized by seven internal disulfide bonds. Crotoxin of the South American rattlesnake *Crotalus durissus terrificus* is a heterodimer of a basic 13 kDa subunit and an acidic 8.4 kDa subunit [64]. In other cases, the holoproteins are composed of three or four subunits, e.g., in taipoxin of the Australian taipan (*Oxyuranus scutellatus scutellatus*) [98].

Another example of a neurotoxin with a different mechanism of action is latrotoxin of black widow spiders (*Latrodectus* sp.). When spiders inject their venom into a vertebrate, the endogenous neuronal proteins neurexin, latrophilin, and protein tyrosine phosphatase σ serve as docking sites on the presynaptic surface of motor neurons. Alpha-latrotoxin creates Ca^{2+}-permeable channels in the plasma membranes of neurons connecting cytosol and extracellular space [176]. Ca^{2+} influx into the cells through these channels along the electrochemical gradient for Ca^{2+}-ions induces uncontrolled transmitter release which may result in spastic paralysis in the target organism.

Venom components of other toxic animals interfere directly with neuronal ion channels in the target organisms. Protein toxins in scorpion venoms (e.g., that of the North African scorpion *Leiurus quinquestriatus*) cause release of copious amounts of neurotransmitters from nerve endings by inhibiting the intrinsic inactivation of activated voltage-dependent sodium channels in the neuronal plasma membrane [22]. Cone snails (Conidae; see Section 3.2.3) generate many different toxic proteins in their venom glands that they use to hunt and to defend themselves. Among these are the ω-conotoxins which are efficient inhibitors of N-type voltage-activated calcium channels (Cav2.2) in mammalian neurons [75]. These channels occur in presynaptic terminals of nerve cells and are required for transmitter release from the respective synapses. They are involved in processing of pain signals in the brain stem. Thus, ω-conotoxins exert analgesic effects in target animals or affected humans. Chemically synthesized ω-conotoxin (ziconotide) is approved for patients with severe chronic pain, including neuropathic pain refractory to opioids [73]. While ω-conotoxins block Ca^{2+} channels, voltage-gated K^+ channels are blocked by another kind of conotoxin, the κ-conotoxins. Voltage-gated Na^+ channels, however, are affected by δ-conotoxins, which, like the abovementioned scorpion toxins, inhibit the autonomous inactivation of these channels which results in overexcitation of the respective neurons [63]. μ-Conotoxins are inhibitors of voltage-gated sodium channels. A newly discovered isoform, SxIIIC, has a strong selectivity toward the human subtype NaV1.7 [106].

Several animal toxins contain components that affect the lifetime of transmitter molecules within the synaptic cleft. Acetylcholine esterases are components of some snake toxins, e.g., that of the Asian Sind krait, *Bungarus sindanus* [1]. These enzymes rapidly cleave acetylcholine released from motor neuron terminals into the synaptic cleft. They generate acetate and choline which are inefficient in activating the nicotinic acetylcholine receptors in the postsynaptic muscle cell surface. This results in flaccid paralysis in affected animals. Sponges and corals, on the other hand, contain small diterpene molecules like the cembranoids of the soft corals *Eunicea knighti* or *Pseudoplexaura flagellosa*. These are inhibitors of endogenous acetylcholine esterases in the extracellular matrix within synaptic clefts in animals [40]. Inhibition of the esterase results in accumulation of acetylcholine within the synaptic cleft and overexcitation of the postsynaptic muscle cell. Cramps and spastic paralysis in the respiratory muscles may be fatal.

That animal toxins are evolutionarily optimized for fitting to highly specific target molecules is illustrated by the fact that different snake toxins (α- or κ-neurotoxins) interact selectively with nicotinic acetylcholine receptors (AChR) in postsynaptic muscle (α) or in presynaptic neuronal cells (κ). Such toxins are being tested for potential medical applications [159].

2.12.5 Myotoxins

Toxins of venomous predators may also block neuromuscular signal transmission by blocking the transmitter receptors in the postsynaptic (muscle cell) plasma membrane. An example is α-bungarotoxin of the Taiwanese many-banded krait (*Bungarus multicinctus*). It belongs to the three-finger toxin family and irreversibly binds to nicotinic acetylcholine receptors (nAChR) in skeletal muscle cells (K_d approx. 10^{-10} mol/L) [80]. This toxin blocks signal transmission from motor neurons to skeletal muscle cells by suppressing the generation of action potentials in the muscle cells. This results in rapid onset of flaccid paralysis of animals bitten by this snake.

Animals embedded between the trophic levels must simultaneously balance pressures to deter predators and acquire resources. Examples are scorpions which use their toxins to hunt for insect prey, but may use them also to deter vertebrate predators (birds and mammals). Recent studies have shown that the venom composition used for defensive purposes depends on the kind of predator in the respective habitat. After 6 weeks of exposure to different predator scents, scorpions exhibited significantly different venom chemistry compared with naive scorpions. This change also included a change in the relative proportions of compounds toxic to vertebrates and compounds effective against invertebrate prey [53]. This, in turn, shows that scorpions have evolutionarily developed different sets of protein toxins. One set is directed against arthropod prey, and the other set is directed against vertebrate predators.

Some animal toxin mixtures contain agents that disturb normal functions of smooth muscle tissue in affected animals. Smooth muscle cells are present in inner organs like the gastrointestinal tract where they mediate the peristalsis moving the gut contents and in blood vessels where they control blood pressure and local blood flow. Protein toxins containing repetitions of the amino acids alanine, valine, isoleucine, and threonine (so-called AVIT scaffold peptides) are found in the saliva of the Komodo dragon (*Varanus komodoensis*) and in skin secretions of toads (*Bombina* sp.). These proteins are potent agonists of mammalian prokineticin receptors which are highly abundant in gastrointestinal smooth muscle cells surrounding the gut. Binding of the toxic proteins to these receptors mimics the effect of an overdose of endogenous prokineticin and elicits gastrointestinal cramps and intense abdominal pain sensations [116].

Vascular smooth muscle cells are targets in prey animals of some toxic predators as well. The glycoprotein gilatoxin [162] that is a constituent of the toxin mixture of the Gila monster (*Heloderma suspectum*) is functionally equivalent with kallikrein, a protease that cleaves circulating kininogens producing kinins, e.g., bradykinin, which induce severe pain sensations [154] and relaxation of vascular smooth muscle cells [14]. This may induce hypotonic shock in animals exposed to these toxins.

Other animals use components of their own saliva to inhibit endogenous kinin production in host animals. The saliva of blood feeding leeches (*Hirudo medicinalis*) contains a heat-labile substance, i.e., a protein, that inhibits plasma kallikrein in vertebrates and minimizes kinin production which otherwise would be a consequence of wounding the body surface of the host animal [9]. This has several benefits for the leech. The host does not feel any pain during the leech's bite and the vascular tone in the host is maintained which suits the leech in that it has a steady supply of blood to the wound that it can ingest.

2.12.6 Cardiotoxins

Cardiotoxins (CTX) in animal toxin mixtures erratically affect the cardiovascular system of target animals. Many snake venoms contain proteins that display cardiotoxic effects, especially those of cobras. The venom of the Caspian cobra (*Naja oxiana*) contains S-type (serine residue at position 28) and P-type (proline residue at position 30) cardiotoxins with 59–61 amino acid residues and 4 internal disulfide bonds. Both types of toxins induced contracture in papillary muscle preparations and strong contractions in aortic rings prepared from rats. The P-type toxin CTX-2 was more potent compared with the S-type toxin CTX-1. The effects of these toxins were significantly attenuated by nifedipine, a blocker of L-type Ca^{2+}-channels, indicating that the toxins may induce an overload of heart muscle cells with calcium ions [5]. In addition,

cobra cardiotoxins seem to be able to pass the plasma membrane of cardiomyocytes and damage the mitochondria which results in cell death [96].

The venom of the Chinese scorpion (*Buthus martensii*) contains bukatoxin [153], a
peptide comprising 65 amino acids and 4 disulfide bonds. Bukatoxin shares 78% and
72% structural similarity with neurotoxin X of *Mesobuthus eupeus* [61] and neurotoxin
IV from *Leiurus quinquestriatus* [84], respectively. Such toxins inhibit the intrinsic inactivation of neuronal voltage-activated Na^+ channels and promote membrane depolarization. This triggers the release of catecholamines from sympathetic nerve fibers
and may play a role in initiating diastolic depolarization in cardiac fibers, which
leads to tachycardia or tachyarrhythmia and hypertension [31]. Desensitization of β-
adrenergic receptors may subsequently induce bradycardia and hypotension [131].

The sarafotoxins (SRTXs) are 21 amino acid peptides which are components of the
venom produced by the burrowing vipers of the genus *Atractaspis* [13]. They consti-
tute approximately 30–40% of the venom protein content of *Atractaspis engaddensis*.
The amino acid sequences are similar to those of endothelins, most effective vasocon-
strictor peptides. Functional assays have revealed that SRTXs compete with endogenous endothelins in binding to the endothelin receptors on vascular smooth muscle
cells. Depending on the subtype of SRTX present in the venom, this may result in muscle relaxation and a drop in blood pressure, initial hypotension followed by a hypertensive period, or irregularities in the electrocardiogram and cardiac arrest in the
intoxicated vertebrate animal.

Shrews are very small mammals and must constantly feed to maintain their high
metabolic rate and their constant body temperature. Many shrew species have toxins
in their saliva to efficiently overwhelm prey organisms, mainly insects, worms, and
frogs. Venomous saliva of the Eurasian water shrew (*Neomys fodiens*) contains cardioactive compounds that decrease contractile activity in the heart muscle of frogs and
beetles (*Tenebrio molitor*) [85] and may serve as prey-immobilizing agents.

2.12.7 Hemotoxins

Hemotoxins in animal venoms generally destroy blood cells or interfere with hemostasis in their victims. Usually, the initial phase of stopping blood flow from injured
blood vessels in vertebrates (primary hemostasis) is mediated by blood platelets
(thrombocytes) that interact via von Willebrand factor (vWF) with collagen molecules
that lie exposed in the injured vessel wall and with each other to form platelet aggregates at the site of injury. Subsequently, a cascade of proteases (blood coagulation
cascade) is activated which ultimately leads to the activation of thrombin. This calcium-dependent protease cleaves circulating fibrinogen to yield fibrin molecules
that form a firm molecular network that stops blood from leaving the vessel (secondary hemostasis) [165].

Hemotoxins may target platelet aggregation and/or the blood coagulation cascade in target organisms which results in severe bleeding or in rapid blood coagulation. Both strategies have been realized in nature and a few examples are described here. Hemophagic animals like leeches, ticks, bed bugs, mosquitos, or vampire bats use salivary components (subsumed under the term 'sialome') to prevent platelet aggregation as well as blood clotting at the sites of bites or stings in host animals [167]. This allows feeding on a continuous flow of liquid blood that is running into the wound.

Leeches have evolutionarily developed salivary compounds that inhibit platelet aggregation as well as those that inhibit blood clotting. Saratin, a protein isolated from the saliva of the medicinal leech, *Hirudo medicinalis*, is an inhibitor of von Willebrand factor-dependent platelet adhesion to collagen [7, 169]. Calin, another sialome component of the medicinal leech, has similar effects by blocking the interaction of platelets with collagen [113] via the platelet glycoprotein IIb/IIIa, i.e., integrin αIIbβ3 [32]. Some snake venoms contain platelet aggregation inhibitors [3]. An example is botrocetin from the South American pit viper *Bothrops jararaca*. This toxin enhances the affinity of the A1 domain in the von Willebrand factor (vWF) for the platelet receptor, glycoprotein Ibα (GPIbα), which prevents regular platelet aggregation [52]. This results in prolongation of bleeding in the affected animal or human.

Another family of proteins, the disintegrins, are important components of animal toxin mixtures interfering with tissue integrity and hemostasis in prey animals. The central molecular core of disintegrins is present in enzymes as diverse as snake venom metalloproteases (SVMP [21, 56]) and ADAMs (*a d*isintegrin *a*nd *m*etalloprotease [168]) which indicates that the members of the disintegrin family are evolutionarily homologous. Some disintegrins interact with αvβ3-, α5β1- or α4β1-integrins expressed on platelets, endothelial cells, fibroblasts, or phagocytes, and affect cell–matrix and cell–cell interactions. Disintegrins in animal toxins may assist in detaching cells from the extracellular matrix or from neighboring cells, thereby damaging the tissue integrity. In other cases, disintegrins may prevent cells from interacting and forming cell aggregates. Inhibition of αvβ3-mediated platelet aggregation may serve as an example. Disintegrins have been discovered in the salivary gland secretions of ticks and leeches but also in hookworms and horseflies [4, 108]. However, they are components of snake venoms as well. Trigramin has been isolated from venom of the bamboo pit viper *Craspedocephalus gramineus* [71]. This disintegrin inhibits binding of von Willebrand factor to activated platelets. Its binding affinity to the relevant integrins in platelets (αIIbβ3) is quite high ($K_d = 10$ nmol/L).

Specific inhibitors of proteases of the blood coagulation cascade (factor X, prothrombin, thrombin) in vertebrates are also components of many animal toxin mixtures. Hirudin is such a highly potent inhibitor of thrombin, a key enzyme of the secondary hemostasis pathway. It is present in several isoforms in the sialome of several leech species of the genus *Hirudo* [104, 112]. Hirudin is a high-affinity, bivalent thrombin inhibitor ($IC_{50} = 180$ pmol/L [172]). The N-terminal portion masks the active site of the protease and the acidic C-terminal covers the fibrinogen binding site of thrombin, the exosite I [138]. Recombinant variants of this natural anticoagulant are used in medical applications [59, 120].

Many snake species carry effective anticoagulants in their venoms [185]. An anticoagulatory protein with the ability to inhibit prothrombin has been purified from the venom of the bamboo pit viper, *Craspedocephalus gramineus* from the south of India. The venom of the abovementioned South American pit viper *Bothrops jararaca* contains a direct thrombin inhibitor (K_d = 0.6 nmol/L), a dimeric protein called bothrojaracin [109]. The two subunits are connected by a disulfide bond, but there are three further disulfide bonds within each of the two subunits. Bothrojaracin does not interact with the active site of thrombin but seems to inhibit the enzyme by interfering with fibrinogen binding at the exosite I.

Rhodniin (11 kDa) from the blood-sucking bug *Rhodnius prolixus* (Heteroptera) belongs to the Kazal-type serine protease inhibitors. It specifically inhibits thrombin with a very high potency (K_i = 0.2 pmol/L). The N-terminal Kazal domain of rhodniin binds the active site of thrombin, while the C-terminal Kazal domain binds to exosite I [48].

Cross-linkage of fibrin molecules in a blood clot occurs via isopeptide bonds formed over the side chains of lysine residues. Factor XIIIa of the blood coagulation cascade is mediating this specific way of cross-linking fibrin molecules [115]. Isopeptidases in animal toxin mixtures counteract hemostasis and thrombus formation by dissolving isopeptide bonds between cross-linked fibrin molecules. Destabilase, a protein in leech saliva that exhibits antimicrobial effects by its muramidase (lysozyme) (see Section 2.12.10) activity, is also able to cleave isopeptide bonds in cross-linked fibrin molecules via its endo-ε-(γ-Glu)-Lys-isopeptidase activity [8, 10]. Compared with streptokinase, a bacterial isopeptidase used in medical thrombolysis, the leech enzyme is clearly more efficient in dissolving blood clots.

Another way of dissolving fibrin clots is activation of plasminogen, an endogenous protein in plasma (concentration in mammals approx. 2 mmol/L) that yields plasmin when proteolytically activated by plasminogen activators that convert the proenzyme plasminogen to the active serine protease plasmin [97]. Plasmin, in turn, is efficiently able to cleave fibrin bundles. In healthy animals, it lyses erratic blood clots within blood vessels. This protects mammals including humans from developing thrombosis. However, when plasminogen is activated in a target animal by ingredients of venoms, the blood becomes incoagulable and keeps flowing out of wounded blood vessels. Such a strategy is used by vampire bats (*Desmodus rotundus*) in Central and South America [62, 141]. Specimens of this species express plasminogen activator in their salivary gland cells and transfer the protein into the bite wound, e.g., in cattle, via their salivary secretions. A number of plasminogen activators have been isolated from venoms of several snake species, e.g., TSV-PA from the Chinese green tree viper (*Trimeresurus stejnegeri*) and LV-PA from the South American bushmaster (*Lachesis muta*) [94]. Bites of such snakes induce profuse bleeding from the bite wounds and formation of large bruises.

Other snake species deliver venom proteins to their prey that are procoagulants and induce blood clotting in the vascular system. Such a snake venom component was isolated from the venom of the European horned viper (*Vipera ammodytes*). This 34 kDa glycoprotein turned out to be a chymotrypsin-like serine protease that specifically activates the blood coagulation factors X and V [92]. In some cases of snakes that routinely deliver procoagulants to their prey there was a paradoxical effect observed that these procoagulants appear to render the blood of target animals incoagulable. A potential explanation for these observations may be that certain snake venom procoagulants seem to induce the release of endogenous plasminogen activators in the prey animals. Venom components of Crotalinae seem to stimulate secretion of cellular plasminogen activators from endothelial cells in bitten prey animals. An example is batroxobin, a thrombin-like enzyme in the venom of the common lancehead, *Bothrops atrox*, which induces the release of tissue plasminogen activator in vivo [83]. This process renders the blood of the victim incoagulable and induces severe bleeding. A similar mechanism has been proposed for a venom component of the pit viper *Protobothrops flavoviridis*, habutobin, which releases t- and u-type plasminogen activators from cultured endothelial cells isolated from pulmonary artery [156].

Moreover, snake venoms of some viperid snakes contain metalloproteases which affect fibrin cross-linking and hemostasis and may provoke hemorrhage in target animals. These enzymes have fibrin(ogen)olytic activities. The mechanisms, however, are different from those of destabilases and independent of the activation of plasminogen. Instead, the Aα-chain of fibrin(ogen) is degraded, which prevents the fibrin clot formation [139]. An example of such enzymes is barnettlysin-I which occurs in the venom of the South American Barnett's pit viper, *Bothrops barnetti*. This venom component may be a potential therapeutic agent to treat major thrombotic disorders in humans.

2.12.8 Allergens

Virtually, all proteins among the animal toxins have the potential to sensitize the immune system of mammals and to induce allergic reactions upon repeated exposure. However, there are differences in the allergic potencies of different proteins. In humans, the members of three protein families have high allergic potentials, namely serum albumins, lipocalins, and secretoglobins [87]. While serum albumins are not generally occurring in animal toxin mixtures, lipocalins are present in insect, tick, and reptile toxins, but secreted globular proteins are regularly present in animal toxin mixtures [50].

The lipocalins are a family of proteins which allow the transmembrane transport of small hydrophobic molecules such as steroids, bilins, retinoids, and lipids between the extracellular space and the cytosol in animals and humans. Their primary sequences are rather diverse, but they all share a common tertiary structure comprising eight strands of antiparallel beta-sheets forming a barrel with an internal ligand-binding site [45]. The caterpillars of the giant silkworm moth (*Lonomia obliqua*) possess urticating bristles for self-defense which are covered with a toxin mixture containing several lipocalin-scaffold variants [135]. Among them is Lopap (*Lonomia obliqua* prothrombin activator protease), which has a unique serine protease-like activity and activates prothrombin [132]. Accidental contact with *Lonomia obliqua* caterpillars may result in acute renal failure and intracranial hemorrhage.

A few of the many secreted globular proteins with allergic potential should be mentioned. A prominent example is melittin, a 26 amino acid peptide in bee venom (e.g., of the honeybee, *Apis mellifera*) that amounts to approximately 50% of the venom dry weight. Besides forming ion-permeable pores in the plasma membranes of affected animal or human cells [158], it is an allergen that activates the immune system of animals and humans upon repeatedly receiving stings as indicated by the presence of high titers of IgE antibodies against melittin in bee venom-sensitive individuals [122]. Other proteins in venoms of hymenopteran insects have allergenic potential as well. Hyaluronidases and phospholipases are present in venoms of bees, bumblebees, wasps, and hornets and may induce anaphylaxis in sensitive persons [70, 121, 125].

2.12.9 Lectins

Lectins are soluble or membrane-associated proteins or glycoproteins that have high affinities for certain monosaccharide, amino sugar, uronic acid, or oligosaccharide residues of glycoproteins or glycolipids. They form strong bonds with other molecules (that are mostly associated with cell surfaces) which carry such residues and may cross-link such molecules on different cells. However, carbohydrate-specific immunoglobulins and enzymes that have similar abilities are not considered to be lectins.

Lectins were initially discovered in extracts of plant seeds, in this case of castor beans (*Ricinus communis*), that were able to agglutinate erythrocytes [155]. Peter Hermann Stillmark named the active component 'ricin' and suggested that it might be a protein. The lectin protein family was later termed 'agglutinins' or, since most of these molecules were isolated from plants, 'phytohemagglutinins'. The toxicity of ricin, however, is more related to its ability to block ribosomal protein synthesis in animal and human cells than to its ability to agglutinate erythrocytes [47].

Lectins occur in venoms of snakes but some of them are also found in fish. The snake venom lectins are calcium-dependent and thus termed 'C-type-lectins' or 'snaclecs' (for 'snake venom C-type lectins') [26]. The functions of the C-type lectins in snake venoms are not well understood. They have been implicated in exerting a variety of biological effects such as hemagglutination, antibacterial functions, blood coagulation, mitogenic activity, platelet aggregation, edema formation, altering blood pressure, cytotoxicity, or modulation of calcium release from skeletal muscle sarcoplasmic reticulum. Mannose-binding C-type lectins have been identified in the venoms of 13 Australian elapid snakes including the common taipan *Oxyuranus scutellatus* [41]. Galactoside-binding lectins recognize and interact with terminal galactoside residues of glycans. Examples are thrombolectin of *Bothrops atrox* [55] and several other lectins isolated from venoms of Viperidae and Elapidae.

Several lectins have been isolated from fish. Nattectin is a galactose-specific lectin produced in the venom gland at the base of the dorsal fin of the Brazilian toadfish, *Thalassophryne nattereri*. It is a basic, non-glycosylated, 15 kDa monomeric protein that functions as a hemagglutinin in a Ca^{2+}-independent manner [102].

An unusual combination of a lectin with a pore forming protein has been detected in eggs of the aquatic golden apple snail, *Pomacea canaliculata* [38]. PcPV2 is a highly stable protein that survives the passage through the intestinal system of egg predators, passes the intestinal lining, and unfolds severe neurotoxicity and lethality in rodents. The toxin is an efficient biochemical defense against rats and mice that otherwise would consume these aerial egg clutches.

2.12.10 Antimicrobial Peptides and Proteins

Lysozyme is an antibacterial enzyme that is produced by epithelial cells, macrophages, and polymorphonuclear neutrophils in animals and humans. Lysozyme is highly abundant in secretions including tears, saliva, human milk, and airway mucus [44]. The enzyme is also known as muramidase or *N*-acetylmuramide glycan hydrolase (E.C. 3.2.1.17). It cleaves glycosidic bonds between *N*-acetylmuramic acid and *N*-acetyl-D-glucosamine residues in bacterial peptidoglycans. Peptidoglycans are components of the bacterial cell wall, especially in Gram-positive bacteria, that are essential for cell stability.

Expression of lysozymes or functionally equivalent defense molecules occurs in virtually all animal species [20]. Three major lysozyme types have been identified in animals, the c-type (chicken or conventional type), the g-type (goose type), and the i-type (invertebrate type). However, family members of the lysozyme types c and g occur also in some invertebrate groups. These lysozyme types differ in gene structures and primary sequences but show remarkable similarities in the 3D structures of the proteins indicating that the evolution of these defense molecules was convergent. Chicken egg white contains a lot of lysozyme [2]. A destabilase that has, among other functions, lysozyme activity has been found in leech saliva [182]. An i-type lysozyme was isolated from tissues of the American oyster *Crassostrea virginica* [175].

Gene-encoded antimicrobial peptides (AMPs) are a group of structurally diverse molecules that are produced by different types of tissues and cells in a variety of animals (invertebrates as well as vertebrates) and in humans. AMPs form a first line of host defense against pathogens and are involved in innate immunity. General features of AMPs are their amphipathicity, overall cationic charge, and relatively small molecular masses (12 to 50 amino acids). These properties allow the AMPs to associate with plasma membranes of microbial cells and to form pores. Although there are several models trying to explain the mechanisms as to how AMPs damage and kill microorganisms, there is still a lot of research necessary. Besides forming pores in the surface of microbial cells, AMPs may inhibit cell wall synthesis, nucleic acid-, or protein synthesis in bacteria [18]. Using structural criteria, three groups of AMPs have been defined, namely, peptides with an α-helical conformation (insect cecropins, magainins, etc.), cyclic peptides with pairs of cysteine residues (defensins, protegrin, etc.) [43, 134], and peptides enriched in certain amino acids (proline, histidine, etc.). Some of these molecules show some promise for therapeutic use in humans [19].

Defensins are the most widely distributed types of antimicrobial peptides [54, 173, 181]. This wide distribution of defensins and defensin-like peptides across all animal taxa and humans testifies to their crucial function in protecting these organisms against microbial pathogens. Vertebrates express three structurally related (and probably homologous) defensin-subtypes, α-, β-, and θ-defensins [107]. Cysteine-stabilized α/β (CSαβ) defensins have been isolated from nematodes, mollusks, insects, and arachnids [184].

Hematophagous insects like mosquitoes inject saliva into the wound inflicted to the host animal. In addition to anticoagulants the saliva contains AMPs which are partially reingested into the crop with the blood meal. AMPs stop the growth of microorganisms and prevent them from degrading the stored blood in the crop. Defensin A1 is such a compound in the salivary gland secretion of the mosquito *Aedes aegypti* [133].

The maggots of the common green bottle fly, *Lucilia sericata*, express different proteins with antimicrobial activities in the midgut, hindgut, salivary glands, crop, and fat body. Injection of Gram-negative bacteria (e.g., *Pseudomonas aeruginosa*) into these larvae results in substantial upregulation of some of these AMPs in selected tissues. Defensin-1 was upregulated in salivary glands, crop, and fat body. Attacin-2, another AMP directed against Gram-negative bacteria, was upregulated over 50,000-fold in the fat body [11].

Similar bioactive peptides with antimicrobial and other functions are found in the defensive skin secretions of *Bombina* toads (Bm8, Bo8, and Bv8) [24] and *Xenopus laevis* (magainins) [180].

Defensins and other AMPs are expressed in cells lining the airways in human lungs [67]. They are secreted into the airway surface liquid and attack inhaled bacteria before they can get in touch with the apical surface of the epithelium.

Other examples of antimicrobial functions of secreted peptides and proteins have been identified in invertebrates. Cnidocyte peptides in the fire coral *Millepora complanata* (Cnidaria) showed inhibitory activity against both Gram-positive and Gram-negative bacteria [65]. The fire coral AMPs showed sequence similarities to histones. Two AMPs, termicin and spinigerin, have been isolated from the fungus-growing termite *Pseudacanthotermes spiniger* (Isoptera). Termicin has 36 amino acid residues with 6 cysteines and is C-terminally amidated which increases its stability. It has antifungal properties. Spinigerin consists of 25 amino acids and has no cysteine. It is active against bacteria and fungi [90].

Mammals have developed yet another group of AMPs, the cathelicidins. The amino acid sequences of cathelicidins are not well conserved between species, but the evolutionary relationship between these peptides can be inferred from the fact that all of these AMPs originally carry a highly conserved cathelin proregion which is proteolytically cleaved off to release the active peptide. Cathelicidins have been isolated from cows (BMAP-27, indolicidin, and bac-tenecin), pigs (protegrins), mice (CRAMP), rabbits (CAP18), and humans (hCAP-18/LL-37) [179]. Such AMPs are present in mucus layers on the surface of mucous epithelia (airways, eyes, etc.) at concentrations of approximately 2 µg/mL. However, upon exposure of such surfaces to pathogens or to endogenous proinflammatory mediators, the expression of these AMPs is substantially upregulated.

Antileukoprotease (ALP), or secretory leukocyte protease inhibitor, is an endogenous inhibitor of serine proteases that is present in the airway surface liquid in human lungs. Besides inhibiting certain proteases, ALP is an antibacterial agent and kills *Escherichia coli* or *Staphylococcus aureus* [68]. In a protein extract of equine neutrophils, an antimicrobial polypeptide (eNAP-2) was discovered that shares some functional features with ALP [29].

References

[1]   Ahmed M et al. (2012) J Venom Anim Toxins Incl Trop Dis 18(2): 236
[2]   Alderton G, Fevold HL (1946) J Biol Chem 164: 1
[3]   Andrews RK et al. (2004) Snake venom toxins affecting platelet function. In: Gibbins JM, Ma-haut-smith MP (eds.) Platelets Megakaryocytes – Methods in Molecular Biology, Vol. 273. Humana Press, p. 335
[4]   Assumpcao TCF et al. (2012) Toxins 4(5): 296
[5]   Averin AS et al. (2022) Toxins 14(2): 88
[6]   Badawi JK (2021) Toxins 13(5): 337
[7]   Barnes CS et al. (2001) Semin Thromb Hemost 27(4): 337
[8]   Baskova IP et al. (2018) Thromb Res 165: 18
[9]   Baskova IP et al. (1992) Thromb Res 67(6): 721
[10] Baskova IP, Nikonov GI (1991) Blood Coagul Fibrinolysis 2(1): 167
[11] Baumann A et al. (2015) PLoS ONE 10(8): e0135093
[12] Betz SF (1993) Protein Sci 2(10): 1551
[13] Bdolah A (2010) Hypertensive and hypotensive snake venom components. In: Kini RM, Clemetson KJ, Markland FS, McLane MA, Morita T (eds.) Toxins and Hemostasis – From Bench to Bedside. Springer, Dordrecht, Heidelberg, London, New York, Chapt. 37, p. 655
[14] Bhoola KD et al. (1992) Pharmacol Rev 44(1): 1
[15] Bickler EP (2020) Toxins 12(2): 68
[16] Bordon KCF et al. (2015) J Venom Anim Toxins Incl Trop Dis 21(1): 43
[17] Bottrall J et al. (2010) J Venom Res 1: 18
[18] Brogden KA (2005) Nat Rev Microbiol 3(3): 238
[19] Bulet P et al. (2004) Immunol Rev 198: 169
[20] Callewaert L, Michiels CW (2010) J Biosci 35(1): 127
[21] Casewell NR et al. (2012) Toxicon 60(2): 119
[22] Catterall WA (1980) Annu Rev Pharmacol Toxicol 20: 15

[23] Chatterjee S (1999) Chem Phys Lipids 102(1): 79
[24] Chen T et al. (2003) Biochem J 371(1): 125
[25] Chu YY et al. (2020) Toxins 12(4): 230
[26] Clemetson KJ et al. (2009) J Thromb Haemost 7(2): 360
[27] Cooke IR et al. (2017) Toxicity in cephalopods. In: Malhotra A, Gopalakrishnakone P (eds.) Evolution of Venomous Animals and Their Toxins. Springer, Dordrecht, Netherlands, p. 125
[28] Costa H, Palma MS (2000) Toxicon 38(10): 1367
[29] Couto MA et al. (1992) Infect Immun 60(12): 5042
[30] Currie BJ, Jacups SP (2005) Med J Aust 183(11–12): 631
[31] Das B et al. (2021) Front Pharmacol 12: 710680
[32] Deckmyn H et al. (1995) Blood 85(3): 712
[33] De Colibus L et al. (2012) Structure 20(9): 1498
[34] Dhananjaya BL, D'Souza CJM (2010) Cell Biochem Funct 28(3): 171
[35] Dobson JS et al. (2021) Toxins 13(8): 549
[36] Doery HM, Pearson JE (1964) Biochem J 92(3): 599
[37] Doorduijn DJ et al. (2019) Bioessays 41: 1900074
[38] Dreon MS et al. (2013) PLoS ONE 8(5): e63782
[39] Dutta S et al. (2017) J Proteom 156: 29
[40] Dvir H et al. (2010) Chem Biol Interact 187(1–3): 10
[41] Earl STH et al. (2011) Biochimie 93(3): 519
[42] El-Safory NS et al. (2010) Carbohydr Polym 81(2): 165
[43] Falanga A et al. (2017) Molecules 22(7): 1217
[44] Fleming A (1922) Proc Roy Soc London B 93(653): 306
[45] Flower DR et al. (1993) Protein Sci 2(5): 753
[46] Fox JW (2013) Toxicon 62: 75
[47] Franke H et al. (2019) Naunyn-Schmiedeberg's Archives of Pharmacology 392(10): 1181
[48] Friedrich T et al. (1993) J Biol Chem 268(22): 16216
[49] Frobert Y et al. (1997) Biochim Biophys Acta, Protein Struct Mol Enzymol 1339(2): 253
[50] Fry BG et al. (2009) Ann Rev Genomics Hum Genet 10: 483
[51] Fry BG et al. (2009) Proc Natl Acad Sci U S A 106(22): 8969
[52] Fukuda K et al. (2005) Nat Struct Mol Biol 12(2): 152
[53] Gangur AN et al. (2017) Proc R Soc B 284(1863): 20171364
[54] Ganz T (2003) Nat Rev Immunol 3(9): 710
[55] Gartner TK et al. (1980) FEBS Lett 117(1): 13
[56] Giorgianni MW et al. (2020) Proc Natl Acad Sci U S A 117(20): 10911
[57] Girish KS et al. (2002) Mol Cell Biochem 240(1–2): 105
[58] Gouaux E (1998) J Struct Biol 121(2): 110
[59] Greinacher A, Warkentin TE (2008) Thrombosis Haemostasis 99(5): 819
[60] Gremski LH et al. (2020) Toxins 12(3): 164
[61] Grishin EV et al. (1979) Toxicon 17(Suppl. 1): 60
[62] Hawkey C (1966) Nature 211: 434
[63] Heinemann SH, Leipold E (2007) Cell Mol Life Sci 64(11): 1329
[64] Hendon RA, Fraenkel-Conrat, H (1971) Proc Natl Acad Sci U S A 68(7): 1560
[65] Hernández-Elizárraga VH et al. (2022) Toxins 14(3): 206
[66] Hershko A et al. (1984) Proc Natl Acad Sci U S A 81(22): 7021
[67] Hiemstra PS (2007) Exp Lung Res 33(10): 537
[68] Hiemstra PS et al. (1996) Infect Immun 64(11): 4520
[69] Hildebrandt J-P (2015) AIMS Microbiol 1(1): 11
[70] Hoffman DR et al. (2001) J Allergy Clin Immunol 108(5): 855

[71]   Huang TF et al. (1989) Biochemistry 28(2): 661
[72]   Iacovache I et al. (2008) Biochim Biophys (BBA) – Acta Biomembr 1778(7–8): 1611
[73]   Jain KK (2000) Expert Opin Investig Drugs 9(10): 2403
[74]   Jared C et al. (2015) Curr Biol 25(16): 2166
[75]   Jin AH et al. (2019) Chem Rev 119: 11510
[76]   Julius D, Basbaum AI (2001) Nature 413: 203
[77]   Kang TS et al. (2011) FEBS J 278(23): 4544
[78]   Khoury GA et al. (2011) Sci Rep 1(1): 90
[79]   Kini RM (2003) Toxicon 42(8): 827
[80]   Kini RM (2019) Biochem J 476(10): 1515
[81]   Kita M et al. (2004) Proc Natl Acad Sci U S A 101(20): 7542
[82]   Kleizen B, Braakman I (2004) Curr Opin Cell Biol 16(4): 343
[83]   Klöcking HP et al. (1987) Haemostasis 17(4): 235
[84]   Kopeyan C et al. (1985) FEBS Lett 181(2): 211
[85]   Kowalski K et al. (2017) Front Zool 14(1): 46
[86]   Krayem N, Gargouri Y (2020) Toxicon 184: 48
[87]   Kuehn A, Hilger C (2015) Front Immunol 6: 40
[88]   Kutsukake M et al. (2008) Mol Biol Evol 25(12): 2627
[89]   Kutsukake M et al. (2004) Proc Natl Acad Sci U S A 101(31): 11338
[90]   Lamberty M et al. (2001) J Biol Chem 276(6): 4085
[91]   Lange S et al. (1999) Eur J Biochem 262: 547
[92]   Latinović Z et al. (2020) Toxins 12(6): 358
[93]   Lavergne V et al. (2015) The structural universe of disulfide-rich venom peptides. In: King GF (ed.) Venoms to Drugs: Venom as a Source for the Development of Human Therapeutics. The Royal Society of Chemistry, London, UK, Chapt. 2, p. 37
[94]   Le Bonniec BF, Libraire J (2010) Plasminogen activators from snake venoms. In: Kini RM, Clemetson KJ, Markland FS, McLane MA, Morita T (eds.) Toxins and Hemostasis – From Bench to Bedside. Springer, Dordrecht, Heidelberg, London, New York, Chapt. 22, p. 371
[95]   Lee G, Bae H (2016) Toxins 8(2): 48
[96]   Li F et al. (2020) Toxins 12(7): 425
[97]   Lijnen HR, Collen D (1988) Enzyme 40: 90
[98]   Lind P, Eaker D (1982) Eur J Biochem 124(3): 441
[99]   Lingappa VR, Blobel G (1980) Recent Prog Horm Res 36: 451
[100] Linker A et al. (1957) Nature 180(4590): 810
[101] Linker A et al. (1960) J Biol Chem 235: 924
[102] Lopes-Ferreira M et al. (2011) Biochimie 93(6): 971
[103] Marino G et al. (2015) ACS Chem Biol 10(8): 1754
[104] Markwardt F (1994) Thromb Res 74(1): 1
[105] Marsh N, Williams V (2005) Toxicon 45(8): 1171
[106] McMahon KL et al. (2022) Toxins 14(9): 600
[107] Menendez A, Brett Finlay B (2007) Curr Opin Immunol 19(4): 385
[108] Min GS et al. (2010) J Parasitol 96(6): 1211
[109] Monteiro RQ et al. (1999) Biochem Biophys Res Commun 262(3): 819
[110] Morante K et al. (2019) Toxins 11(7): 401
[111] Morgan BP et al. (2017) Semin Cell Dev Biol 72: 124
[112] Müller C et al. (2020) FEBS Lett 594(5): 841
[113] Munro R et al. (1991) Blood Coagul Fibrinolysis 2(1): 179
[114] Murakami M, Kudo I (2002) J Biochem 131(3): 285
[115] Muszbek L et al. (2011) Physiol Rev 91(3): 931

[116] Negri L et al. (2007) Life Sci 81(14): 1103
[117] Nguyen TV, Osipov AV (2017) J Anim Sci Technol 59: 20
[118] Niermann NC et al. (2020) Toxins 12(4): 260
[119] Norment BR et al. (1979) Toxicon 17(6): 539
[120] Nowak G, Schrör K (2007) Thrombosis Haemostasis 98: 1) 116
[121] Palm NW et al. (2013) Immunity 39(5): 976
[122] Paull BR et al. (1977) J Allergy Clin Immunol 59(4): 334
[123] Pawson T et al. (2002) FEBS Lett 513: 2
[124] Peraro MD, van der Goot FG (2016) Nat Rev Microbiol 14(2): 77
[125] Perez-Riverol A et al. (2019) Insect Biochem Mol Biol 105: 10
[126] Piot N et al. (2021) Toxins 13(1): 36
[127] Postic G et al. (2018) Nucl Acid Res 46(D1): D454
[128] Ramazi S, Zahiri J (2021) Database (Oxford) 2021: 1
[129] Rawlings ND et al. (2010) Nucl Acid Res 38(Suppl. 1): D227
[130] Rawlings ND et al. (2018) Nucl Acid Res 46(D1): D624
[131] Reddy CR et al. (2017) BMJ Case Rep 2017: 221606
[132] Reis CV et al. (2006) Biochem J 398(2): 295
[133] Ribeiro JM et al. (2007) BMC Genomics 8: 6
[134] Ribeiro R et al. (2022) Marine Drugs 20(6): 397
[135] Ricci-Silva ME et al. (2008) Toxicon 51(6): 1017
[136] Richmond GS, Smith TK (2011) Int J Mol Sci 12(1): 588
[137] Rigoni M et al. (2008) J Biol Chem 283(49): 34013
[138] Rydel TJ et al. (1991) J Mol Biol 221(2): 583
[139] Sanchez EF et al. (2017) Toxins 9(12): 392
[140] Sandoval K, McCormack GP (2022) Marine Drugs 20(1): 74
[141] Schleuning WD (2001) Haemostasis 31(3–6): 118
[142] Schwanhäusser B et al. (2013) Bioessays 35(7): 597
[143] Schweitz H, Moinier D (1999) Perspect Drug Discovery Des 15-16(0): 83
[144] Scieuzo C et al. (2021) Sci Rep 11(1): 5032
[145] Sharpe HJ et al. (2010) Cell 142(1): 158
[146] Sherman RG et al. (1976) Comp Biochem Physiol Part A 53(3): 227
[147] Shiomi K et al. (2004) Toxicon 44(5): 499
[148] Shogomori H, Kobayashi T (2008) Biochim Biophys Acta (BBA) Gen Subj 1780(3): 612
[149] Smith LA et al. (1995) Toxicon 33(4): 459
[150] Smith TM et al. (2002) Arch Biochem Biophys 406(1): 105
[151] Sobotka AK et al. (1976) J Allergy Clin Immunol 57(1): 29
[152] Spirin AS (2002) FEBS Lett 514: 2
[153] Srinivasan KN et al. (2001) FEBS Lett 494(3): 145
[154] Steranka LR et al. (1988) Proc Natl Acad Sci U S A 85(9): 3245
[155] Stillmark H (1889) Arbeiten des Pharmakologischen Institutes zu Dorpat 3: 59
[156] Sunagawa M et al. (1996) Toxicon 34(6): 691
[157] Tan NH, Ponnudurai G (1992) Comp Biochem Physiol C – Comp Pharmacol Toxicol 103(2): 299
[158] Terwilliger TC, Eisenberg D (1982) J Biol Chem 257(11): 6010
[159] Tsetlin VI, Hucho F (2004) FEBS Lett 557(1–3): 9
[160] Tu AT, Hendon RR (1983) Comp Biochem Physiol B 76(2): 377
[161] Undheim EAB et al. (2016) Bioessays 38(6): 539
[162] Utaisincharoen P et al. (1993) J Biol Chem 268(29): 21975
[163] Valdez-Cruz NA et al. (2004) Eur J Biochem 271(23–24): 4753
[164] Vardjan N et al. (2013) Commun Integr Biol 6(3): e23600

[165] Versteeg HH et al. (2013) Physiol Rev 93(1): 327
[166] Vigetti D et al. (2014) Biochim Biophys Acta (BBA) Gen Subj 1840(8): 2452
[167] Ware FL, Luck MR (2017) Biosci Horiz: Int J Student Res 10: 1
[168] White JM (2003) Curr Opin Cell Biol 15(5): 598
[169] White TC et al. (2007) FEBS J 274(6): 1481
[170] Willard NK et al. (2021) Toxins 13(9): 654
[171] Williams EJ et al. (1961) J Biol Chem 236: 1130
[172] Wirsching F et al. (2003) Mol Genet Metab 80(4): 451
[173] Wong JH et al. (2007) Curr Protein Pept Sci 8(5): 446
[174] Wright RP et al. (1973) Arch Biochem Biophys 159(1): 415
[175] Xue Q et al. (2010) BMC Evol Biol 10: 213
[176] Yan S, Wang X (2015) Toxins 7(12): 5055
[177] Yang CC, Chang LS (1991) Biochem J 280(3): 739
[178] Zambelli V et al. (2017) Toxins 9(12): 406
[179] Zanetti M (2004) J Leukocyte Biol 75(1): 39
[180] Zasloff M (1987) Proc Natl Acad Sci U S A 84(15): 5449
[181] Zasloff M (2002) Nature 415(6870): 389
[182] Zavalova LL et al. (2000) Biochim Biophys Acta, Protein Struct Mol Enzymol 1478(1): 69
[183] Zeldin DC (2001) J Biol Chem 276(39): 36059
[184] Zhu S et al. (2005) Cell Mol Life Sci 62(19–20): 2257
[185] Zingali RB, Nogueira ACF (2010) Bothrojaracin – A potent thrombin inhibitor. In: Kini RM, Clemetson KJ, Markland FS, McLane MA, Morita T (eds.) Toxins and Hemostasis – From Bench to Bedside. Springer, Dordrecht, Heidelberg, London, New York, Chapt. 12, p. 179
[186] Zuliani JP et al. (2009) Protein Pept Lett 16(8): 908

3 Venomous and Poisonous Animals

3.1 Synergism of Venom Components

Animal venoms are rarely based on one substance, but mostly composed of several toxic compounds which may represent entirely different classes of molecules. In many cases, one venom constituent supports the effects of other compounds which results in synergistic actions of all venom components. Examples are bee-, scorpion-, or spider venoms which contain hyaluronidases besides cyto-, neuro-, or myotoxins. Hyaluronidases improve the diffusion of the toxins from the site of injection to wide tissue areas which results in rapid whole body paralysis or large-scale tissue destruction in the target animal.

3.2 Animals Producing Poisons or Venoms and Modes of Their Actions

The purpose of these chapters is to review the many species of toxic animals and their toxin mixtures and to describe the modes of application and the modes of action in target animals where possible. We follow the current systematics of the animal kingdom as outlined by Giribet and Edgecombe [1] to allow the reader to quickly find the description of a toxic animal of interest (Fig. 1.4).

Reference

[1]   Giribet G, Edgecombe GD (2020) The Invertebrate Tree of Life. Princeton University Press, Princeton, New Jersey, USA

3.2.1 Sponges (Porifera)

Sponges are aquatic multicellular organisms built of two layers of surface cells which are connected by a layer of extracellular gelatinous matrix called mesohyl which, in conjunction with embedded sklerotizing minerals (silica or calcium carbonate spicules), functions as an endoskeleton and gives the sponge body its outer shape (Fig. 3.1). The presence of pores and channels through these cell layers and the activities of ciliated cells (choanocytes) lining the inner surface of the sponge body enable the sponge to actively drive currents of ambient water along the surface of virtually every single cell of the sponge body. Sponges do not have nervous, digestive, respiratory, circulatory, or excretory systems. Instead, oxygen and carbon dioxide enter or leave the individual

https://doi.org/10.1515/9783110728552-003

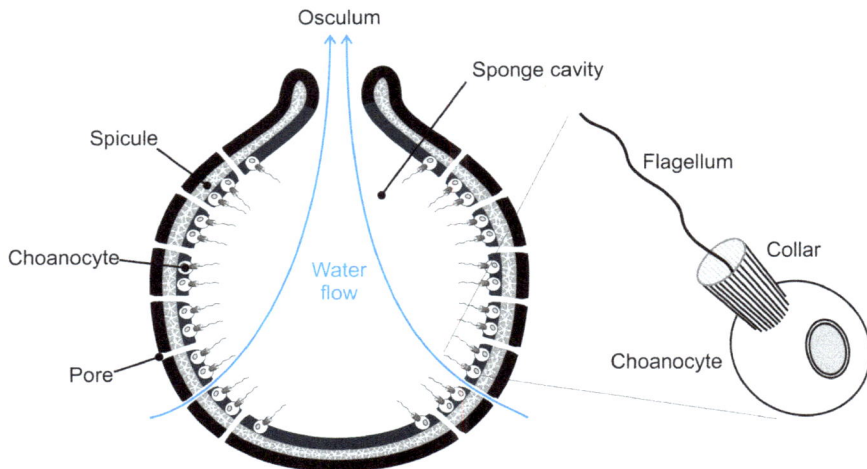

Fig. 3.1: Scheme of a sponge. Its body is built from two layers of surface cells (black) which are connected by a layer of gelatinous extracellular matrix (gray) with embedded spicules and spongin fibers. Small openings in the surface are connected to pores through these layers connecting the ambient medium to the internal cavity. The pores are formed by tubelike cells, so-called porocytes. The inner layer of cells contains many choanocytes which are characterized by an apical collar and a flagellum. They create water currents through the pores and the internal cavity. Food particles are endocytosed and digested intracellularly. Water and waste products leave the cavity through the osculum, whose diameter is regulated to adjust water flow according to the needs of the sponge.

cells by diffusion. Nutrients (detritus, microorganisms, planktonic organisms) are taken up by endocytosis and digested intracellularly. The digestive products are shared between all cells. Indigestible material is excreted by exocytosis.

Sponges occur in marine and in limnic ecosystems. Many marine species host photosynthesizing organisms as endogenous symbionts, mostly cyanobacteria, but also eubacteria. These symbionts may make up to one third of the total mass in certain sponge species. Sponges may gain more than 50% of their total energy supply from these symbionts [7, 30]. Freshwater sponges host green algae as endosymbionts, and they benefit from nutrients produced by the algae.

Sponge larvae are freely swimming but adult sponges are sessile organisms that cannot move away when threatened by predatory animals or microorganisms. Thus, evolution has provided these animals with a huge arsenal of chemical weapons that are present within the sponge cells and directed against predators or grazers [8, 22] or secreted to the outer surface of the sponge body to inhibit the growth of microorganisms like bacteria or fungi [12, 18].

Chemical defense in sponges is provided by a large repertoire of toxic secondary metabolites, including polyynes (see Section 2.2), but also polyketides (see Section 2.3), monoterpenes, sesquiterpenes, diterpenes, sesterterpenes (see Section 2.4), steroid saponins (see Section 2.6), steroids (see Section 2.5), alkaloids (see Section 2.9), and peptides (see Section 2.12). Such substances are used to impregnate the sponge tissue which becomes practically inedible for potential predators.

The chemical defense is very effective and works reliably against fish. Nudibranch gastropods (e.g., sea hares), however, may consume sponge tissue without any adverse effects and may even use the sponge secondary metabolites to protect themselves from being predated on. Sponges use polyyne impregnation of tissues and surfaces (see Section 2.2) not only for defense against predators but also to defend themselves against colonization by other organisms (algae, bacteria, fungi, larvae of invertebrates) and against attacks by bacteria and viruses. Balancing symbiontic relationships with microorganisms is another function of these substances [22].

Since many shallow-water sponges live in symbiosis with phototrophic bacteria, microalgae, cyanobacteria, dinophyceae, or with heterotrophic but highly biosynthetically active fungi, they profit from the biosynthetic activities of their partners. It is likely, and in some cases also proven, that these symbionts are the producers of some of the toxins found in sponge tissue or are the producers of precursors for the biosynthesis of sponge secondary substances [5].

All secondary metabolites in sponges seem to have cytotoxic effects either by damaging cell membranes of epithelial cells in the mucous surfaces of the attacker's integument or by interfering with DNA integrity (intercalation and topoisomerase inhibition) and/or through inhibition of key enzymes.

Divers, snorkelers, and sponge collectors who come in close contact with sponges may be in danger of intoxication, especially if sponges are torn off the substrate or if they are squeezed. Potentially problematic sponge species are the 'touch-me-not' sponge (*Neofibularia nolitangere*) from the Caribbean (Ph. 3.1), the fire sponge (*Tedania ignis*; Ph. 3.2) which occurs in the Caribbean and in the Pacific, or *Tedania anhelans* (around Australia and Indonesia). The skeletal needles may penetrate the skin, break off, and allow bacteria or the sponge toxins to interact with cell surfaces. Severe dermatitis may occur. Eye contact with the slimy liquid that escapes when sponges are squeezed is particularly dangerous [6, 33].

> ☠ Symptoms after intensive contact with sponges can include reddening of the skin, a stinging sensation, erythema, swelling, blistering, and stiffness of the finger joints. The symptoms last 2–3 days, but the tingling and stabbing symptoms may persist much longer.

> 🩹 Affected skin areas should be washed off intensively with water. Skeletal needles stuck in the skin should be removed out using an adhesive strip or sticking plaster if possible. Itching may be treated with antihistamines. Inflammatory responses may be attenuated by applying creams or lotions containing glucocorticoids.

Research on the use of sponge secondary products in medicinal applications is still in its infancy due to the fact that only minor amounts of these substances are available from natural sources and that synthesis pathways for these substances are not known.

Ph. 3.1: The Caribbean sponge, *Neofibularia nolitangere* (Source: Joseph Pawlik, from [35] with permission).

Ph. 3.2: The fire sponge, *Tedania ignis* (Source: Joseph Pawlik, from [35] with permission).

The Okinawan marine sponge *Xestospongia* sp. generates nepheliosyne A, an acetylenic acid (polyyne) with 47 C atoms [14]. Polyacetylenic alcohols with 22 C atoms, the callyspongenoles, have been isolated from a Red Sea sponge, *Callyspongia* sp. [34]. These substances have mild cytotoxic properties with respect to human cells. Other polyacetylenic alcohols isolated from the marine sponge *Petrosia* sp. inhibited DNA replication in mitotic mammalian cells [20].

An important polyketide present in high amounts in sponge tissues is okadaic acid (see Section 2.3.9). Okadaic acid has been isolated from sponges of the genus *Halichondria* (Ph. 3.3) and from *Suberites domuncula* [27, 32]. The latter species is often found on snail shells occupied by hermit crabs. Okadaic acid efficiently binds and inhibits protein serine-threonine phosphatases and disturbs the balance of phosphorylation and dephosphorylation of cellular proteins. Thus, okadaic acid is an important cytotoxin to animal and human cells [10].

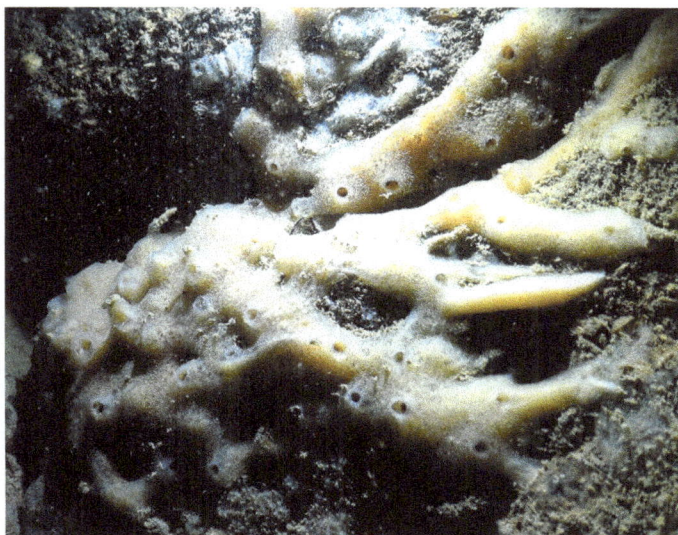

Ph. 3.3: The marine cave sponge, *Halichondria bowerbanki* (Source: Oleg A. Kovtun; from [11], with permission).

Bromine-containing pyrrole compounds like hymenidin and hymenin have been isolated from the Okinawan marine sponge *Hymeniacidon* sp. They are potent antagonists of serotonergic receptors or β-adrenergic receptors, respectively, in animal and human cells [15, 16].

Furanoterpenes (see Section 2.4.5) identified in tissues of the Indonesian marine sponges, *Ircinia* sp. and *Spongia* sp., are inhibitors of protein tyrosine phosphatase 1B [2]. Another furanoterpene, furospongin-1, from marine sponges (the bath sponge *Spongia officinalis* and the honeycomb bath sponge *Hippospongia communis*) collected in the bay of Naples caused inhibition of contractions of guinea-pig ileum segments by interfering with mitochondrial ATP generation [3]. Diterpene alkaloids, the agelasines, from Indonesian marine sponges of the genus *Agelas* were identified as potent inhibitors of mycobacterial growth [4]. Bisnorditerpenes, the gracilins, were isolated from the sponge *Spongionella pulchella* [23]. One of the gracilins, gracilin B, mediates inhibition of integrin-mediated adhesion of cultured cells to their substrates. Derivatives of gracilin A are currently tested as potential drugs with neuroprotective or immunosuppressive activities [1]. Triterpenoid saponins with moderate cytotoxicity against human cells were isolated from the sponge *Erylus nobilis* collected near Jaeju Island in Korea [24].

Steroidal alkaloids (see Section 2.9.16) from the marine sponge *Corticium niger* collected in the Philippines, the plakinamines, inhibit growth of human colon carcinoma cells with IC_{50} values in the lower micromolar range [26].

Diketopiperazines (e.g., cyclo(L-Pro-L-thioPro); Fig. 3.2) [9] occur in fire sponges (*Tedania ignis*). They are the smallest cyclic peptides known. It is likely that they are generated by sponge symbiontic bacteria of the genus *Micrococcus*. The sponges use them as defense substances as diketopiperazines are efficient inhibitors of plasminogen activator inhibitor (PAI-1) in animals and humans, which is the main physiological inhibitor of the serine proteases urokinase plasminogen activator (uPA) and tissue plasminogen activator (tPA) [21]. Contact of humans with fire sponges may result in thromboembolic disease.

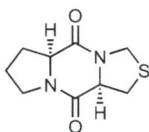

Fig. 3.2: Cyclo(L-Pro-L-thioPro), a sulfur-containing diketopiperazine of the fire sponge, *Tedania ignis*. The compound is assumed to be synthesized from the amino acid proline and 3-thioproline.

Halogenated cyclic peptides, corticiamide A and cyclocinamide B, have been detected in sponges of the genus *Corticium*. Due to their amphiphilic character they may disrupt plasma membranes of animal and human cells and function as cytotoxins [17].

A perforin-like protein was cloned from the sponge *S. domuncula* that displayed strong antibacterial activity by formation of pores in the surface of bacterial cells [28]. The expression of this antibacterial protein in the sponge is inducible by the presence of sponge pathogens [31]. Accumulation of autonomously produced brominated alkaloids, e.g., in *Pseudoceratina durissima*, provides protection of this and other verongiid sponge species against pathogenic microbes as well [19]. Moreover, symbiontic α-and γ-proteobacteria in sponges secrete as yet unknown substances which seem to protect sponge tissue from microbial overgrowth [28].

It is interesting that some marine nudibranchs (Gastropoda) feed on the outer cell layers of sponges on purpose. By doing so, they acquire toxins which have been synthesized by the sponges or by their symbiontic cyanobacteria. The gastropods are resistant against these toxins and accumulate them in their own tissues to repel potential predators. An example is the sea lemon, *Peltodoris nobilis* (Ph. 3.4), which accumulates an *N*-methylpurine riboside termed doridosine in its digestive gland [13]. When tested on anesthetized rodents, doridosine induces prolonged hypotension and bradycardia. Feeding on sponge tissue was also observed in the yellow umbrella slug, the Mediterranean opisthobranch *Tylodina perversa*. These animals sequester the brominated isoxazoline alkaloids of their food sponge *Aplysina aerophoba* (Ph. 3.5) [29].

Ph. 3.4: The yellow dorid nudibranch, *Peltodoris nobilis* (Source: Brocken Inaglory, Wikimedia, GNU-CC-BY-SA).

Ph. 3.5: The yellow tube sponge, *Aplysina aerophoba* from the eastern Atlantic Ocean (Source: Yoruno, Wikimedia, GNU-CC-BY-SA).

Whether these substances provide benefits for the slugs, potentially as feeding deterrents against marine fish, is not known.

Marine sponges contain a lot of chemically diverse secondary metabolites which serve as defense substances, and most of them are still unknown. Some of the already identified substances or derivatives are considered as potential drugs because they interfere with important cell physiological processes in animal and human cells [25]. Isolating even more natural products from different sponge species increases the chance of developing selective drugs for specific targets.

References

[1]   Abbasov ME et al. (2019) Nat Chem 11(4): 342
[2]   Abdjul DB et al. (2017) Bioorganic Med Chem Lett 27(5): 1159
[3]   Anderson AP et al. (1994) Clin Exp Pharmacol Physiol 21(12): 945
[4]   Arai M et al. (2014) Chem Bio Chem 15(1): 117
[5]   Biabani MAF, Laatsch H (1998) J für Praktische Chem – Chemiker-Zeitung 340(7): 589
[6]   Bonamonte D et al. (2016) Dermatitis caused by sponges. In: Bonamonte D, Angelini G (eds.) Aquatic Dermatology: Biotic, Chemical and Physical Agents. Springer International Publishing, Cham, p. 121
[7]   Burgsdorf I et al. (2022) ISME J 16(4): 1163
[8]   Carroll AR et al. (2019) Nat Prod Rep 36(1): 122
[9]   Dillman RL, Cardellina JH (1991) J Nat Prod 54(4): 1159
[10]   Dounay AB, Forsyth CJ (2002) Curr Med Chem 9(22): 1939
[11]   Ereskovsky A et al. (2018) PeerJ 6 e4596
[12]   Esposito R et al. (2022) Mar Drugs 20(4): 244
[13]   Fuhrman FA et al. (1980) Science 207(4427): 193
[14]   Kobayashi J et al. (1994) J Nat Prod 57(9): 1300
[15]   Kobayashi J et al. (1986) Experientia 42(10): 1176
[16]   Kobayashi J et al. (1986) Experientia 42(9): 1064
[17]   Laird DW et al. (2007) J Nat Prod 70(5): 741
[18]   Lenz KD et al. (2021) Toxins 13(5): 347
[19]   Lever J et al. (2022) Mar Drugs 20 554
[20]   Lim YJ et al. (2001) J Nat Prod 64(1): 46
[21]   Martins MB, Carvalho I (2007) Tetrahedron 63(40): 9923
[22]   Proksch P (1994) Toxicon 32(6): 639
[23]   Rueda A et al. (2006) Lett Drug Des Discov 3(10): 753
[24]   Shin J et al. (2001) J Nat Prod 64(6): 767
[25]   Sipkema D et al. (2005) Mar Biotechnol 7(3): 142
[26]   Sunassee SN et al. (2004) J Nat Prod 77(11): 2475
[27]   Tachibana K et al. (1981) J Am Chem Soc 103(9): 2469
[28]   Thakur NL et al. (2003) Aquat Microb Ecol 31(1): 77
[29]   Thoms C et al. (2003) Zeitschrift für Naturforschung C 58(5–6): 426
[30]   Usher KM (2008) Mar Ecol 29(2): 178
[31]   Wiens M et al. (2005) J Biol Chem 280(30): 27949
[32]   Wiens M et al. (2003) Mar Biol 142 213
[33]   Yaffee HS, Stargardter F (1963) Arch Dermatol 87 601
[34]   Youssef DTA et al. (2003) J Nat Prod 66(5): 679
[35]   Zea S et al. (2014) The Sponge Guide: A Picture Guide to Caribbean Sponges, 3rd ed. Available online at www.spongeguide.org, accessed on: January 19, 2023

3.2.2 Cnidaria

Cnidarians (or Medusozoa [13]) are water-dwelling animals. Most species live in marine habitats, and some live in freshwater. Their common feature is the presence of cnidocytes in the outer cell layer, the ectoderm. The cnidocytes are specialized on the application of toxins to target organisms (Fig. 3.3). Thus, cnidarians are venomous animals. The cnidocytes are preferentially used for capturing prey but may also be used for defense purposes.

The body wall of cnidarians is build from two layers of epithelial cells (diploblastic organisms), the entoderm or gastroderm which covers the gastric lumen, and the ectoderm which covers the outer body surface (Fig. 3.4). The two cell layers are attached to each other by the mesoglea, a jelly-like extracellular matrix. Two differently organized life stages occur in cnidarians, the pelagic (freely swimming) medusae, and the sessile polyps both of which are radially symmetrical with mouths surrounded by tentacles that bear the cnidocytes.

The cnidocytes are embedded in the ectodermal cell layer and especially concentrated in the surface of the tentacles. The tentacles grap or engulf prey animals (from planktonic animals to fish) and the cnidocytes are used for stunning or paralyzing the victims and to inject enzymes (hyaluronidase, collagenase, elastase, and DNAses) that facilitate toxin distribution and start the digestive processes [11]. Each cnidocyte is embedded in a nematocyte and has a sclerotized capsule. The inner apical membrane of the cnidocyte is infolded and forms a long hollow tube. The lumen of the cnidocyte is filled with venom and pressurized. Upon contact of a sensillum in the apical membrane of the nematocyte the cnidocyte bursts, releases its harpoon-like structure that penetrates the victim's integument. Driven by the high pressure within the cnidocyte the hollow tube unfolds turning its inside outward while advancing into the tissues of the victim and releasing the venom (Fig. 3.3). Major venom components are cytotoxins, neurotoxins, and cardiotoxins. The venom composition of the cnidocytes varies depending on which cnidarian species is inspected and where the cnidocytes are exactly located in the animals [11, 15].

3.2.2.1 Anthozoa

The polyps of species in the taxon Anthozoa (Fig. 3.5) may live solitarily (sea anemones, Actiniaria) or build large colonies (stony corals, Scleractinia, as well as the Octocorallia). Due to the hexagonal anatomy of the polyps, the Actiniaria and the Scleractinia are joined together in the class of Hexacorallia, while the species with octagonal polyps are members of the class of Octocorallia. The colony-building species (especially the stony corals) use calcium carbonate as mineral material for sklerotizing their colonies. They are mainly responsible for the formation of coral reefs. The Staurozoa are a small group of stalked jellyfishes which have no major toxicological significance.

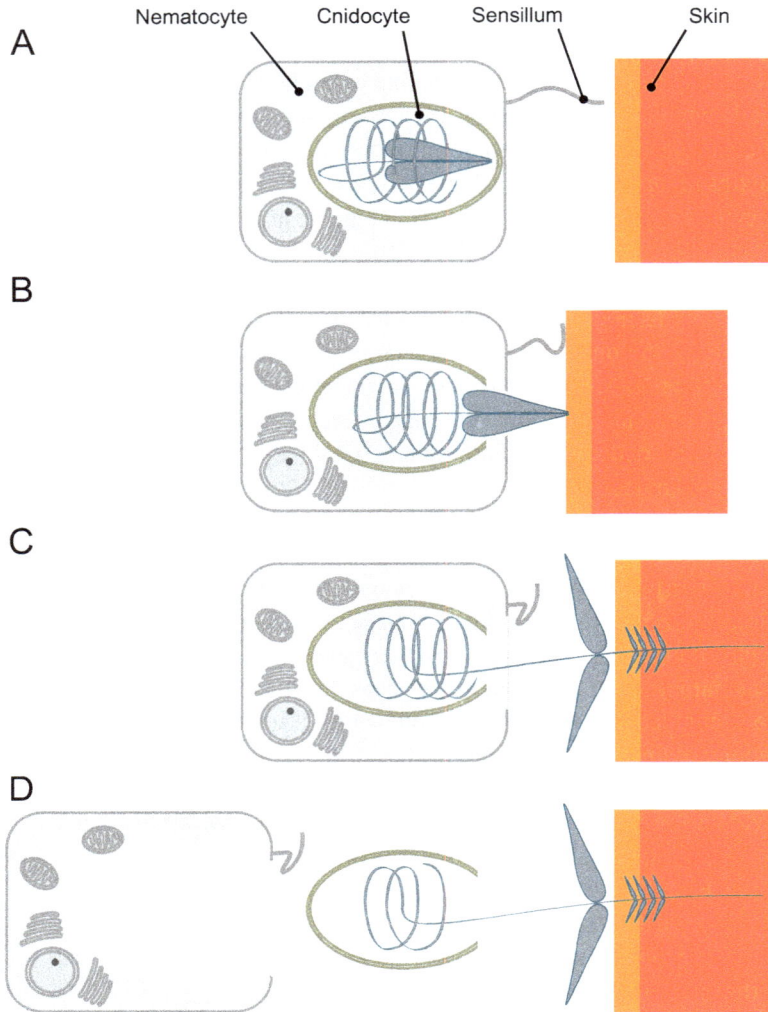

Fig. 3.3: Cnidocyte function in Cnidaria. Each cnidocyte is embedded in an ectodermal epithelial cell, the nematocyte. Nematocytes occur in very high densities on the tentacles of cnidarians. The cnidocyte is full of venom and contains the venom apparatus which is maintained in a standby mode under high internal pressure as long as the sensillum of the nematocyte is not stimulated by prey or by an attacker (A). When there is such a contact, the apical surface of the cnidocyte ruptures and a harpoon-like device is shot into the integument of the victim (B). An elastic protein is suspected to store the energy needed for the rapid ejection of the cnidocyte harpoon [5]. The tip of the harpoon unfolds and anchors itself in the surface tissues of the victom using small barbs. Simultaneously, a fine hose that is connected to the harpoon is unrolled and turned inside out while advancing into the victim's tissue due to the high pressure in the cnidocyte (C). If the victim tries to get rid of the tentacles the cnidocyte may rip out of the nematocyte and continues to pump venom into the wound (D).

Medusa

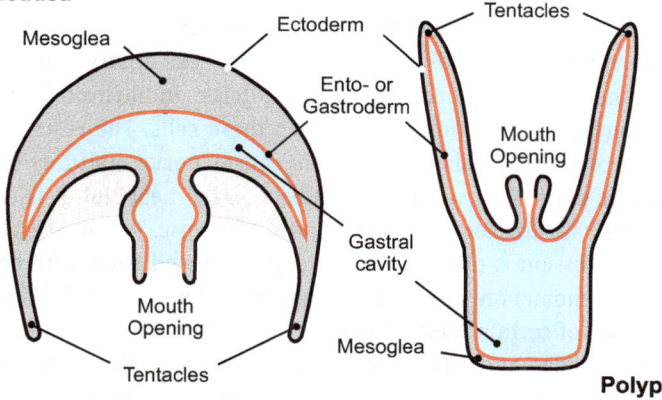

Fig. 3.4: Scheme Cnidaria. The two life forms of Cnidaria, the swimming medusae (left) and the sessile polyps (right), show radial symmetry and a similar body plan. They are built from two layers of epithelial cells, the ectoderm covering the body surface and the entoderm (or gastroderm) lining the gastral cavity. The two cell layers are connected by the mesoglea, a jelly-like extracellular matrix which is more prominent in medusae compared with that in polyps. The ectoderm of the tentacles contains a lot of nematocytes, each carrying a cnidocyte that contains a venom reservoir and a venom application apparatus. Thus, cnidarians are venomous animals. The cnidocytes are used to acquire prey organisms or for defense purposes.

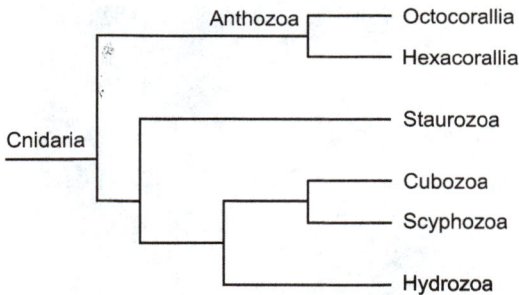

Fig. 3.5: Simplified phylogenetic tree of the Cnidaria according to Collins et al. [13]. The Anthozoa are the sister group to all other taxa of cnidarians. This evolutionarily ancient group encompasses the corals in which only the sessile polyps are present. They use their toxins for obtaining planktonic prey or for defense against coral eating fish or grazers like echinoderms or mollusks. The highly toxic Australian sea wasp, *Chironex fleckeri*, is a member of the taxon Cubozoa. Among the Scyphozoa are all major groups of freely swimming (pelagic) medusae (jellyfish) like the toxic lion's mane jellyfish (*Cyanea capillata*). The polyps of these animals are small and live solitarily. They generate new medusae by a vegetative fragmentation process called strobilation. Different species of Hydrozoa occur in salt or in fresh water. Most species show a regular switch between asexual reproduction (strobilation) and sexual reproduction in the medusa stage. *Hydra*, the freshwater polyp, is an exception as there is no medusa stage and sexual reproduction occurs in the polyp.

The sessile sea anemone (*Anemonia sulcata*; Ph. 3.6) contains several homologous peptide toxins (ATXs, 24–46 amino acid residues; Fig. 3.6). ATXs induce the elongation of open times of different voltage-gated sodium channels in the cells of target organisms including the $Na_v1.1$ and $Na_v1.2$ channels. The resulting delays in intrinsic inactivation of these channels prolong the action potentials in these cells. The resulting sodium overload of nerve or muscle cells in the victim causes nerve malfunction, muscle cramps (even spastic paralysis), or inotropic effects on the heart [6]. It is assumed that these toxins are used for acquiring prey and for defense. Recent studies have revealed that toxin expression is different in cnidocytes isolated from different portions of the tentacles, a phenomenon termed 'regionalisation of venom production' [3]. The specific toxin profiles of certain tentacle portions in individuals belonging to the sea anemone families Aliciidae and Thalassianthidae indicate that these venoms may function to protect structures that harbor large numbers of photosynthetic symbionts. As sea anemones are sessile organisms that have to compete for space on the solid supports in their habitats, they generate pore-forming toxins with structural similarities to bacterial porins [24], the actinoporins, to deter attackers or competitors that may try to settle too close [1, 32].

Ph. 3.6: The cnidarian, *Anemonia sulcata* (Source: Liné1, Wikimedia, GNU-CC-BY-SA).

Small molecule and protein toxins are present in coral polyps to deter grazers or predators that try to feed on corals. Some of these toxins are not even products of the polyps but obtained from their food or from symbionts present in the polyp cells. An example for such a case is the polyketide palytoxin (see Section 2.3.7), which is one of the most poisonous nonprotein substances known (LD_{50} (mice, i.v.) is approx. 0.15 µg/kg BW [31]). It is originally synthesized by certain Dinoflagellata (*Ostreopsis* sp.) and

```
        10           20          30          40
GAACLCKSDGPNTRGNSMSGTIWVFGCPSGWNNCEGRAIIGYCCKQ
```

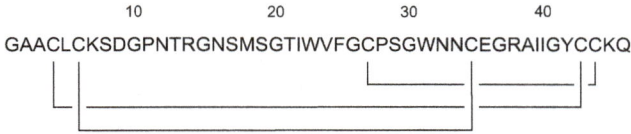

Fig. 3.6: Peptide toxin ATX-1 of the sea anemone, *Anemonia sulcata*. The amino acid sequence of the toxin is shown in one-letter code. The brackets indicate the location of disulfide bridges between cysteine residues stabilizing the 3D structure of the peptide.

accumulates in crust anemones of the genus *Palythoa* which feed on planktonic organisms. Feeding on these crust animals exposes animal intestinal cells to the toxin. Handling of crust anemones in aquaria may result in exposure of external mucous cells in humans. Palytoxin interacts with the Na^+/K^+-ATPase which provides the ion gradients for secondary transport processes in animal and human cells. It converts the transporter into an open channel that is permeable to sodium and potassium ions [20, 22]. This results in the loss of the electrical plasma membrane potential and cell death. Severe illnesses like rhabdomyolysis have been described in humans upon eating fish or other seafood that had accumulated palytoxin through the food web [33].

⚕ Human cases of exposure to aerolized palytoxin have been described by Sud et al. [39]: Four adults and two children were exposed to palytoxin aerosolized from *Palythoa* corals during cleaning of seawater aquaria or fish tanks. Their symptoms included respiratory problems (shortness of breath, chest tightness, and dry cough), fatigue, myalgias, paresthesias of the upper extremities, low-grade fevers, and gastrointestinal symptoms (nausea and vomiting). All patients survived without sequelae.

3.2.2.2 Cubozoa and Scyphozoa

The Cubozoa and the Scyphozoa encompass all species that have pelagic medusae as prominent life stages. Of toxicological significance for humans are only a few species. Only a fraction of these species has been investigated with respect of venom composition using modern techniques [18].

Among the Cubozoa, these are the sea wasps of the genus *Chironex*, the Moreton Bay stinger *Morbakka fenneri*, and the Irukandji jellyfish, *Carukia barnesi* [11]. The Irukandji syndrome occurs upon contact of humans with the tentacles of the latter species. It encompasses symptoms ranging from headaches, severe pain, nausea and vomiting to pulmonary edema, severe hypertension, and cardiac failure resulting in death [12].

Box jellyfishes of the genus *Chironex* (Ph. 3.7) occur in coastal areas of Southeast Asia and Northern Australia. The bell of the 'sea wasp', *Chironex fleckeri*, can reach the size of a man's head, and each of its multiple (up to 60) tentacles, which are densely covered with cnidocytes, may be as long as 3 m. *Chironex* captures prawns and small fish, but seems to avoid contact with larger organisms. Accidental contact of humans with the tentacles results in excruciating pain and lasting burning sensations. Depending

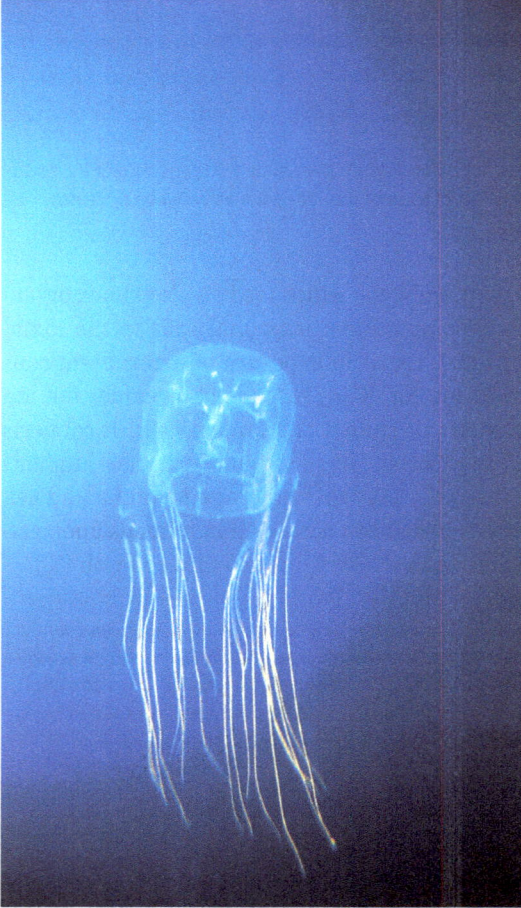

Ph. 3.7: The box jellyfish, *Chironex* sp. (Source: Guido Gautsch, Wikimedia, CC-BY-SA).

on the intensity of the tentacle contact, an untreated human victim may die within a few minutes [14]. Since this boxjelly species has been initially described in 1883, *C. fleckeri* has caused approximately 70 human deaths, but most encounters result in milder envenomation [14, 17].

The venom of *Chironex* contains several proteins (10–60 kDa) with neurotoxic, myotoxic, cardiotoxic, and cytotoxic properties as well as pain-inducing histamine and kinins [9]. Many of these protein toxins belong to a family of toxins that are only found in Cnidaria. However, some of the components of *C. fleckeri* venom are potential homologs of proteins in venoms of other cnidarian species. This applies to two of the most abundant proteins found in the cnidocytes of *C. fleckeri*, the toxins CfTX-1 and CfTX-2, with molecular masses of 43 or 45 kDa, respectively. Highly conserved regions of these proteins encode transmembrane regions indicating that these proteins are potentially able to interfere with the integrity of cell membranes in target organisms [7]. Their function as pore-forming toxins was later confirmed as these proteins assemble to form multisubunit complexes of up to 370 kDa and induce lysis in animal red blood cells [8]. Additionally, two other pore-forming toxins which are structurally similar to CfTX-1 or CfTX-2 have been found in the venom of *C. fleckeri*. CfTX-A has a molecular mass of approximately 40 kDa and CfTX-B has approximately 42 kDa. Combinations of these subunits have even higher hemolytic potential than those of CfTX-1 and CfTX-2 [10].

> 🦠 Rapid inactivation of cnidocytes can be achieved using washing of affected limbs with acetic acid (4–6%, v/v) or commercial vinegar [21]. In case the patient becomes unconscious or if there is life-threatening cardiac or respiratory arrest, cardiopulmonary resuscitation is required. Three vials of antivenom (20,000 units, diluted 1:10 with isotonic saline) should be given i.v. The antivenom has been shown to effectively neutralize jellyfish toxin proteins in cell-based assays [2]. If necessary, up to six vials may be given consecutively together with 0.2 mmol/kg magnesium sulfate (up to 10 mmol in adults) as an intravenous bolus over 5–15 min. In conscious patients without life-threatening conditions, ice packs may be applied for pain relief, together with oral, intravenous, or intramuscular analgesia using 0.1 mg/kg morphine (up to 5 mg in adults) [14].

> ⚕ A fatal case of *Chironex fleckeri* envenomation has been reported by Currie and Jacups [14]: 'In February 1996, a young girl (3 years old, healthy) was playing in shallow water at the seashore in a remote area of Australia's Northern Territory. Suddenly, she screamed out in pain and was brought to the beach with several jellyfish tentacles dangling from one arm and one leg. It was estimated that she had contact with approximately 1.2 m of tentacles. Tape samples from the skin confirmed that the tentacles belonged to *Chironex fleckeri*. Only several minutes after the incident, the girl suffered from cardiorespiratory arrest. She was taken to a nearby healthcare facility where she arrived with no detectable pulse or spontaneous ventilation. While cardiopulmonary resuscitation commenced over the next 20 min she was treated with three vials (20,000 units each) of *C. fleckeri* antivenom as well as adrenaline and atropine (i.m.). Additional adrenaline and bicarbonate, and two doses of 1.5 mg of verapamil were given intravenously during the following 60 min. A medical evacuation plane arrived 90 min after the incident with a retrieval doctor on board who intubated the patient and gave her two further vials of antivenom intravenously. Ongoing resuscitation, further cardioactive drugs, and six attempts of electrical cardioversion could not save the life of the patient and she died 110 min after receiving the jellyfish stings.'

The taxon Scyphozoa encompasses the typical jellyfish species with freely swimming (pelagic) medusae as the prominent life stages. Scyphozoa that are dangerous for humans, to name just a few, are the sea nettle, *Chrysaora quinquecirrha*, the cabbage head- or cannonball jellyfish, *Stomolophus meleagris*, Nomura's jellyfish, *Nemopilema nomurai*, the lion's mane jellyfish, *Cyanea capillata* (Ph. 3.8), and the mauve stinger, *Pelagia noctiluca* [11].

Extracts of sea nettle (*Chrysaora quinquecirrha*) cnidocytes induced release of lactate dehydrogenase from cultured primary rat liver cells [23] indicating that the toxins of this species may be hepatotoxic. The mechanism of action, however, is still unkown. A proteomic study on the cnidocyte venom composition of *Chrysaora fuscescens* revealed the presence of pore-forming toxins and proteases besides other components [34].

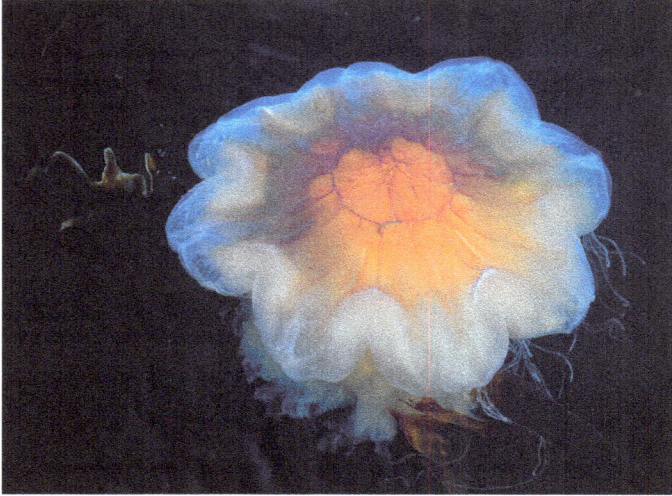

Ph. 3.8: Lion's mane jellyfish, *Cyanea capillata* (Source: Arnstein Rønning, Wikimedia, CC-BY).

The cannonball jellyfish (*Stomolophus meleagris*) has a dome-shaped bell which can reach 25 cm. A cluster of bushy tentacles carrying cnidocytes surround the mouth opening. The toxins in the cnidocytes may harm fishes that try to prey on these jellyfish and drives away most predators with the exception of certain crustaceans and leatherback turtles whose integuments cannot be penetrated by the cnidocyte harpoons. Stings by this jellyfish in humans are rare events. In most cases, stings cause local reactions such as pain, redness, and swelling. In some cases, stings cause problems like cardiac arrhythmia or disturbance of electrical conduction between cardiac myocytes [41].

Nomura's jellyfish, *Nemopilema nomurai* (Ph. 3.9), is a very large rhizostome jellyfish (bell diameter up to 2 m) in the Japanese Sea. Contact of humans with the tentacles may be fatal. As potentially lethal components of *N. nomurai* venom were identified a phospholipase A_2, a potassium channel inhibitor, a pore-forming hemolysin, and the procoagulant thrombin [28].

Comparative proteomic analyses of the toxin mixtures of *N. nomurai* and *Cyanea capillata* revealed the presence of approximately 60 proteins that could be grouped in 10 functional categories. The venoms contain proteases, phospholipases, neurotoxins, cysteine-rich secretory proteins (CRISPs), lectins, pore-forming toxins (PFTs), protease inhibitors, ion channel inhibitors, and other toxins. The relative composition, however, was different in the two species with metalloproteases, proteases, and pore-forming toxins being predominant in *N. nomurai*, while phospholipases, neurotoxins, and proteases were more abundant in venom of *C. capillata* [43].

Previous de novo transcriptome sequencing of the tentacle tissue of the jellyfish *Cyanea capillata* had revealed that the venom is composed of hemolytic proteins resembling those in other cnidarians, phospholipases, metalloproteases, serine proteases, and

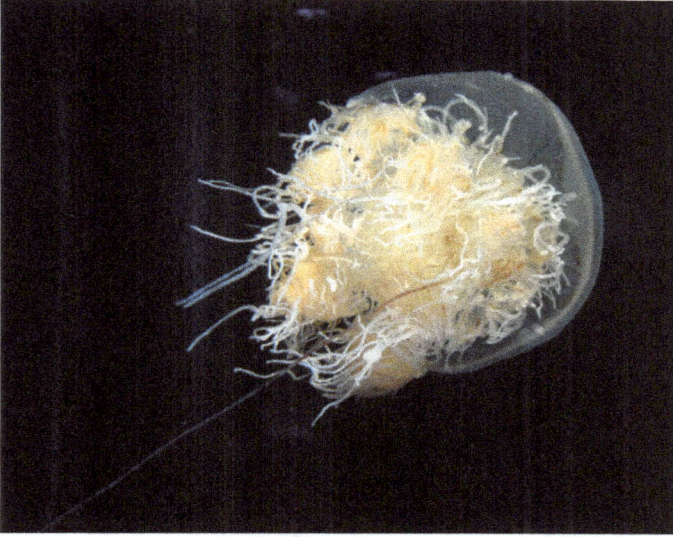

Ph. 3.9: Giant jellyfish (*Nemopilema nomurai*) from the Sea of Japan. Its bell may reach 2 m in diameter (Source: KENPEI, Wikimedia, GNU-CC-BY-SA).

serine protease inhibitors, among many other unknown proteins [29]. A basic 70 kDa protein isolated from cnidocytes of *C. capillata* induced irreversible spasmogenic actions on isolated cells of smooth muscle, mammalian heart, or frog rectus muscle [42]. A neurotoxin inhibiting voltage-activated sodium channels in excitable cells of target animals was isolated from the tentacles of *C. capillata* medusae [27].

Pelagia noctiluca is the most venomous jellyfish in the Mediterranean Sea. Due to the fact that these medusae form dense blooms during summer, exposure of humans to the *Pelagia* tentacles occurs frequently. Crude venom extracts have neurotoxic, haemolytic, and cytotoxic effects on test cell systems of human, rabbit, chicken, or fish origin [4]. Neurotoxic effects of *Pelagia* venom fractions were observed in the Atlantic ghost crab, *Ocypode quadrata* [37]. Injection of aliquots of purified fractions of *Pelagia* venom into the walking leg muscle of the crab resulted in convulsions, paralysis, and even death. A homolog of a zinc metalloprotease that is also present in the venoms of other jellyfish species was recently discovered to be an important constituent of *P. noctiluca* venom [19].

🚑 Rapid rinsing of the sting area with vinegar (5% acetic acid) and the application of heat (hot pack/ immersion in hot water) or lidocaine have been shown to be appropriate first aid measures for humans who have had contact with tentacles of scyphozoan medusae [35].

3.2.2.3 Hydrozoa

Hydrozoa are predatory animals using their cnidocyte toxins for obtaining prey and for defense. Some live solitary and others colonial. Many hydrozoans have a medusoid stage, but this is not always free-living. In many species it exists solely in the form of buds on the surface of the hydroid colony. These buds are the sites of sexual reproduction as they produce eggs and sperm. Most species inhabit saline water, but there are some species that live in limnic habitats (e.g., the freshwater polyps of the genus *Hydra*).

Hydra polyps are solitary and have a tubular, radially symmetric body up to 10 mm in total length including the trunk and the 12 tentacles around the mouth opening forming the 'head'. The base of the polyp's trunk, the 'foot', is attached to underwater stones or wood. Gland cells in the foot area secrete a sticky fluid that is used to glue the polyp to the substrate. An individual of *Hydra* is composed of up to 100,000 cells which are continually renewed by cell divisions in three stem-cell populations. The outer layer of ectodermal cells accumulates hydramacin-1 which is a bactericidal peptide containing a knottin protein fold [25]. It protects the surfaces of the polyps from microbial infection and overgrowth (EC_{50} = 0.5 µmol/L) by mediating aggregation and clumping of the bacteria. Another defensive protein, the 27 kDa protein hydralysin, is present in surface cells of hydrozoan polyps and has initially been isolated from the green hydra *Chlorohydra viridissima* [38]. This toxin belongs to the aerolysin type of pore-forming toxins [26] that occur in bacteria as well as in animals. The 27 kDa monomers liberated from the polyp cells when the polyp is under attack by a predator assemble into multimeric complexes that form transmembrane pores in the plasma membranes of oral or intestinal cells in the predator followed by cell death through lysis. Due to this defense chemistry polyps are not considered to be proper food by potential predators.

Some of the colony-building hydrozoa build highly differentiated structures. An example is the Portuguese man o' war (*Physalia physalis*; Ph. 3.10) which is a marine species and occurs in the Atlantic and the Indian Oceans. The community of individual organisms keeps itself swimming at the ocean surface by forming an air-filled balloon on a sclerotic platform that is used as a floating device. The rest of the colony forms many long threats with thousands of polyps per threat dangling from the floating balloon into the upper layers of the water body. Using their cnidocytes and the toxins therein they jointly capture fish or small other marine animals. Fatal outcomes of accidental contacts between swimmers, divers, or shipwrecked sailors with these polyp colonies have been reported. Most encounters, however, are less harmful, but patients experience *Physalia* stings as extremely painful, and some have severe systemic effects (like tremors of cramps) lasting for several days. Skin lesions usually heal within several weeks.

Ph. 3.10: Portuguese man o'war, *Physalia physalis* (Source: US Dept. of Commerce, National Oceanic and Atmospheric Administration, Wikimedia, public domain).

⚲ A 35-year-old biologist, Scuba diving off Crandon Beach, Miami, Florida, was stung by the cnidarian, *Physalia physalis*. While swimming toward shore, in approximately six feet of water, the victim surfaced. He immediately experienced a 'blinding flash of pain' across his entire body and realized that he had been stung by a *Physalia*. In pain, he swam the remaining distance to shore, but as he arose from the surf, he saw that he was dragging the animal along with him. Its body, approximately 10 inch in diameter, trailed about 5 feet behind him; its tentacles were wrapped about his neck, chest, abdomen, arms and legs. As soon as he struggled free of his diving gear, he began to tear the gelatinous tentacles from his body. He was aware of severe pain, anxiety, and slight respiratory distress. A passer-by attempted to help him remove some of the tentacles from his back but was stung immediately. Finally, the patient rolled in the wet sand and rubbed sand over his arms and legs to dislodge the remaining tentacles, which covered most of his body except for his head, and those areas protected by the Scuba tank and his swimming trunks. The patient walked to the nearest Life Guard Station, approximately 50 yd away. He noted that he could 'hardly catch his breath'. His chest was expanded and his breathing was superficial and laborious, as if he had 'run out of oxygen'. He noted that his breathing was almost entirely intercostal; diaphragmatic excursions were almost absent. He felt as though his chest was in a tetanic spasm; and he feared suffocation.

According to the Life Guard the patient was in acute distress when first seen. He was given a fresh water shower and following this sprayed with prednisolone-isopropyl myristate (Meti-Derm). Then, because of his 'paleness', and his perhaps slightly cyanotic appearance, he was taken to a physician. On arrival at the physician's office, he was still in acute pain and respiratory distress. He also noted a board-like rigidity of his abdomen. He was given 100 mg meperidine hydrochloride (Demerol) and an intramuscular antihistamine. This was followed by 16 mg of morphine. Oxygen was administered by mask. Questioned later about his sensations at this time, he recalled no unusual paresthesia, but did remember the presence of a fine tremor over his entire body. The symptoms and signs receded over the next 10 hr. The only residual discomfort was a soreness in the large muscle masses, particularly of the abdomen and chest. Numerous wheels, approximately 1–2 cm were observed over his arms and legs. The center of each wheel was brown and was surrounded by an erythematous zone of approximately 0 ~ 5 cm in diameter. These the patient first noted approximately 20 min following the injury, although he thought they probably appeared earlier than this. The brownish areas of hyperpigmentation were evident for about 3 weeks after the stinging. The patient states that there were at least several hundred of these. – Cited from Russell [36].

Cnidocytes isolated from the tentacles of *Physalia physalis* contain venom composed of several proteins [40]. One of the most abundant ingredients (approx. 28% of total venom protein) is physalitoxin, a 240 kDa protein that has hemolytic properties and kills injected mice with an EC_{50} of approximately 1 µmol/L. The hemolysin appears to be composed of three glycosylated subunits.

Two other cnidocyte toxins, PpV9.4 and PpV19.3, were purified from *P. physalis*. These are small proteins with molecular masses of 551 or 4,720 Da, respectively. Both toxins have structural similarities to acylpolyamines that are ingredients of the venoms of certain spiders. Both toxins induced elevations in intracellular free calcium ion concentrations in isolated rat pancreatic β-cells and stimulated insulin secretion from these cells [16]. It has been speculated that these toxins modify the gating behavior of calcium channels in the plasma membrane of pancreatic β-cells.

A high-molecular-mass toxin in *P. physalis* venom has been labeled P3 [30]. When the toxin was applied to neurons of the central nervous system in in-situ preparations of the snail *Zachrysia guanesis* it reversibly blocked the activation of glutamate receptors by exogenously added glutamate. Suppression of glutamate-evoked membrane potential changes was also observed when the toxin was applied to neuromuscular junctions of the crayfish *Procambarus clarkii*. It has been concluded that this toxin is an effective reversible glutamate antagonist [30] and may serve the hydrozoan as a neuro- and a myotoxin-inducing paralysis in prey organisms.

Besides these toxins which are directed against specific molecular mechanisms in target organisms there are generally small molecules like histamine, serotonin, prostaglandins, and kinin-like proteins in cnidocytes of cnidarian species [11]. These substances are supposedly responsible for the intense pain sensations that are induced by touching the tentacles of problematic cnidarian species.

Tab. 3.1: Toxins of pelagic Cnidaria and their biological effects.

Taxon	Toxin	Molecular mass (kDa)	Biological Effects
Cubozoa			
Chironex fleckeri	CfTX-1	43	cardiotoxic cytotoxic hemolytic
	CfTX-2	45	cardiotoxic cytotoxic hemolytic
	CfTX-A	40	cardiotoxic cytotoxic hemolytic
	CfTX-B	42	cardiotoxic cytotoxic hemolytic
Chiropsoides sp.	CqTX-A	44	hemolytic lethal to Crustacea
Carybdea sp.	CrTX-A	43	hemolytic lethal to Crustacea
	CrTX-B	45	hemolytic lethal to Crustacea
	Cytolysin	102-107	pore forming hemolytic
	Cytolysins	220 139 36	cytolytic cytolytic cytolytic
	CmNT	120	neurotoxic
Alatina sp.	CaTX-A	43	hemolytic lethal to Crustacea
	CaTX-B	45	hemolytic lethal to Crustacea
Carukia barnesi	CbTX1	21	neurotoxic

Tab. 3.1 (continued)

Taxon	Toxin	Molecular mass (kDa)	Biological Effects
Scyphozoa			
Cyanea sp.	ClGP-1	~27	cytotoxic
	CcTX-1	~31	cytotoxic
	CcNT	~8	neurotoxic
	Cc basic protein	70	myotoxic cardiotoxic
Pelagia noctiluca	Zinc Metalloprotease nas15-like	34	cytotoxic
Rhopilema nomadica	Phospholipase A_2		cytotoxic
Nemopilema nomurai	Phospholipase A_2		cytotoxic
Stomolophus meleagris	SmP90	90	ROS scavenging
Rhizostoma pulmo	Rhizolysin	260	pore forming hemolytic
	Metalloprotease	95	cytotoxic
Hydrozoa			
Hydra vulgaris	hydramacin-1	7	bacteriocidal
Chlorohydra viridissima	hydralysin	27	pore forming cytolytic
Physalia physalis	Physalitoxin	240	pore forming hemolytic
	P1	220	neurotoxic
	P3	85	neurotoxic
	PpV9.4	~5.5	secretagogue in pancreatic β-cells
Olindias sambaquiensis	Oshem 1	~3	hemolytic
	Oshem 2	~3.4	hemolytic

Note: Data were collected from different sources. The toxin names refer to the species in which they have been discovered. The names are also used for homologs in other species.

References

[1]   Alvarez C et al. (2021) Toxins 13(8): 567
[2]   Andreosso A et al. (2014) J Venom Anim Toxins Incl Trop Dis 20: 34
[3]   Ashwood LM et al. (2021) Toxins 13(7): 452
[4]   Badré S (2014) Toxicon 91: 114
[5]   Beckmann A et al. (2015) BMC Biol 13(1): 3
[6]   Beress L et al. (1975) Toxicon 13(5): 359
[7]   Brinkman D, Burnell J (2007) Toxicon 50(6): 850
[8]   Brinkman D, Burnell J (2008) Toxicon 51(5): 853
[9]   Brinkman DL et al. (2012) PLoS ONE 7(12): e47866
[10]   Brinkman DL et al. (2014) J Biol Chem 289(8): 4798
[11]   Burnett JW, Calton GJ (1987) Toxicon 25(6): 581
[12]   Carrette TJ et al. (2012) Diving Hyperb Med 42(4): 214
[13]   Collins AG et al. (2006) Syst Biol 55(1): 97
[14]   Currie BJ, Jacups SP (2005) Med J Aust 183(11–12): 631
[15]   D'Ambra I, Lauritano C (2020) Mar Drugs 18(10): 507
[16]   Diaz-Garcia CM et al. (2012) Curr Med Chem 19(31): 5414
[17]   Fenner PJ, Williamson JA (1996) Med J Aust 165(11–12): 658
[18]   Frazão B, Antunes A (2016) Mar Drugs 14(4): 75
[19]   Frazão B et al. (2017) Prot J 36(2): 77
[20]   Habermann E (1989) Toxicon 27(11): 1171
[21]   Hartwick R et al. (1980) Med J Aust 1(1): 15
[22]   Hilgemann DW (2003) Proc Natl Acad Sci U S A 100(2): 386
[23]   Houck HE et al. (1996) Toxicon 34(7): 771
[24]   Iacovache I et al. (2008) Biochim Biophys (BBA) – Acta Biomembr 1778(7–8): 1611
[25]   Jung S et al. (2009) J Biol Chem 284(3): 1896
[26]   Knapp O et al. (2010) Open Toxinol J 3: 53
[27]   Lassen S et al. (2012) Toxicon 59(6): 610
[28]   Li R et al. (2020) J Proteome Res 19(6): 2491
[29]   Liu G et al. (2015) PLoS ONE 10(11): e0142680
[30]   Mas R et al. (1989) Neuroscience 33(2): 269
[31]   Moore RE, Scheuer PJ (1971) Science 172(3982): 495
[32]   Morante K et al. (2019) Toxins 11(7): 401
[33]   Okano H et al. (1998) Int Med 37(3): 330
[34]   Ponce D et al. (2016) Toxins 8(4): 102
[35]   Remigante A et al. (2018) Toxins 10(4): 133
[36]   Russell FE (1966) Toxicon 4(1): 65
[37]   Sánchez-Rodríguez J, Lucio-Martínez NL (2011) Cienc Mar 37(3): 369
[38]   Sher D et al. (2005) J Biol Chem 280(24): 22847
[39]   Sud P et al. (2013) J Med Toxicol 9(3): 282
[40]   Tamkun MM, Hessinger DA (1981) Biochim Biophys Acta (BBA) – Prot Struct 667(1): 87
[41]   Toom PM et al. (1975) Toxicon 13(3): 159
[42]   Walker MJA (1977) Toxicon 15(1): 3
[43]   Wang C et al. (2019) J Proteome Res 18(1): 436

3.2.3 Mollusca

Universal features of the body structure of mollusks are a mantle forming a mantle cavity containing environmental medium that is used for breathing and for expelling excretory material (urine and feces) and the double-stranded organization of the nervous system. Many mollusks have a calcareous shell. Of toxicological relevance, especially for humans, are three major groups, the Gastropoda (snails and slugs), the Bivalvia (mussels), and the Cephalopoda (squid, octopus, and cuttlefish).

3.2.3.1 Snails and Slugs (Gastropoda)

Among the gastropoda there are some marine snail species belonging to the families Conidae (cone snails), Turridae (turrid snails), or Terebridae (auger snails) that are venomous [11, 13, 14]. They are carnivors and use their toxins to obtain prey, usually polychaete worms. *Conus geographus* is a fish-hunting species. These snails may, however, use their venom also for defense when threatened. Because all species of cone snails are highly venomous and potentially capable of attacking humans for defense, live snails should be handled with care or not at all.

As the Conidae are the most dangerous ones for animals and humans among the venomous snails we will focus on this group. In a fish hunting species (*Conus geographus*; Ph. 3.11) of this group of tropic and subtropic marine snails, the radula, which is used for grazing in other snail taxa, is transformed to a production site of elongated, hollow, arrow-like radula teeth that detach from the radula surface and are stored in a pouch, the radular sheath (Fig. 3.7). If needed for attacking prey or for defense, one of these radula teeth is filled with venom, which is generated and stored in the venom gland and transported to the radula sheath via a venom duct. The end of the radula tooth is placed in front of the opening of the venom duct, completely filled with venom, and transported to the tip of the proboscis. As soon as a fish comes close enough, the snail shoots the radula tooth toward the fish, but holds on to the back of the tooth using ring muscles of the proboscis. The radula tooth penetrates the fish integument, the venom is released, and paralyzes the victim instantly. Within a few minutes, the snail swallows the paralyzed fish.

Each *Conus* species has a specific toxin mixture that is specifically tailored to its prey, e.g., fish or marine worms. A toxin mixture is necessary to ensure that the toxins attack as many molecular targets as possible and to achieve a rapid onset of the paralytic effect since the snail has no way of pursuing escaped prey animals. The toxins of the piscivorous *Conus* species are therefore the most effective ones and the most dangerous ones for humans.

Various types of C-terminally amidated peptides are present in the toxin mixtures of *Conus* snails, and the assortment is specific for each snail species. The venom of *Conus geographus* contains a multitude of toxic peptides belonging to different toxin families. The four toxins G I, G IA, G II, and M I are representatives of one family, and two other peptides (Geo I and Geo II) represent another family (Fig. 3.8). Within each

Ph. 3.11: The cone snail, *Conus geographus* (Source: Kerry Matz, US National Institute of General Medical Services, Wikimedia, public domain).

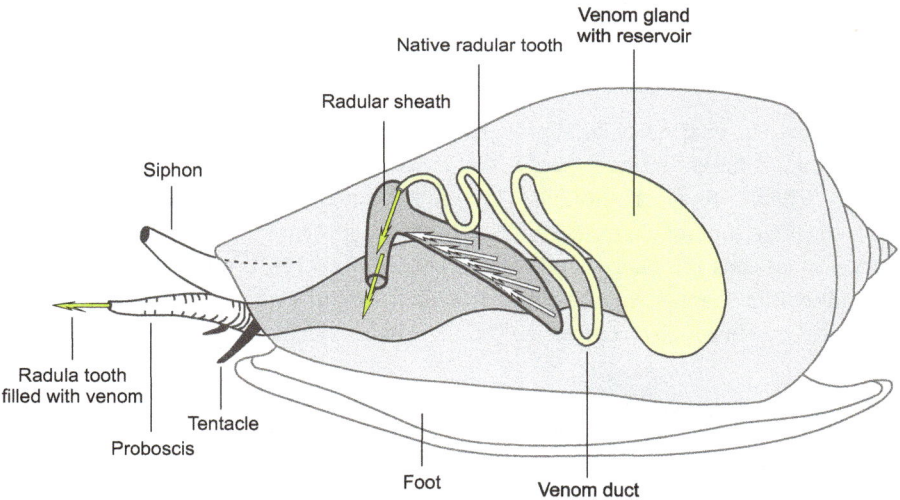

Fig. 3.7: Scheme of the body plan of a cone snail (e.g., *Conus geographus*). The shell is shown in grey and transparent to visualize some of the inner organs. The venom apparatus is shown in light green, the venom-filled radula teeth used as harpoons to catch fish are shown in dark green.

family these peptides are homologs (paralogs) as more than 50% of all amino acid positions are filled with identical amino acid residues. The former family of *Conus geographus* toxins belong to the framework I (CC-C-C) in the nomenclature introduced by Jin et al. [13] as there are two cysteine residues following each other in positions 2 and

3 and two additional cysteine residues following in a distance with several other amino acid residues in between. There are generally two disulfide bridges between the cysteine residues in this type of toxins. The latter family of toxins of *Conus geographus* belongs to the framework III (CC-C-C-CC) which allows the formation of three disulfide bridges.

Alpha-Conotoxins in *Conus geographus*:

```
G I:    Glu-Cys-Cys-Asn-Pro-Ala-Cys-Gly-Arg-His-Tyr-Ser-Cys-NH₂
G IA:   Glu-Cys-Cys-Asn-Pro-Ala-Cys-Gly-Arg-His-Tyr-Ser-Cys-Lys-NH₂
G II:   Glu-Cys-Cys-His-Pro-Ala-Cys-Gly-Lys-His-Phe-Ser-Cys-NH₂
M I: Gly-Arg-Cys-Cys-His-Pro-Ala-Cys-Gly-Lys-Asn-Tyr-Ser-Cys-NH₂
          ↑   ↑         ↑                ↑
```

Micro-Conotoxins in *Conus geographus*:

```
G IIIA: Arg-Asp-Cys-Cys-Thr-Hyp-Hyp-Lys-Lys-Cys-Lys-Asp-Arg-Gln-Cys-Lys-Hyp-Gln-Arg-Cys-Cys-Ala-NH₂
G IIIB: Arg-Asp-Cys-Cys-Thr-Hyp-Hyp-Arg-Lys-Cys-Lys-Asp-Arg-Arg-Cys-Lys-Hyp-Met-Lys-Cys-Cys-Ala-NH₂
              ↑   ↑                     ↑                   ↑                 ↑   ↑
```

Fig. 3.8: Conotoxins of *Conus geographus*. Amino acid sequences of selected conotoxins of *Conus geographus*. The C-termini of these toxins are amidated, a posttranslational modification which increases protein longevity. The sequences are presented in three-letter code. Hyp (blue lettering) stands for hydroxyproline. Hydroxylation of proline residues in proteins is a posttranslational modification. The peptides in one species are homologs (paralogs) as more than 50% of all amino acid positions are occupied by the same amino acid. Positions in which different amino acids occur are labeled in red. The arrows point to cysteine residues which form disulfide bridges within the peptide.

According to their pharmacological properties in target organisms the conotoxins can be arranged in a different scheme (see Section 2.12.4). α-Conotoxins (Fig. 3.8) inhibit nicotinic acetylcholine receptors (nAChRs) [10] while δ-conotoxins delay the inactivation of voltage-sensitive sodium channels resulting in overexcitation in muscle and nerve cells of target organisms [8]. Micro-conotoxins (Fig. 3.8), however, are blockers of voltage-activated sodium channels suppressing action potentials in such cells [4]. Other conotoxins affect G protein-coupled receptor activation or block calcium channels (Tab. 3.2).

🐌 After being stung by a dangerous *Conus* snail, a painful punctate wound develops. In exceptional cases, the pain is so weak that the sting is hardly noticed. After 20–30 min whole body paresthesia sets in, followed by vomiting, impaired vision, difficulty swallowing, speech disorders, shortness of breath, and in severe cases, muscle spasms, muscle paralysis (also of the respiratory muscles), coma, and cardiac arrest. These symptoms appear between 40 min and 5 h after the sting.

🩹 For treatment of a cone snail sting medical help should be sought as soon as possible so that the person can be given artificial ventilation if necessary. The affected limb should be immobilized. Whether applying a pressure bandage has a beneficial effect or not is controversial.

Certain sea slugs, also called sea hares, e.g., *Aplysia brasiliana* (Ph. 3.12), *Stylocheilus longicauda* or *Dolabella auricularia*, contain polyketide or amine toxins that are originally

Tab. 3.2: List of conotoxin families and representative members according to their pharmacological actions in target organisms.

Conotoxin family	Target mechanism	Action	Representative toxin	Reference
Alpha, α	Nicotinic acetylcholine receptor (nAChR)	Inhibition	G I	[10]
Beta, β	Voltage-activated sodium channels	Delay in intrinsic inactivation	PVIA5	[22]
Gamma, γ	Cation channels in neuronal pacemaker cells	Activation	PnVIIA, TxVIIA	[9]
Delta, δ	Voltage-activated sodium channels	Delay in intrinsic inactivation	TxVIA	[8]
Epsilon, ε	Presynaptic calcium channels or presynaptic G protein-coupled receptors	Inhibition	TxIX	[18]
Iota, ι	Voltage-activated sodium channels	Activation	RXIA	[2]
Kappa, κ	Voltage-activated Shaker potassium channels	Inhibition	PVIIA	[22]
Mu, μ	Voltage-activated sodium channels in skeletal muscle cells	Inhibition	GIIIA	[4]
Rho, ρ	α1-Adrenergic receptor	Inhibition	TIA	[20]
Sigma, σ	Serotonin-gated ion channels, 5-HT3-receptor	Inhibition	GVIIIA	[7]
Tau, τ	Somatostatin receptor	Inhibition	CnVA	[17]
Chi, χ	Noradrenalin reuptake transporter in neurons	Inhibition	MrIA, MrIB	[20]
Omega, ω	Voltage-activated calcium channels	Inhibition	CVIA	[15]

synthesized by cyanobacteria such as *Lyngbya gracilis*. These toxins, aplysiatoxin (see Section 2.3.3), acutiphycin, scytophycin B, polycavernoside A, or malyngamides are taken up by the animals during feeding and accumulate in the slug tissues. They are efficient feeding deterrents in potential slug predators [16]. When humans handle these slugs with bare hands aplysiatoxin may induce skin irritation, called 'seaweed dermatitis'. Aplysiatoxin acts as a tumor promoter similar as plant phorbol esters. The LD_{50} (mouse, i.p.) is 0.2 mg/kg BW.

3.2.3.2 Bivalvia

Eating blue mussels (*Mytilus edulis*; Ph. 3.13) or other marine animals such as clams, oysters, geoduck, or scallops collected at the wrong time of the year (during summer

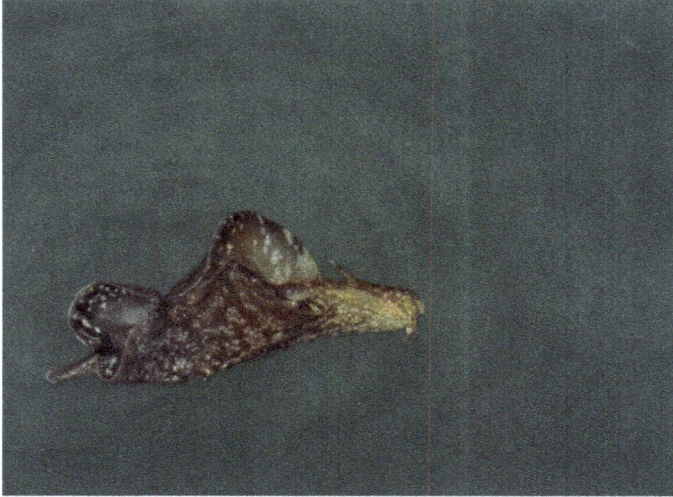

Ph. 3.12: The sea hare, *Aplysia brasiliana* (Source: SongayeNovell/iStock/Getty Images Plus).

when algal or cyanobacterial blooms occur) may result in severe gastrointestinal problems. The disease is called 'diarrhetic shellfish poisoning' (DSP). The symptoms are related to the uptake of okadaic acid (see Section 2.3.9) and saxitoxin (see Section 2.9.14) which accumulate in the bivalves during filter feeding on unicellular algae, cyanobacteria, and dinoflagellates [5]. Especially dinoflagellates like *Dinophysis acuminata* or the 'red tide'-dinoflagellate *Gonyaulax catanella* are rich in these toxins. Okadaic acid is a potent inhibitor of protein phosphatases [6]. Saxitoxin is a potent blocker of voltage-gated sodium channels and inhibits the generation of action potentials in excitable cells (neurons and muscle cells) in target animals and humans [3].

🐾 Intoxications may result from eating marine bivalves or fish that have consumed great amounts of these organisms. Severe cases of saxitoxin poisoning are known as 'paralytic shellfish poisoning' (PSP). The symptoms, which begin with a tingling or burning sensation, then numbness of the lips, gums, tongue, and face, gradually spread over the entire body. Gastrointestinal symptoms as nausea, vomiting, abdominal pain, and diarrhea may develop. Other symptoms including weakness, joint aches, and muscular paralysis may occur as well. Fatalities due to this illness have been reported in humans and animals. There is no specific treatment or antidote.

3.2.3.3 Cephalopoda

Many cephalopods have the ability to release a cloud of melanin-containing ink to the surrounding medium from an ink sac which is an extension of the hindgut. Inking is used for camouflage and to confuse attacking predators [1]. Cephalopods feed by capturing prey (fish, crabs, etc.) with their tentacles, drawing it into their mouth, and taking bites from it using their beaks. They produce saliva-containing enzymes and other

Ph. 3.13: The blue mussel, *Mytilus edulis* (Source: eddyfish/iStock/Getty Images Plus).

toxic ingredients (that may be produced from symbiontic microorganisms) which is injected into prey and used for extraintestinal predigestion. The posterior venom glands produce a variety of enzymes that may support digestion of prey [19].

The blue-ringed octopus (*Hapalochlaena maculosa*; Ph. 3.14) is found in tidal rock pools along the south coast of Australia. Adults grow up to 20 cm in overall length. It is a highly venomous species. Agitated individuals display a higher coloration intensity of the blue rings. Although being quite docile animals they may bite when threatened or attacked. Their saliva contains histamine, tyramine, serotonin, and other phenolic amines, a potent hyaluronidase [12] as well as tetrodotoxin [21]. The latter (see Section 2.9.7) is likely being produced by symbiontic bacteria of the genus *Vibrio* that live in the salivary glands. Tetrodotoxin is a highly toxic substance due to its ability to bind with high affinity to voltage-activated sodium channels of animals and humans ($K_d = 10^{-9}$ mol/L). TTX blocks the ion pores of these channels which renders the excitable cells (nerve and muscle cells) unable to generate any action potentials. Bites by the blue-ringed octopus may be fatal even in humans due to paralysis of the ventilatory muscles. It has been calculated that a tiny 25 g octopus possesses enough venom to kill 10 humans of average body weight [24].

☠ Most bites go along with excessive bleeding and moderate local pain for the first few minutes, but then the affected limbs get numb. After approximately 10 min central symptoms become apparent: general numbness, nausea, vomiting, erratic vision, and difficulties in swallowing and breathing. This is followed by paralysis of the entire skeletal muscle system. The victim may require artificial ventilation. If medical care is not provided immediately, respiratory failure and death may occur.

Ph. 3.14: The blue-ringed octopus, *Hapalochlaena maculosa* (Source: Saspotato, Wikimedia, public domain).

🐾 Even wild animals may be victims of envenomation by *Hapalochlaena* octopuses. Two fatal cases of TTX envenomation of green sea turtles (*Chelonia mydas*) grazing in sea grass beds and accidentally ingesting *Hapalochlaena fasciata* have been reported [23].

References

[1]   Boyle P, Rodhouse P (2005) Cephalopods: Ecology and Fisheries. Blackwell Science Ltd, John Wiley & Sons, Hoboken, New Jersey, USA
[2]   Buczek O et al. (2007) Biochemistry 46(35): 9929
[3]   Catterall WA (1980) Annu Rev Pharmacol Toxicol 20: 15
[4]   Cruz LJ et al. (1985) J Biol Chem 260(16): 9280
[5]   Dawson JF, Holmes CF (1999) Front Biosci 4, D646
[6]   Dounay AB, Forsyth CJ (2002) Curr Med Chem 9(22): 1939
[7]   England LJ et al. (1998) Science 281(5376): 575
[8]   Fainzilber M et al. (1991) Eur J Biochem 202(2): 589
[9]   Fainzilber M et al. (1998) Biochemistry 37(6): 1470
[10]   Gray WR et al. (1981) J Biol Chem 256(10): 4734
[11]   Heralde FM et al(2008) Toxicon 51(5): 890
[12]   Jacups SP, Currie BJ (2008) North Territ Nat 20: 50
[13]   Jin AH et al. (2019) Chem Rev 119: 11510
[14]   Kendel Y et al. (2013) Toxins 5(5): 1043
[15]   Kerr LM, Yoshikami D (1984) Nature 308(5956): 282
[16]   Paul VJ, Pennings SC (1991) J Exp Mar Biol Ecol 151(2): 227
[17]   Petrel C et al. (2013) Biochem Pharmacol 85(11): 1663
[18]   Rigby AC et al. (1999) Proc Natl Acad Sci U S A 96(10): 5758

[19]   Ruder T et al. (2013) J Mol Evol 76(4): 192
[20]   Sharpe IA et al. (2001) Nat Neurosci 4(9): 902
[21]   Sheumack DD et al. (1978) Science 199(4325): 188
[22]   Terlau H et al. (1996) Nature 381(6578): 148
[23]   Townsend KA et al. (2012) Mar Biol 159(3): 689
[24]   Williamson JAH (1987) Clin Dermatol 5(3): 127

3.2.4 Echinodermata

Echinodermata (starfish, sea urchins, and sea cucumbers) are marine invertebrates that emerge from bilaterally symmetrical larvae and, at least the former two taxa, are usually radially symmetrical in the adult stage. One can usually differentiate between the ventral side of the animal where the mouth is located (oral side) and the dorsal side (aboral side). Echinoderms are slowly moving animals that rely on their mechanical and chemical defenses. The adult animals are characterized by their endoskeleton composed of calcareous plates or ossicles. As adults, echinoderms are living on the sea bed at different water depths from the intertidal zone to the abyss. The animals can move using their tube feet, but their movements are slow. Nevertheless, many echinoderms (especially the sea stars) are predators that feed on sessile animals (e.g., corals or bivalves). Sea stars use a mixture of digestive enzymes from the midgut gland for extraintestinal digestion of their prey before the products are taken up into the midgut gland tubules and absorbed by the epithelial cells. Others (e.g., the sea urchins) are herbivores that graze on biofilms. Sea cucumbers are mostly detritivores. The sea lilies are feeding on plankton that they catch using a sticky mucus layer on the surface of their long arms.

3.2.4.1 Sea Urchins (Echinoidea)

For passive defense against potential attackers (e.g., predatory fish, crabs, or sea birds), sea urchins use more or less elongated spines sticking out from the body surface. The calcareous spines are covered by epidermis and are surrounded by toxin glands with reservoirs at their bases (Fig. 3.9). If stepped on or contacted by a predator, the spines penetrate the skin of the attacker, the epithelial cell layer around the calcareous spines is ruptured, and the toxin is squeezed into the wound. Various complications such as granuloma, synovitis, arthritis, edema, hyperkeratosis, and even neuromas may occur in humans injured by sea urchin spines.

There is yet another defense mechanism present in sea urchins. Little grasping organs, the pedicellariae, are distributed all over the body and are usually used to remove foreign particles from the body surface [1]. However, they may function as toxin applicators as well. Pedicellariae are complex organs that contain muscles to anchor the three claws to the endoskeleton and move the claws. Each of the claws carries a spike at the tip which is hollow and connected to a venom gland reservoir located at the shaft of the pedicellarium (Fig. 3.10). When grasping another organism with the claws, these spikes penetrate the skin and inject the toxins. There are approximately 80 sea urchin species that potentially pose danger to humans as they are able to penetrate human skin, among them those of the taxa Diadematidae and Echinothuriidae (leather urchins).

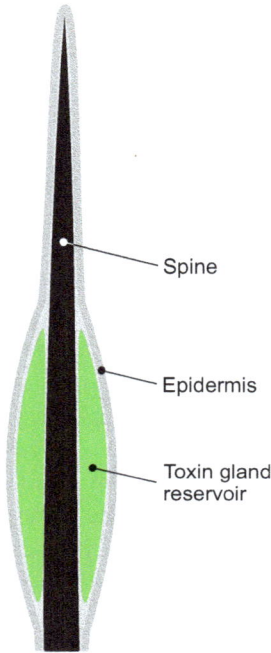

Spine

Epidermis

Toxin gland reservoir

Fig. 3.9: Scheme of a sea urchin spine with toxin reservoir used as a means of passive defense.

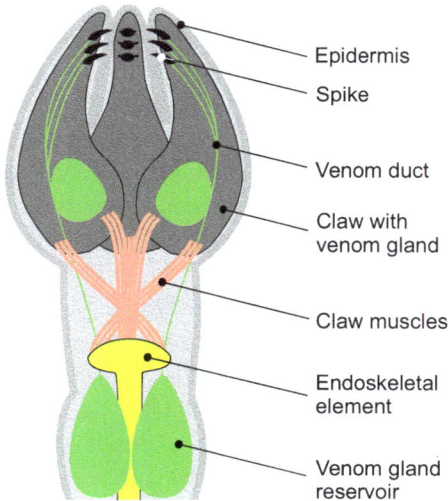

Epidermis
Spike

Venom duct

Claw with venom gland

Claw muscles

Endoskeletal element

Venom gland reservoir

Fig. 3.10: Scheme of a sea urchin pedicellarium which is used to remove particles from the body surface or defensively to pince attackers.

Sea urchin toxins contain pain-inducing serotonin, acetylcholine-like substances, norepinephrine, and steroid derivatives similar to those that occur in sea stars and brittle stars. The most toxic sea urchins, the flower urchin *Toxopneustes pileolus* (Ph. 3.15), the collector urchin *Tripneustes gratilla, Diadema setosum*, and *Asthenosoma* sp. have been more thoroughly investigated with respect to their toxin composition in spines and pedicellariae [9]. Most of the toxin ingredients besides those described above are proteins, among them some proteases (e.g., cathepsins). Several lectins have been found as well, e.g., the rhamnose-binding lectins SUL-I [5], the mannose-containing glycoprotein contractin A (18 kDa) which is a lectin that causes contractions in isolated guinea pig tracheal smooth muscle [10], and the phospholipase A_2-like substance, UT841 which inhibits Ca^{2+} uptake in synaptosomes prepared from chick brain [16]. Pedicillariae of *Toxopneustes pileolus* also contain another protein toxin, termed peditoxin, which carries a prosthetic group called pedoxin. Pedoxin is a low-molecular-mass substance (approx. 0.2 kDa) composed of a heterocyclic lactone structure formed from pyridoxal and glycine. When pedoxin is injected into mice in sublethal doses, it reduces basal body temperature (hypothermia) and leads to sedation and muscle relaxation [8].

Ph. 3.15: The sea urchin *Toxopneustes pileolus* (Source: François Michonneau, Wikimedia, CC-BY).

♟ Skin penetration of poisonous sea urchin spines results in severe pain that may last for a few hours. Further symptoms are erythema and swelling at the injection site. Fragments of spines which are not removed will be encapsulated and foreign body granulomas form. These may be permanently painful when pressure is applied. The spines of tiara sea urchins are hollow and may bring a blue dye into the tissue that may be visible for several weeks. Systemic symptoms such as nausea, fatigue, and circulatory problems are occasionally observed. When envenoming occurs by *Toxopneustes* species systemic effects are prevailing. In such cases, pain lasts only for 15–60 min. In addition, edema and hemorrhages, local paralysis, e.g., in the affected limb, and breathing difficulties may occur. Death from drowning of a pearl diver due to paralysis has been reported. Injuries by pedicellariae in humans are rare but may cause severe pain, local edema, bleeding, general lethargy, and dizziness, tingling sensations in the skin, joint pain, muscle paralysis, respiratory distress, and hypotension. However, fatalities are very rare.

🐾 When injured by a sea urchin, spines and broken tips of spines should be carefully removed from the wound to prevent the development of chronic pain or secondary bacterial infections and to avoid the transfer of additional toxin material. Poisoning symptoms are treated symptomatically. Bathing the puncture site in water as hot as possible will relieve the pain, but poses danger of scalding. Antibiotic prophylaxis is necessary if the immune system is compromised. Resting the affected limb is not beneficial.

3.2.4.2 Sea Stars (Asteroidea)

Many organs of echinoderms contain different kinds of toxins and bioactive compounds to deter potential predators. The tissues of most sea stars are loaded with saponins (asterosaponins [7]) that are potent antifeedants so that sea stars are usually not predated on by other animals. The common sea star, *Asterias rubens*, produces steroid glycosides (see Section 2.6.4) which bear a sulfate group attached to the aglycone [2]. The blue bat star, *Patiria pectinifera*, contains a variety of polyhydroxysterol esters together with steroid derivatives. These substances have been shown to have antiviral properties besides being cytotoxic to human and animal cells [11]. The crown-of-thorns starfish, *Acanthaster planci* (Ph. 3.16), is a large starfish (up to 35 cm in diameter) that preys upon coral polyps and contributes to the degradation of coral reefs in the Indo-Pacific region due to its mass reproduction. It carries poisonous thorn-like spines on its backside which may get dangerous to humans when they step on such a sea star. The poison of these spines contains at least two different variants

Ph. 3.16: Crown-of-thorns sea star, *Acanthaster planci* (Source: Jon Hanson, Wikimedia, CC-BY-SA).

of phospholipases A_2 which have hemolytic properties [12] and the hepatotoxic planci-toxins I and II which have DNase activities [13].

☠ Injuries caused by spines of *Acanthaster planci* are very painful and may result in the formation of granulomas if the broken tips of the spines are not completely removed from the wounds. The affected limb swells rapidly. Other symptoms that can last for several days include headache, nausea, vomiting, and fever. Circulatory problems also occur, but less frequently.

3.2.4.3 Sea Cucumbers (Holothuroidea)

More than 100 different triterpene glycosides (saponins) (see Section 2.6.4) have been found in sea cucumbers (Holothuroidea; Ph. 3.17) which protect these slow-moving invertebrates against predators, even against sharks [15] because they seem to exert cytotoxic and hemolytic effects [3, 6, 14].

Ph. 3.17: The sea cucumber *Holothuria tubulosa* (Source: Nachosan, Wikimedia, CC-BY-SA).

Some species of sea cucumbers have developed a peculiar way of actively defending themselves. They eject their Cuvierian tubules from the gut in response to stress. These tubules are caeca, elongated invaginations of the hindgut epithelium, which are usually located within the posterior part of the intestines. The mass of expelled tubules transforms into sticky white threads that entangle the attacker and brings its body surface in intense contact with toxic saponins which are concentrated in these tubules [4].

☠ Some humans experience skin irritations or even severe allergic reactions upon skin contract with holothurians.

⬛ Skin areas that have come in contact with holothurian tissues may be treated with seawater washes to dilute the toxin. Using freshwater may increase pain sensations. Upon eye contact the eyes have to be rinsed with copious amounts of saline. Wounds should be left uncovered to heal. Checks may be required to monitor for any signs of infection. In such cases, the use of antibiotics should be considered by a physician.

Sea cucumbers are used as food in Southeast Asia. Care must be taken to remove all internal organs before thoroughly cooking the animals to destroy residual toxin molecules.

References

[1]   Coppard SE et al. (2012) Acta Zoologica 93(2): 125
[2]   Demeyer M et al. (2014) Comp Biochem Physiol B Biochem Mol Biol 168: 1
[3]   Eeckhaut I et al. (2015) Biol Bull 228(3): 253
[4]   Hamel JF, Mercier A (2000) Mar Freshw Behav Physiol 33(2): 115
[5]   Hatakeyama T et al. (2017) Protein Sci 26(8): 1574
[6]   Kalinin VI et al. (1996) Toxicon 34(4): 475
[7]   Komori T (1997) Toxicon 35(10): 1537
[8]   Kuwabara S (1994) J Biol Chem 269(43): 26734
[9]   Nakagawa H et al. (2003) J Toxicol 22(4): 633
[10]   Nakagawa H et al. (1991) Arch Biochem Biophys 284(2): 279
[11]   Peng Y et al. (2010) Chem Pharm Bull 58(6): 856
[12]   Shiomi K et al. (1998) Toxicon 36(4): 589
[13]   Shiomi K et al. (2004) Toxicon 44(5): 499
[14]   Thron CD et al. (1964) Toxicol Appl Pharmacol 6(2): 182
[15]   Van Dyck S et al. (2011) J Exp Biol 214(8): 1347
[16]   Zhang Y et al. (2001) Toxicon 39(8): 1223

3.2.5 Worms

3.2.5.1 Flat Worms (Plathelminthes)

The phylum of Plathelminthes (flat worms) encompasses many species of relatively simple worm-like, bilaterian, unsegmented, and soft-bodied invertebrates. Their flat-tend body shape allows them to absorb sufficient amounts of oxygen and nutrients through their integuments by diffusion or by facilitated diffusion. Some flatworm species are free living, while others have adapted a parasitic life style. The gastrointestinal parasitic worms, e.g., the tape worm *Taenia saginata,* defend themselves against digestion by the intestinal enzymes of their hosts by secretion of copious amounts of mucus and protease inhibitors. Moreover, these parasitic worms have developed sophisticated strategies to evade the defensive actions of the host immune systems [6].

In some cases, however, toxins are used by plathelminths to efficiently hunt for prey. A marine flatworm from Guam has been described to release stored tetrodo-toxin (TTX) to the environment to kill gastropods that are its preferred prey [44]. The limnic predaceous flatworm *Mesostoma lingua* uses a neurotoxin diffusing into the water from webs of mucus produced by the worm to make its hunt for water fleas (*Daphnia magna*) more efficient [10].

3.2.5.2 Ribbon Worms (Nemertea)

The animal phylum Nemertea contains mostly marine worms that are also known as ribbon worms or proboscis worms. Some of these animals have an eversible proboscis carrying a venomous stiletto that is used to inject prey animals (annelids, mollusks, crustaceans, or even fish) with venom. The venoms as well as the mucus cover of the animal's integument contain pyridine alkaloids (see Section 2.9.8) like anabaseine and nemertilline [14, 22, 24]. These substances interact with the nicotinic acetylcholine re-ceptor (nAChR) in target animals and humans. Activation of this receptor in the ab-sence of endogenous transmitter results in muscle spasms and paralysis in animals and man. The LD_{50}s of anabaseine in the mouse or the red swamp crayfish *Procamba-rus clarkii* (i.v.) are 85 or 70 µg/kg BW, respectively. Proteo-transcriptomic analyses of toxin mixtures of the predatory ribbon worm *Amphiporus lactifloreus* (Ph. 3.18) re-vealed that the substance mix in Nemertea may be highly complex with respect to their protein compounds [54]. A comparative transcriptomic study of peptide toxins in different Nemertea species revealed the presence of ion channel inhibitors (knot-tins), pore-forming toxins, and neurotoxins in the venoms [55].

Ph. 3.18: The ribbon worm, *Amphiporus lactifloreus* (Source: Diego Delso, Wikimedia, CC-BY-SA).

Some Nemertea species (e.g., *Cephalothrix* cf. *simula* from the Sea of Japan) accumulate tetrodotoxin of bacterial origin in their tissues in extremely high concentrations to deter potential predators and to kill prey [56]. Other species (e.g., the milky nemertean *Cerebratulus lacteus*) use other hunting techniques. They wrap the long anterior body around their victims and cover them with skin secretions, which contain so-called rhabdites (toxin granules). These contain neurotoxins (B toxins) that act on voltage-gated sodium channels in the target organism. Furthermore, cytolytically active polypeptides are present in these granules. Polypeptide A-III consists of 95 amino acid residues and contains 3 disulfide bridges [3]. It attaches to cell membranes in the target organism and forms large transmembrane pores that allow even small proteins to pass through.

🐾 An interesting case of an invasive nemertean driving native population of soil arthropods (e.g., isopods and amphipods) to the verge of extinction has recently been reported [47]. Predation of the land nemertean *Geonemertes pelaensis* on such animals has led to substantial declines in the abundances of soil arthropods in the Ogasawara Islands in the Pacific Ocean, a UNESCO World Heritage site.

3.2.5.3 Ringed Worms (Annelida)
The annelids (Annelida), also known as the ringed worms or segmented worms, are an animal phylum including the ragworms (Nereididae, Polychaeta), the earthworms (Oligochaeta, Clitellata), and the leeches (Hirudinea, Clitellata). Ragworms and earthworms use certain toxins present in their body fluids, tissues, or in the mucus covering the body surface to deter potential predators. The ectoparasitic leeches, in turn, have evolved a multitude of salivary venom compounds that allow them to feed on body fluids of warm-blooded animals upon creating an integumental wound using either saw-like jaws (Arhynchobdellidae) or a proboscis (Rhynchobdellidae).

Some polychaetes like *Kuwaita heteropoda* harbor nereistoxin in the mucus covering the body surface [18]. Nereistoxin is a sulfur-containing amine (1-dimethylamino -3,4-dithiolane) (Fig. 2.28) and blocks the nAChRs in excitable cells of target animals and humans in a dose-dependent manner [23, 38, 45]. Handling of these worms with bare hands may cause increased saliva production, miosis, headaches, and vomiting. The LD_{50} (rabbit, s.c.) is 1.8 mg/kg BW. An analog of nereistoxin, thiocyclam, affects the nAChRs in the insect central nervous system and had been used for a while as a broad-spectrum insecticide until the observation of adverse effects in domestic animals resulted in a ban in many countries.

Polychaetes belonging to the taxon Glyceridae are commonly known as bloodworms. These venomous annelids have an eversible proboscis that carries four cross-arranged claws, each of which is associated with a venom gland surrounded by several layers of muscle cells. When a worm grabs a prey organism using its claws, contraction of these muscles drives venom from the gland reservoir through ducts and pores in the claws into the wound. Humans who have been stung report

Ph. 3.19: The bloodworm, *Glycera* sp. (Source: US National Oceanic and Atmospheric Administration, Wikimedia, public domain).

local effects of envenomation by *Glycera* sp. (Ph. 3.19) including rapid onset of pain and swelling followed by numbness [12].

Transcriptomic profiling of the venom glands of three *Glycera* species indicated that the venoms contain pore-forming toxins (PFTs), membrane-disrupting toxins, neurotoxins, protease inhibitors, and CAP domain toxins [53]. Proteomic studies of venom gland extracts of *Glycera tridactyla* revealed the presence of proteins of low as well as of high molecular masses. Injecting test animals with venom protein fractions demonstrated the ability of components of the high-molecular-mass fraction to induce neurotransmitter release from nerve terminals and paralysis in the test animals [4, 29]. Subsequent studies revealed that the active ingredient of this protein fraction was a glycoprotein of approximately 300 kDa which was named α-glycerotoxin (GLTx). This toxin activates Cav2.2 channels (N-type Ca^{2+} channels) in presynaptic plasma membranes of nerve cells in target organisms [31] in a reversible manner [39]. In vitro, GLTx potently stimulates transmitter release from isolated synaptosomes with an EC_{50} of 50 pmol/L [46].

Mediterranean specimens of the bearded fireworm, *Hermodice carunculata* (Ph. 3.20), contain quaternary ammonium compounds derived from betaine, the carunculines, in all tissues including the dorsal setae. When an animal is approached by a predator, these setae puncture the integument of the attacker, break off, and release the toxins into the wound [43]. The carunculines may induce neurotoxic effects and edema in target organisms.

Earthworms (Oligochaeta) are prey to many different animals like birds or moles. Thus, some have evolved defensive molecules that at least partially protect them. An example is the manure worm (*Eisenia fetida*; Ph. 3.21) which lives in great densities in compost heaps. When researchers analyzed the coelomic fluid of these animals they

found a protein that was able to aggregate when in contact with foreign cells and to form transmembrane pores which results in cell lysis. This 33 kDa protein is produced by coelomocytes and free chloragocytes and was termed lysenin [25]. The protein has structural similarities to other pore-forming proteins even of bacteria, requires the presence of sphingolipids in the target membranes [27], and shares many mechanistic aspects of pore formation with the bacterial toxins [26] indicating that these proteins may have a common evolutionary origin [9]. Lysenin forms transmembrane pores in cell membranes of eukaryotic organisms and has antibacterial properties [5]. Another antibacterial defense molecule present in the body fluids of *Eisenia* sp. is lysozyme, an enzyme that destabilizes bacterial cell walls [28].

Ph. 3.21: The manure worm, *Eisenia fetida* (Source: Mihai Duguleana, Wikimedia, public domain).

Leeches are hematophagic animals. Upon attachment of a jawed leech to a suitable host using its anterior sucker, the leech starts periodic tilting movements of its saw-like jaws that carry rows of calcified teeth. This creates a wound in the host skin and induces bleeding through severing small blood vessels. Blood and lymph are then sucked up by a muscular pharynx pump into the crop, a large storage compartment from which small portions are transferred to the short intestine for digestion over the following weeks and months. Leeches belonging to the genera *Hirudo* (Eurasian leeches), *Macrobdella* (American leeches), or *Haemadipsa* (Asian land leeches) are large (5–20 cm in length) and strong enough to suck blood from mammals including humans [13, 57].

Simultaneously with the start of jaw movements, the leech initiates secretion of saliva from unicellular gland cells lying dispersed in the connective tissue and the musculature in the anterior body part [7]. Saliva is constantly secreted from the gland cell reservoirs into the wound during the feeding period. The ingredients fulfil important functions for the leech. To stay undetected by the host during the feeding period (30 min to 2 h) the leech has to introduce analgesics or anesthetics into the wound. Substances that suppress inflammatory responses in the host are transferred as well as substances (matrix-degrading enzymes) that help to distribute the salivary compounds to adjacent tissues around the feeding site. To ensure a steady and sustained blood flow to the feeding site the leech has to introduce substances into the wound that relax vascular smooth muscle cells and others that prevent blood platelet aggregation at the feeding site. Anticoagulants that suppress the proteolysis of plasma fibrinogen by inhibiting thrombin or other factors of the blood coagulation cascade effectively keep the blood in a liquid state. Last but not the least, the leech may enrich the ingested blood with antimicrobial substances to minimize loss of stored material and potential harm from microbial toxins in its crop. Although not all of the many salivary compounds (potentially >200 [58]) have been functionally characterized yet, the leech seems to provide one or more active ingredients to achieve each of these goals [19].

Saliva of the medicinal leech, *Hirudo* sp. (Ph. 3.22), contains antistasin, a 15 kDa protein [48], which has, besides other functions, the potency to inhibit tissue kallikrein. Tissue kallikreins are proteases which cleave inactive preproteins (kininogens) to generate the biologically active kinins which are proinflammatory and pain-inducing agents [50, 59].

Another salivary protein in the leech, eglin C, has been shown to inhibit *N*-formyl-Met-Leu-Phe-induced chemotaxis of rat neutrophils [20] and the generation of reactive oxygen species (ROS) in human neutrophils [40, 52]. These observations may indicate that release of eglin C into host tissue at the feeding site may impair neutrophils from entering the surrounding tissue and inducing defense reactions which promote inflammation. Moreover, leech saliva contains a Kazal-type inhibitor of mast cell tryptase (leech-derived tryptase inhibitor, LDTI) [49, 51]. LDTI may suppress inflammation and immune reactions in the host as elevated levels of tryptase in human serum are associated with inflammatory conditions and anaphylactic reactions [37]. Host defense reactions related to the

Ph. 3.22: The medicinal leech, *Hirudo verbana* (Source: Jan-Peter Hildebrandt).

complement system may also be attenuated by ingredients of leech saliva. An inhibitor of the C1 complement factor with a molecular mass between 60 and 70 kDa has been discovered that blocks activation of classical as well as alternative pathways of complement activation [2].

Leech saliva contains a hyaluronidase [21, 41] that has a β-endoglucuronidase activity and is able to digest hyaluronic acid within the extracellular matrix of host tissues. This mobilizes the water molecules originally bound to hyaluronic acid and facilitates diffusion of other ingredients of leech saliva through the tissues.

Leeches secrete several salivary factors which attenuate platelet adhesion to injured vessel walls. One of these factors, saratin, is a 12 kDa protein which binds to exposed collagen I and II in the walls of injured blood vessels [16] and inhibits binding of von Willebrand factor (vWF) as well as platelet aggregation in a competitive manner. Another factor in leech saliva, calin, is a 65 kDa protein that binds to collagen I [35] thereby preventing vWF binding to exposed collagen and platelet aggregation [8, 17]. The activation of platelets by purinergic signaling is inhibited by the presence of a 45 kDa adenosine 5′-diphosphate diphosphohydrolase (apyrase) in leech saliva [42]. This enzyme dephosphorylates ADP preventing ADP-mediated activation of purinergic receptors in platelets.

Leech saliva contains several factors that interfere with elements of the blood coagulation cascade in warm-blooded host animals, thus preventing blood clotting during feeding and during storage of blood in the crop. Antistasin is a potent inhibitor of factor Xa in the coagulation cascade [11]. A similar factor with the same mode of action, bdellastasin, has been isolated from the saliva of *Hirudo medicinalis* [32]. A direct thrombin inhibitor has been found in saliva of different leech species. Hirudin [36] is a 7.1 kDa protein in *Hirudo* sp. containing a sulfated Tyr63 seems to support high-affinity binding to thrombin. Hirudin is a bivalent thrombin inhibitor. Its N-terminus interacts with the active site of thrombin and the C-terminus with the fibrinogen-binding domain of thrombin, the exosite II (Fig. 3.11) which explains the low K_d of this interaction of 10^{-14} mol/L [1]. Natural or recombinant hirudin has been successfully used for antithrombotic therapy [15, 30]. Hirudins and similarly build hirudin-like factors (HLFs) occur in several isoforms in each leech species indicating that such agents have important roles as anticoagulants or in other functions [34].

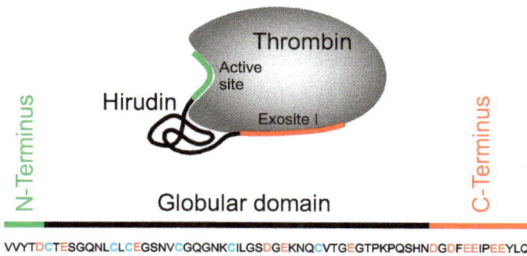

VVYTDCTESGQNLCLCEGSNVCGQGNKCILGSDGEKNQCVTGEGTPKPQSHNDGDFEEIPEEYLQ

Fig. 3.11: Inhibition of the protease thrombin by the salivary anticoagulant hirudin of *Hirudo medicinalis*. Hirudin is a bivalent thrombin inhibitor: The N-terminus of hirudin (green) blocks the active site of thrombin while the C-terminus (red) which contains an accumulation of acidic amino acids (red lettering in the hirudin sequence) blocks the exosite I, the binding site of the thrombin substrate, fibrinogen. The central globular domain of hirudin contains six cysteine residues (blue lettering) which form three disulfide bridges. The pI of hirudin as calculated from its amino acid sequence is 4.41 [33].

A destabilase gene has been identified in *Hirudo medicinalis* that encodes an enzyme, the destabilase, which specifically cleaves endo-epsilon (γ-Glu)-Lys isopeptide bonds [60]. These bonds are a hallmark of crosslinked fibrin molecules mediating secondary hemostasis. Thus, this enzyme is not only able to avoid the formation of fibrin meshworks but also to dissolve blot clots that have already been formed. The destabilase molecule contains a domain that has antimicrobial potency by lysozyme-like as well as nonenzymatic mechanisms [61].

References

[1]   Ascenzi P et al. (1992) J Mol Biol 225(1): 177
[2]   Baskova IP, Zavalova LL (2001) Biochemistry (Mosc) 66(7): 703
[3]   Blumenthal KM, Kem WR (1980) J Biol Chem 255(17): 8266
[4]   Bon C et al. (1985) Neurochem Intl 7(1): 63
[5]   Bruhn H et al. (2006) Dev Comparat Immunol 30(7): 597
[6]   Chulanetra M, Chaicumpa W (2021) Front Cell Infect Microbiol 11: 702125
[7]   Damas D (1974) Arch Zool Exp Gen 115: 279
[8]   Deckmyn H et al. (1995) Blood 85(3): 712
[9]   De Colibus L et al. (2012) Structure 20(9): 1498

[10]   Dumont HJ, Carels I (1987) Limnol Oceanogr 32(3): 699
[11]   Dunwiddie C et al. (1989) J Biol Chem 264(28): 16694
[12]   Durkin DM et al. (2022) Toxins 14(7): 495
[13]   Elliott JM (2008) Freshwater Biol 53(8): 1502
[14]   Göransson U et al. (2019) Toxins 11(2): 120
[15]   Greinacher A, Warkentin TE (2008) Thrombosis Haemostasis 99(5): 819
[16]   Gronwald W et al. (2008) J Mol Biol 381(4): 913
[17]   Harsfalvi J et al. (1995) Blood 85(3): 705
[18]   Hashimoto Y, Okaichi T (1960) Ann NY Acad Sci 90: 667
[19]   Hildebrandt J-P, Lemke S (2011) Naturwissenschaften 98: 995
[20]   Hornebeck W et al. (1987) Cell Biochem Funct 5(2): 113
[21]   Hovingh P, Linker A (1999) Comp Biochem Physiol B Biochem Mol Biol 124(3): 319
[22]   Kem W et al. (2006) Mar Drugs 4(3): 255
[23]   Kem WR et al. (2022) Mar Drugs 20(1): 49
[24]   Kem WR et al. (1976) Experientia 32(6): 684
[25]   Kobayashi H et al. (2004) Int Rev Cytol 236: 45
[26]   Kulma M et al. (2019) Toxins 11(8): 462
[27]   Lange S et al. (1997) J Biol Chem 272(33): 20884
[28]   Lassalle F et al(1988) Comp Biochem Physiol Part B 91(1): 187
[29]   Manaranche R et al. (1980) J Cell Biol 85(2): 446
[30]   Markwardt F et al. (1988) Thromb Res 52(5): 393
[31]   Meunier FA et al. (2002) EMBO J 21(24): 6733
[32]   Moser M et al. (1998) Eur J Biochem 253(1): 212
[33]   Müller C et al. (2020) FEBS Lett 594(5): 841
[34]   Müller C et al. (2016) Mol Genet Genomic 291(1): 227
[35]   Munro R et al. (1991) Blood Coagul Fibrinolysis 2(1): 179
[36]   Nowak G, Schrör K (2007) Thrombosis Haemostasis 98(1): 116
[37]   Payne V, Kam PC (2004) Anaesthesia 59(7): 695
[38]   Raymond Delpech V et al. (2003) Invert Neurosci 5(1): 29
[39]   Richter S et al. (2017) BMC Evol Biol 17(1): 64
[40]   Rigbi M et al. (1987a) Comp Biochem Physiol C 88(1): 95
[41]   Rigbi M et al. (1987b) Comp Biochem Physiol B 87(3): 567
[42]   Rigbi M et al. (1996) Seminars in Thromb Hemost 22(3): 273
[43]   Righi S et al. (2022) Mar Drugs 20: 585
[44]   Ritson-Williams R et al. (2006) Proc Natl Acad Sci U S A 103(9): 3176
[45]   Sattelle DB et al. (1985) J Exp Biol 118(1): 37
[46]   Schenning M et al. (2006) J Neurochem 98(3): 894
[47]   Shinobe S et al. (2017) Sci Rep 7(1): 12400
[48]   Söllner C et al. (1994) Eur J Biochem 219(3): 937
[49]   Sommerhoff CP et al. (1994) Biol Chem Hoppe-Seyler 375(10): 685
[50]   Steranka LR et al. (1988) Proc Natl Acad Sci U S A 85(9): 3245
[51]   Stubbs MT et al. (1997) J Biol Chem 272(32): 19931
[52]   Suter S, Chevallier I (1988) Biol Chem Hoppe-Seyler 369(7): 573
[53]   von Reumont BM et al. (2014) Genome Biol Evol 6(9): 2406
[54]   von Reumont BM et al. (2020) Mar Drugs 18(8): 407
[55]   Vlasenko AE et al. (2022) Toxins 14(8): 542
[56]   Vlasenko AE, Magarlamov TY (2020) Toxins 12(12): 745
[57]   Wilkin PJ, Scofield AM (1990) Freshwater Biol 23(2): 165
[58]   Yanes O et al. (2005) Mol Cell Proteomics 4(10): 1602

[59] Yousef GM, Diamandis EP (2001) Endocrine Rev 22(2): 184
[60] Zavalova L et al. (1996) Mol Gen Genet 253(1–2): 20
[61] Zavalova LL et al. (2006) Chemotherapy 52(3): 158

3.2.6 Tunicata

Most Tunicata (also called urochordates) are sessile animals attached to rocks or other hard substrate when they have reached the adult stage. Some live as solitary individuals, but others replicate by budding and form colonies of clonal animals. Their body consists of a sac-like water-filled lumen with two tubular openings, the siphons, through which environmental water is drawn into and out of the lumen.

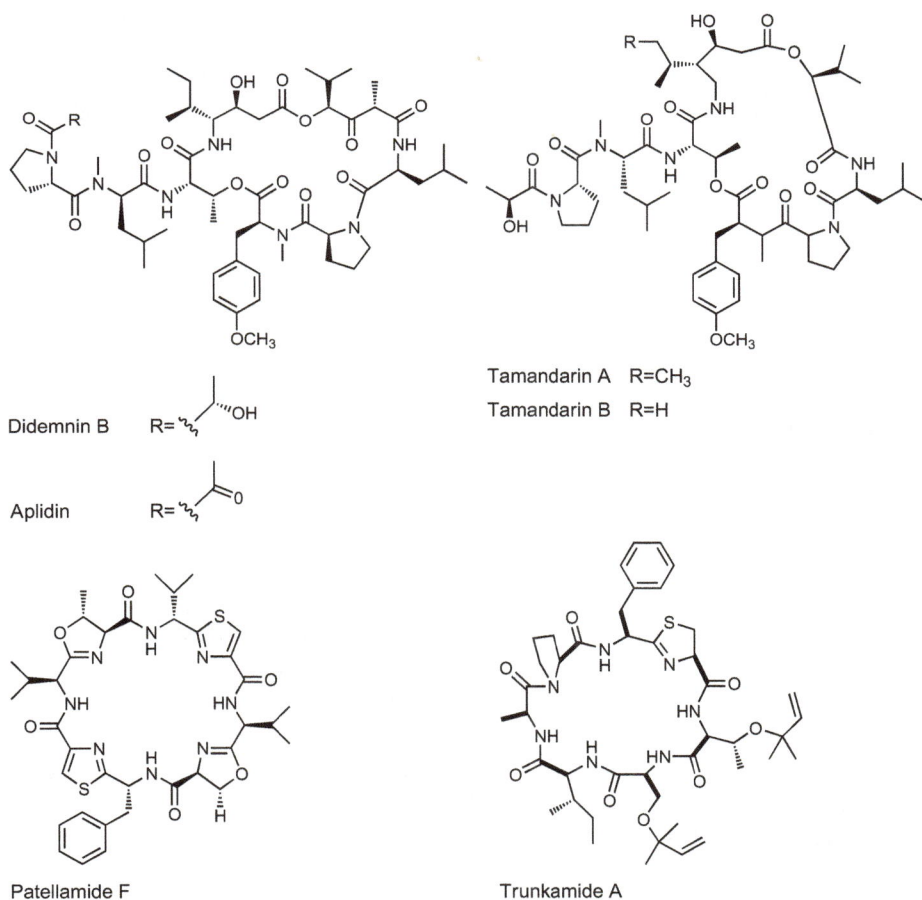

Tamandarin A   R=CH₃
Tamandarin B   R=H

Didemnin B   R=

Aplidin   R=

Patellamide F

Trunkamide A

Fig. 3.12: Cyclic peptides of tunicates.

Tunicates are marine filter feeders that are in close contact with the microorganisms in the environment and cannot run away from potential predators. Thus, they have developed a range of chemical defenses. Antimicrobials prevent them from being overgrown or consumed by bacteria or fungi, repellent substances deter other animals that may try to prey on them. Virtually all of these substances are potent cytotoxins.

Tetracyclic aromatic alkaloids have been isolated from the Okinawan tunicate *Cystodytes dellechiajei* [2]. These cystodytines (see Section 2.9.5) induce the release of calcium ions from intracellular stores which may disturb normal cell signaling function in epithelial cells of predatory animals that get in contact with these sea squirts. Brominated β-carbolines, eudistomin E, and eudistalbin A have been isolated from the marine tunicate *Eudistoma album* [1]. Cyclic peptides with cytotoxic potencies like the tamandarins, didemnin B, patellamide F, or trunkamide A (Fig. 3.12) occur in different tunicate species and have been tested as anticancer drugs. The ecteinascidins of the mangrove tunicate *Ecteinascidia turbinata* are biologically active compounds containing two to three tetrahydroisoquinoline subunits. They are not only cytotoxins but also potent antimicrobials. One of these molecules, trabectedin, is in use for the treatment of soft tissue cancer [3]. Many tunicate-derived compounds are currently screened for their potential usefulness in medical applications [4, 5].

References

[1] Adesanya SA et al. (1992) J Nat Prod 55(4): 525
[2] Kobayashi J et al. (1988) J Org Chem 53(8): 1800
[3] Le VH et al. (2015) Nat Prod Rep 32(2): 328
[4] Palanisamy SK et al. (2017) Nat Prod Bioprospecting 7(1): 1
[5] Watters DJ (2018) Mar Drugs 16(5): 162

3.2.7 Arachnida

The class of Arachnida is a subgroup of arthropods comprising spiders, scorpions, ticks, mites, pseudoscorpions, harvestmen, camel spiders, whip spiders, and whip scorpions (vinegaroons). General features of these animals are four pairs of walking legs and a body divided into two tagmata: the cephalothorax and the abdomen. The cephalothorax carries the mouthparts (a pair of chelicerae and a pair of jaw palps, the pedipalps) and the walking legs. Most arachnids are terrestrial animals. Among the Arachnida, the Araneae (spiders), the Scorpiones (scorpions), and the Acari (mites and ticks) are of toxicological interest.

3.2.7.1 Spiders (Araneae)

Web spiders are not able to shred their prey before ingestion. Therefore, they grab their prey, puncture its body surface using their chelicerae, and inject digestive enzymes produced in their midgut diverticula into the victim. After some time of extra-intestinal digestion, they suck up the liquified material that is filtered through oral hairs and grooves on its way into the digestive tract where the final digestion occurs [23]. This mode of feeding explains why it is essential for a spider to immediately immobilize its prey upon catching it. This is achieved by the injection of venoms through the chelicerae right after the prey item has been grabbed and its body surface punctured. The venom glands are located either within the chelicerae or under the carapace. Venom ducts connect the venom gland reservoirs with the tips of the chelicerae. The venom glands probably evolved from accessory digestive glands.

Spider venoms are generally a mixture of more than 100 components. Each component is supposed to have a special function in the target organism. Most venoms act on plasma membrane receptors in muscle or nerve cells, and others help to digest the extracellular matrix and/or are cytotoxic and induce tissue necrosis. Although each of the venom components is a toxin per se, it is the synergistic action of all toxins that mediates the full envenomation effect [63].

Most of the toxins in spider venoms are peptides and proteins. Their concentrations in the venom may reach 150 µg/µL [30]. As mass production of venom components is costly for a spider it adjusts the amount of venom applied to a prey item carefully according to prey size and agility [62]. Spiders may even apply 'dry bites' if they use their chelicerae just for defense purposes [22]. However, defensive use of venom in some spider species may even be dangerous to humans.

Spider venoms have mainly evolved to secure the rapid immobilization of prey. This is optimally achieved by inhibiting pre- and/or postsynaptic mechanisms in the neuromuscular junction. There are peptides that inhibit presynaptic calcium channels. ω-Grammotoxin, a 36-amino-acid peptide of *Grammostola rosea* (Ph. 3.23), is a potent inhibitor of P-, Q-, and N-type calcium channels [40]. Polyamines, e.g., the polyspermine FTX-3.3 of the desert grass spider *Agelenopsis aperta*, block N-type calcium channels [17]. These channels mediate voltage-activated calcium influx into synaptic endings and the formation of action potentials. Block of these channels results in arrest of transmitter release into the synaptic cleft. Other polyamines block receptors on the postsynaptic side of neuronal synapses. Argiopine (Argiotoxin-636) of *Argiope lobata* (Ph. 3.24) or JSTX-3 of *Trichonephila clavata* inhibits different subtypes of ionotropic glutamate receptors in the central nervous system of vertebrates [28, 42], but also in sensory neurons and in skeletal muscle of invertebrates, especially in arthropods [37, 52]. The synergistic effects of pre- and postsynaptic inhibitors favor an immediate paralyzing effect of envenomation in prey animals.

The males of the Australian funnel web spider *Atrax robustus* carry a total of 1.7 mg venom in their gland reservoirs. The LD_{50} of its venom (mouse, s.c.) is 0.16 mg/kg BW (Tab. 3.3.) which indicates that this spider is dangerous for young children.

Ph. 3.23: The Chilean rose hair tarantula, *Grammostola rosea* (Source: Viki, Wikimedia, GNU-CC-BY-SA).

Even adults may be in danger if attacked by females of the Brazilian wandering spider, *Phoneutria nigriventer* (Ph. 3.25). Its venom gland reservoir may be filled with up to 2 mg venom, and the LD_{50} of this venom (mouse, s.c.) is approximately 0.002 mg/kg BW. However, venomous spiders will typically not bite humans unless they feel threatened or provoked.

Different species of back widow spiders of the genus *Latrodectus* occur almost worldwide. The females of the southern black widow in North America, *Latrodectus mactans* (Ph. 3.26), are well known for their shiny black color with bright red marks. Females can measure 13 mm in their body length. These spiders contain a very potent neurotoxin, α-latrotoxin (120 kDa), which causes a syndrome called 'latrodectism' [39]. This syndrome includes severe muscle pain, abdominal cramps, excessive sweating, tachycardia, hypertension, and muscle spasms. Symptoms usually last for 3–7 days, but may persist for several weeks. Latrotoxin induces neurotransmitter release from nerve terminals by binding to neuronal adhesion molecules (neurexins) and G protein-coupled receptors (latrophilins) in neuronal cells of affected animals or humans. These interactions lead to a nonspecific influx of calcium ions into the cytosol of synaptic terminals. This results in excessive transmitter release into the synaptic cleft. Acetylcholinergic, noradrenergic, dopaminergic, glutamatergic, and enkephalinergic systems in invertebrates and in vertebrates are susceptible to the toxin. Uncontrolled excitation of postsynaptic cells results in muscle cramps or spastic paralysis. Fatalities may occur when ventilatory muscles are affected [55]. Moreover, black widow venom is rich in hydrolases including proteases, hyaluronidase, and alkaline as well as acid

Ph. 3.24: *Argiope lobata* (Source: Javier Virués Ortega, Wikimedia, CC-BY-SA).

Tab. 3.3: Examples of spiders potentially posing a danger to humans and the LD$_{50}$s of their venoms.

Spider species	Distribution	LD$_{50}$ (mouse, s.c.) mg/kg BW
Trechona venosa	South America	0.35
Atrax robustus (males)	Australia	0.16
Latrodectus mactans (females)	North America	0.9
Loxosceles intermedia	The Americas	0.48
Phoneutria nigriventer	The Americas	0.002–0.008

phosphatases. These lytic enzymes are responsible for the more or less extensive tissue necroses around the bite sites in affected animals and humans [64].

The Sydney funnel-web spider (*Atrax robustus*) is native to eastern Australia. Both sexes are glossy and darkly colored. While the females (Ph. 3.27) reside at their webs the males are highly mobile searching for females to mate with. That is the reason why these spiders can get in accidental contact with humans. The bite of the Sydney funnel-web spider can be fatal and requires immediate medical attention. Remarkable about

Ph. 3.25: The Brazilian wandering spider, *Phoneutria nigriventer* (Source: Techuser, Wikimedia, GNU-CC-BY-SA).

Ph. 3.26: A female black widow spider, *Latrodectus mactans* (Source: James Gathany, US Department of Health and Human Services, Wikimedia, public domain).

this spider species is that the venom of the male, which is only up to about 2.5 cm in body length, acts four to six times stronger than that of the female, which is larger (up to about 5 cm). The venom toxicity is primarily due to a neurotoxic protein called δ-atracotoxin. Chemical synthesis and functional testing of the properly refolded δ-atracotoxin of *Atrax robustus* revealed that it inhibits the intrinsic inactivation of voltage-gated sodium channels in activated neurons of target organisms. This results in neuronal overexcitation [4]. Noteworthy is the fact that the venom is particularly

Ph. 3.27: A female Sydney funnel-web spider, *Atrax robustus* (Source: Rosie Steinberg, Wikimedia, CC-BY).

dangerous for humans, primates, and newborn mice (Tab. 3.3), while it has little effect on other mammals (such as dogs and cats).

The brown recluse spider, *Loxosceles reclusa* (Ph. 3.28), occurs in the western areas of North America. *Loxosceles laeta* occurs further south in Chile. The bodies of brown recluse spiders are usually between 6 and 20 mm long. *Loxosceles* venom has several toxins, the most important being the enzyme sphingomyelinase D [29]. This enzyme cleaves sphingomyelin and lysophosphatidylcholine in plasma membranes of exposed cells and results in more or less extensive dermonecrosis around the bite site in affected animals or humans.

Atrax sp. and other spider species (genera *Phoneutria*, *Dugsiella*, *Pterinochilus*, or *Lycosa*) produce 'short neurotoxins' in their venom glands that affect voltage-activated calcium-, sodium-, or potassium channels in target organisms. Examples are the neurotoxins of the Brasilian wandering spider (*Phoneutria nigriventer*), a member of the Ctenidae family that is often erroneously called 'banana spider'.

Phoneutria venom can be separated into five crude fractions according to their molecular sizes by gel filtration [15, 18]. Three of these fractions (P1, P2, and P3) contain proteases. The P4 fraction contains the short neurotoxins with several subtypes (Txs 1–5). The P5 fraction contains low-molecular-mass substances below 1 kDa.

Ph. 3.28: A brown recluse spider, *Loxosceles reclusa* (image reproduced from [56] with permission).

Initially, it was assumed that these substances were of no relevance in the toxicity of spider venoms. However, the discovery of nigriventrine (Fig. 3.13), which accounts for 0.4% of the venom weight, as a potent neurotoxin changed this view. Nigriventrine causes convulsions in rats [25].

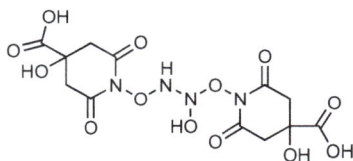

Fig. 3.13: Nigriventrine, a component of *Phoneutria nigriventer* venom.

Among the short neurotoxins in fraction P4 are the toxins 1–3 which are primarily directed against mammalian targets. Tx-4, however, is more toxic to insects than to mammals. Tx-5 shows no lethal effects on mice but acts on vascular smooth muscle by increasing the muscle tone. Tx-1 accounts for about 0.45% of the total mass of proteins in the venom. Injection of the toxin in mice results in general excitation, tail erection, and muscle spasms. The toxin acts on different subtypes of voltage-activated sodium channels, but not on the cardiac isoform Nav1.5.

Tx-2 is actually a collection of nine different peptides which account for the high degree of toxicity of *Phoneutria* venom. The LD_{50} (mice, s.c.) (Tab. 3.3) of this fraction is 0.002 mg/kg BW [13]. These toxins inhibit the intrinsic inactivation of activated voltage-gated sodium channels and are mainly responsible for the neurotoxic effects of the venom. Tx-2-5 and Tx-2-6 are responsible for the penile erection (priapism) observed in envenomated mammals brought about by enhanced nitric oxide (NO) generation in affected cells. Other symptoms of intoxication of animals with Tx-2 toxins are excessive salivation and lacrimation, seizures, and spastic paralysis.

Six toxins could be identified in the Tx-3 fraction. They act in an inhibitory fashion on voltage-gated cation channels and cause flaccid paralysis. Major targets seem to be potassium channels that are important players in spontaneously active pacemaker neurons. Voltage-gated calcium channels, however, are located in the presynaptic membranes of neuronal synapses, and inhibition of these N-type calcium channels results in the suppression of transmitter release into the synaptic cleft.

Due to its high degree of toxicity to insects, the Tx-4 fraction is also referred to as the insecticidal fraction. Envenomation of insects causes motoric hyperactivity in dipterans and beetles. It has been suggested that Tx-4 may activate the glutamate signaling system in insect motor neurons which results in rapid exhaustion of prey animals.

Transcriptomic and proteomic approaches revealed the presence of even more biologically active compounds in *Phoneutria nigriventer* venom [15]. It was found that cysteine-rich peptide toxins are among the most abundant components. These molecules show a highly uniform disulfide scaffold indicating that they may be homologs. Other components are protease inhibitors, metalloproteases, and hyaluronidases.

Bites of adult hobo spiders (*Eratigena* sp., formerly known as *Tegenaria* sp.; Ph. 3.29) native to Europe, Central Asia, and western North America can have similar effects of brown recluse spider bites, but are usually not associated with serious illness.

Ph. 3.29: A giant house spider, *Eratigena* sp. (Source: Jan-Peter Hildebrandt).

Wolf spiders (members of the familiy Lycosidae) are distributed worldwide. They will only bite when they feel threatened. Symptoms of their bites include swelling, mild pain, and itching. Necroses at the bite site have been reported upon bites of wolf spiders in certain areas, e.g., in South America, but these case descriptions may have been wrong and due to misdiagnosed *Loxosceles* spider bites [44]. Wolf spiders of the genus *Lycosa*, however, may apply bites that cause local skin necroses [31]. This spider genus includes the European *Lycosa tarantula* (Ph. 3.30), which has formerly been associated with 'tarantism', a mythical affliction with symptoms as shaking, cold sweats, and high fever. Relief from these symptoms was allegedly achieved by dancing the traditional tarantella dance [47]. However, there has never been scientific proof that *Lycosa tarantula* bites are responsible for such symptoms.

Ph. 3.30: The European wolf spider, *Lycosa tarantula* (Source: David Pérez, Wikimedia, GNU-CC-BY).

☠ Spider bites of Central European species are generally not life-threatening, but may be painful and cause numbness in the affected limb. Nausea can occur in rare cases. Species that are able to penetrate human skin wth their chelicerae are the yellow sac spider, *Cheiracanthium punctorium*, and the water spider, *Argyroneta aquatica* (Ph. 3.31). The bite of the common garden spider, *Araneus diadematus*, feels like a pinprick but has no further complications. In the south of Europe spiders of the genus *Latrodectus* (black widows) pose a thread because their bites are generally followed by more or less extensive tissue necrosis. Medical attention should be sought immediately after being bitten.

The bites of the Australian funnel-web spider *Atrax robustus* (only found in the Sydney area) or spiders of the genus *Hadronyche* (wide distribution in Australia) are very painful and associated with long-lasting additional symptoms. The envenomation syndrome caused by these spiders is known as 'atraxism' and is characterized by nausea, vomiting, abdominal pain, diarrhea, sweating, intense salivation and lacrimation, an increase in blood pressure, and breathing difficulties. These symptoms set in just a few minutes after the bite. Muscle twitching and cramps were also observed. The affected person may spend several hours in a coma. Death (13 cases from 1927 to 1981) may result from respiratory paralysis between 30 min and 30 h after the incident. An antidote is available since 1981 and no further deaths have been reported since its introduction.

North America (especially California, Arizona, and Nevada) does have a couple of spiders whose bite can cause harm to animals and humans. Bites of recluse spiders (*Loxosceles* sp., also known as brown spiders, fiddle-backs, violin spiders, or reapers) cause a characteristic set of symptoms known as 'loxoscelism' [56]. Bites usually cause stinging sensations followed by intense pain that may hold on for 12 h. Tissue necrosis may occur around the bite site. In some cases, fatigue, nausea, and vomiting as well as fever may occur. Hemolysis and thrombocytopenia may occur in affected children.

Bites of the Brazilian wandering spider (*Phoneutria nigriventer*) may lead to neurogenic shock. Symptoms are agitation, somnolence, sweating, nausea, profuse vomiting, excessive salivation and lacrimation, hypertension, tachycardia, tachypnea, tremors, muscle spasms, and priapism. Medical attention should be sought immediately after being bitten.

🦀 If a person is bitten by a spider and experiences allergic reactions, or if the rash or the wound grows bigger, or if tissue becomes necrotic, medical attention is necessary. If reactions to the bite include symptoms as fever, dizziness, or vomiting a physician should be contacted immediately.

Ph. 3.31: A male water spider, *Argyroneta aquatica* (Source: Norbert Schuller Baupi, Wikimedia, GNU-CC-BY-SA).

3.2.7.2 Scorpions (Scorpiones)

Scorpion venoms have been studied for quite some time due to the facts that humans in the tropical and subtropical regions are frequently affected by scorpion stings [1, 10, 57] and that scorpion venoms contain peptides and proteins which are highly specific inhibitors of cellular ion channels not only in prey organisms (mainly arthropods) but also in potential vertebrate predators. The latter are under investigation for potential uses in medicine [3, 14, 38]. In addition to such peptide and proteotoxins, the venoms contain small molecules that cause intense pain, such as serotonin [61] and, in some cases, histamine and noradrenaline. Together with the vertebrate-directed proteotoxins, these ingredients of scorpion venom are interpreted as defense agents against vertebrate predators. As the production of venom is costly, scorpions make economic use of it by adjusting the likelyness of stinging to the thread levels imposed by predators [12]. It takes 2–3 weeks to refill the venom gland reservoir after a sting. The amount of venom released during the sting is between 0.04 and 0.38 mg DW for the South American scorpion, *Tityus bahiensis* (Ph. 3.32) and up to 7 mg for the African cape burrower scorpion, *Opistophthalmus capensis* (Ph. 3.33).

Ph. 3.32: The South American scorpion *Tityus bahiensis* (Source: Limberger, Wikimedia, CC-BY-SA).

Ph. 3.33: The mildly venomous cape burrower scorpion, *Opistophthalmus capensis* (Source: Willem Van Zyl, imgur.com, with permission).

The venom apparatus of scorpions is located in the last segments of the tail. Openly visible is the tail stinger which is the last digit in the tail. The venom gland and the venom reservoir reside within the last segments of the tail. The size of a scorpion is not a good indicator of its toxicity as the small species are often more venomous than the large ones. As a general rule, scorpion species with narrow claws (chelicerae) and thick tails are more toxic than those with thick claws and a slender tail [19]. When scorpions grab a prey animal using their claws they bend their tails over the head and inject venom into the body cavity of the victim upon puncturing its integument. The most potent venoms are produced by scorpions native to The Americas, Northern Africa, and India (Tab. 3.4). Only a relatively few scorpion species are dangerous to humans, e.g., members of the genera *Buthus*, *Androctonus*, *Leiurus*, *Parabuthus*, *Meso-buthus*, *Hemiscorpion*, *Centruroides*, and *Tityus*.

Tab. 3.4: Examples of scorpions potentially posing danger to humans and the LD_{50}s of their venoms.

Scorpion species	Distribution	LD_{50} (mouse, s.c.) mg/kg BW
Leiurus quinquestriatus	North Africa	0.25–0.33
Androctonus australis	North Africa, Asia	0.32
Tityus serrulatus	Brazil	0.43
Centruroides limpidus	Mexico	0.7
Tityus trinitatis	Trinidad and Tobago	2.0

Scorpion peptide toxins are often named using letters derived from the name of the genus, species, subspecies, sometimes including the author's name of the animal's Latin name, and a number, e.g., BotIT6 for a 62 amino acid neurotoxin of *Buthus occitanus* (full name: *Buthus occitanus tunetanus* insect toxin 6) or Aah I for toxin I of *Androctonus australis* Hector. In many cases, however, names were allocated completely randomly, e.g., in case of charybdotoxin of the deathstalker, *Leiurus quinquestriatus* (Ph. 3.34). According to databases of scorpion toxins we know more than 400 scorpion toxins to date. They are relatively heat-stable neurotoxic polypeptides that can be divided into two main groups according to their chemical structures (Fig. 3.14):
- Large (long) peptide toxins – basic single-chain polypeptides of 57–78 amino acids and mostly four disulfide bridges (type 1 toxins)
- Small (short) peptide toxins – basic single-chain polypeptides of 23–39 amino acids and three or four disulfide bridges (type 2 toxins)

The long toxins (type 1) target voltage-gated sodium channels in neuronal cells of invertebrate or vertebrate animals. These toxins delay or inhibit the intrinsic inactivation of activated channels which results in sodium overload in the cells (α-subgroup of type 1 toxins [26]) or induce a direct activation of such channels (β-subgroup of type 1 channels [27]). The small toxins (type 2) inhibit voltage-gated potassium channels which are important carriers of hyperpolarizing potassium currents out of the cytosol into the extracelluar space during an action potential. All of these processes result in prolonged depolarization of the plasma membrane potential in neuronal cells, overexcitation, and

Ph. 3.34: The highly venomous deathstalker scorpion, *Leiurus quinquestriatus* (Source: Danny S., Wikimedia CC-BY-SA).

excessive transmitter release from these cells. Physiological effects of the toxins are anxiety, erratic vision and speech, irregular heartbeat, and muscle cramps up to spastic paralysis. If the ventilatory muscles are involved, envenomation may be fatal.

The Arizona bark scorpion (*Centruroides sculpturatus*; Ph. 3.35) is the most venomous scorpion in North America. Stings in humans cause severe pain and may be associated with numbness of or tingling in the affected limbs, muscle cramps, and vomiting. Such symptoms may occur up to 72 h after the incident. While there are reports of many thousand stings by this scorpion in humans per year, only two fatal cases have been reported during the last 60 years. The related Mexican scorpion *Centruroides noxius* is one of the most venomous scorpions in Mexico and other countries in Latin America. The median lethal dose of its venom in mice is 250 µg/kg BW [50]. One of the two peptide toxins that have been identified in the venom is noxiustoxin (Fig. 3.14), a short neurotoxin blocking neuronal potassium channels [53]. The other one is Cn2, a peptide with 66 amino acid residues that is a very potent and specific ligand for Na^+-channels of subtype Nav1.6 [41].

Charybdotoxin (ChTX) (Fig. 3.14) is a 37-amino-acid neurotoxin in the venom of the North African deathstalker, *Leiurus quinquestriatus hebraeus* [36]. ChTX contains three disulfide bridges. It blocks calcium-activated *Shaker* K^+-channels as well as voltage-dependent potassium channels (Kv1.3 channels) [24]. ChTX occludes the pores of these channels by binding to one of four independent, but overlapping binding sites

```
                          10              20              30            40
                          |               |               |             |
CsE v3    - K E G Y L V K K S D G C K Y G C L K L G E N E G C D T E C K A K N Q G G S Y G ⌐
AaH II    V K D G Y I V D D - V N C T Y F C G R - - - N A Y C N E E C T K L - K G E S - G |
Bot I     G R D A Y I A Q P - E N C V Y E C A Q - - - N S Y C N D L C T K D - G A T S - G |
Lqq V     L K D G Y I V D D - K N C T F F C G R - - - N A Y C N D E C K K K - G G E S - G ⌐

                          50              60              70
                          |               |               |
CsE v3    ⌐ Y C Y A F - - - - - A C W C - E G L P E S T P T Y P L P - N K S C
AaH II    | Y C Q W A S P Y G N A C Y C Y K - L P D H V R T K G - P - G R - C H
Bot I     | Y C Q W L G K Y G N A C W C - K D L P D N V P I R I - P - G K - C H F
Lqq V     | Y C Q W A S P Y G N A C W C Y K - L P D R V S I K E K - - G R - C N
```

CsE v3   of *Centruroides sculpturatus* (Arizona bark scorpion)
AaH II   of *Androctonus australis* (fat-tailed scorpion)
Bot I   of *Buthus occitanus tunetanus* (common yellow scorpion)
Lqq V   of *Leiurus quinquestriatus quinquestriatus* (deathstalker)

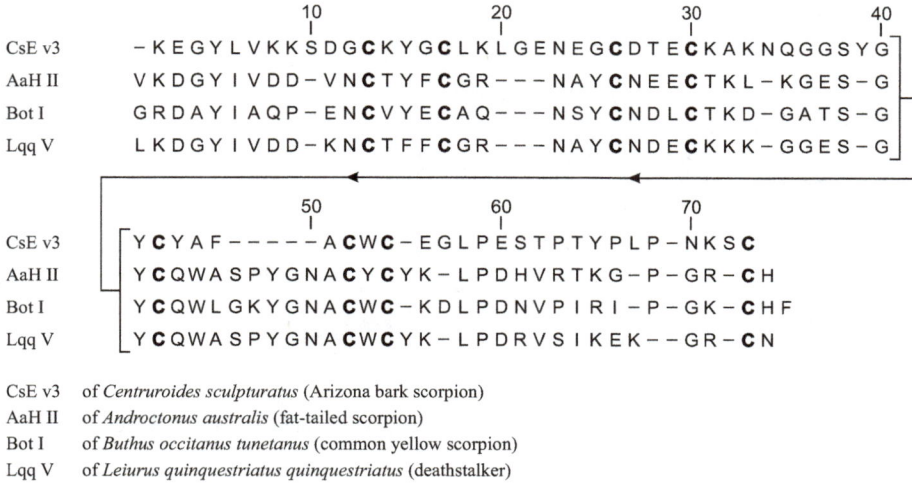

Large peptide toxins (long neurotoxins)

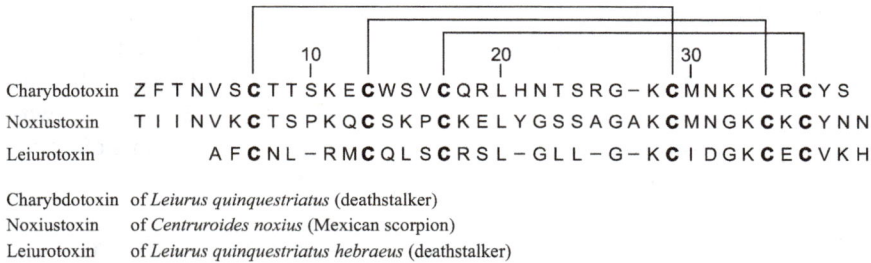

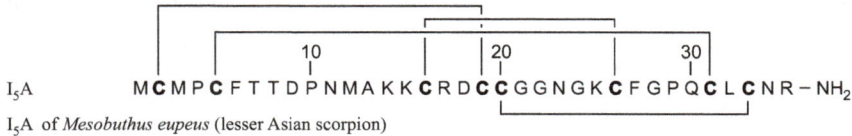

```
                       10              20              30
                       |               |               |
Charybdotoxin  Z F T N V S C T T S K E C W S V C Q R L H N T S R G - K C M N K K C R C Y S
Noxiustoxin      T I I N V K C T S P K Q C S K P C K E L Y G S S A G A K C M N G K C K C Y N N
Leiurotoxin          A F C N L - R M C Q L S C R S L - G L L - G - K C I D G K C E C V K H
```

Charybdotoxin of *Leiurus quinquestriatus* (deathstalker)
Noxiustoxin   of *Centruroides noxius* (Mexican scorpion)
Leiurotoxin   of *Leiurus quinquestriatus hebraeus* (deathstalker)

```
                         10              20              30
                         |               |               |
I₅A    M C M P C F T T D P N M A K K C R D C C G G N G K C F G P Q C L C N R - NH₂
```

I_5A of *Mesobuthus eupeus* (lesser Asian scorpion)

Small peptide toxins (short neurotoxins)

Fig. 3.14: Groups of scorpion toxins: Peptide sequences of large (long) and small (short) scorpion toxins (one-letter amino acid code) and most likely patterns of disulfide bridge formation between cysteine residues (fat print) in the small (short) peptide toxins.

in the channel vestibules [65]. The blockade of these K^+ channels by ChTX causes neuronal hyperexcitability.

The Trinidad thick-tailed scorpion *Tityus trinitatis* is responsible for human deaths because of its severe neurotoxic and cardiotoxic venom. *T. trinitatis* is the only known scorpion species (except *Leiurus quinquestriatus*) that is able to produce a pancreatotoxic venom. Envenomation causes abdominal pain, repeated vomiting, respiration

Ph. 3.35: The Arizona bark scorpion, *Centruroides sculpturatus* (Source: Andrew Meeds, Wikimedia CC-BY).

difficulties, abnormalities in cardiac function, and edematous or hemorrhagic pancreatitis [6]. The LD_{50} of the venom is 2 mg/kg BW.

Amino acid sequencing of neurotoxin I of the North African scorpion *Androctonus australis* (Ph. 3.36) [45] revealed that this toxin, now named AaH I, is a member of the long neurotoxins in scorpions that affect voltage-gated sodium channels. The LD_{50} of the full venom in mice (s.c.) is approximately 0.32 mg/kg BW [60]. Experiments with synthesized AaH I toxin revealed that it selectively prolongs the open time of Nav1.2 sodium channels in target cells [34]. There is also an isoform of this toxin, AaH II, present in *A. australis* venom (Fig. 3.14).

BotIT6 is an insect-directed neurotoxin in *Buthus occitanus* (Ph. 3.37) venom. It consists of 62 amino acid residues with a surplus of basic residues. BotLT6 binds to site 4 of voltage-gated sodium channels in crickets and cockroaches and decreases the amplitude of the action potential which paralyzes the insects. The LD_{50} determined in injected cockroaches is 10 ng/100 mg BW [35].

☠ Scorpion stings cause severe, burning pain at the site of the sting, which progresses to a tingling sensation and eventually to numbness. Agitation, states of anxiety, visual, and speech disorders can occur. Sickness, irregular pulse, fluctuations in blood pressure, and breathing problems are common. In severe cases, death occurs after 20–30 h from respiratory paralysis, pulmonary edema, or cardiovascular failure.

🚑 Basic first aid measures upon receiving a scorpion sting are cleaning of the sting site with soap and water, application of a cool compress (cool cloth), and application of oral painkillers. It is suggested that medical attention is sought as soon as possible, especially if children are affected or if severe symptoms occur. There are antivenoms available in local hospitals for treating stings of the most dangerous scorpions. It is important for proper medical treatment that the patient knows the scorpion species that applied the sting or that the scorpion is brought to the hospital for proper identification.

Ph. 3.36: The yellow fat-tailed scorpion, *Androctonus australis* from Northern Africa (Source: HTO, Wikimedia, public domain).

Ph. 3.37: The common European scorpion, *Buthus occitanus* (Source: Álvaro Rodriguez Alberich, Wikimedia, CC-BY-SA).

⚥ A 65-year-old female was brought to the Toxicological Research Center of the Shahid Beheshti University of Medical Sciences, Tehran, Iran because she had been stung by an Arabian fat-tailed scorpion, *Androctonus crassicauda*. The patient was agitated, experienced tachycardia, and generalized sweating, cold and wet extremities, bilateral diffuse crackle in the base of lungs, tachypnea, and lethargy. She received a low dose of scorpion antivenom. Especially the symptoms of cholinergic and adrenergic activation, pulmonary edema, and the electrocardiographic abnormalities subsided immediately. The patient fully recovered within a few hours [2].

♦ A 54-year-old Indian woman was admitted to a hospital upon being stung by a scorpion. A few hours after admission, the patient developed tachycardia and hypotension. Cardiac evaluation (measurement of creatine kinase isoenzymes and echocardiography) indicated that she developed myocarditis. Over 12 h after the incident her condition worsened. Magnetic resonance imaging of the brain showed indications of massive infarction probably due to cardioembolism. The patient fell into coma and succumbed to her illness the next day.

3.2.7.3 Mites (Acari)

Subgroups of the mites (Acari) are the Oribatida (moss mites) and the Parasitiformes which include the ticks (Ixodida). The former group contains species with narrow ecological niches that have to protect themselves from predators by chemical means. They have evolved the ability to produce cyanogenic substances (see Section 2.10) [7] or complex alkaloids which generally deter predators. In some cases, however, predators have developed the ability to sequester the mite alkaloids and utilize them as defense substances in their own right, e.g., in poison arrow frogs [51].

The Parasitiformes among the Acari include two major families, the Ixodidae or hard ticks, and the Argasidae, or soft ticks. All of them are hematophagous ectoparasites. Ticks use their mouthparts to penetrate the integument of suitable hosts (mostly mammals and birds). The animal tightly glues itself to the host upon insertion of its mouthparts and forms a subcutaneous cavern in which tissue fragments, blood, and lymph are collected. Feeding on material from this pool ('pool feeder') may last for up to 10 days before the tick is fully saturated and drops off. There are several tick species that are able to feed on humans, at least as adults. Among them are species belonging to the genera *Argas, Ornithodos, Dermacentor, Ixodes, Amblyomma, Haemaphysalis, Hyalomma*, and *Rhipicephalus* [16].

Due to this feeding habit the tick needs to inject a variety of substances from the salivary glands into the host. Thus, ticks should be considered as venomous animals [8]. The salivary substances provide several benefits to the tick, namely, to stay undetected by the host (anesthetics), to suppress host inflammatory responses (antiinflammatory agents), and to prevent immune reactions in the host (suppressors of the host's adaptive immune response), thrombocyte aggregation (platelet inhibitors) and clotting of host blood (anticoagulants). As a byproduct of releasing saliva into the wound ticks are able to transfer infectious agents like virus particles (e.g., FSME virus), protists (e.g., *Babesia*), or bacteria (e.g., *Rickettsia, Ehrlichia, Borrelia*) to the host to name just a few. Tick-borne diseases in domestic animals and humans become more widespread due to the geographical expansion of their tick vectors in the course of the current climate change [46]. These infectious agents, however, are not obligatory components of tick saliva and are therefore not considered any further.

Some types of ticks can cause so-called 'tick paralysis' in the host which is the only tick-borne disease that is not caused by an infectious organism. It is due to the presence of neurotoxins in the salivary gland secretions which are transmitted during prolonged feeding sessions when the tick has already engorged large volumes of blood. Within a few days after the sting the toxin causes weakness in the arms or legs of the host. The paralysis ascends to the trunk and head within hours and may lead to death by respiratory failure. The causative neurotoxins are in the low-molecular-mass fraction of salivary proteins. The genes encoding these peptides have been cloned from the salivary glands of the Australian tick *Ixodes holocyclus*. The gene products have been called holocyclotoxins and include at least three isoforms (HT-1, HT-3, and HT-12) (Fig. 3.15) [9] which are all members of the cysteine knot peptide family [58]. These toxins act through presynaptic inhibition of neurotransmitter (acetylcholine) release from synaptic terminals of motor neurons.

```
HT-1:    MSKVTTVFIGALVLLLLIENGF---SCTNPGKKRCNAKCSTHCDCKDGPTHNFGAGPVQCKKCTYQ-FKGEAYCKQ
HT-3:    MVKATATLVCALIILAIVHEGFPSSSCSTPGRRNCNQDCYTHCDCVGGKEYNNGAGMVLCKTCTYPLGKKVGFCKFAP
HT-12:   MAKFTAALFFALIILAIVQEG--SAGCSNPGKKNCNADCYTHCDCSGGEPHDFGAGPKLCTSCTYQPFKSVGYCK
```

Fig. 3.15: Alignment of neurotoxic holocyclotoxins of *Ixodes holocyclus* (one-letter amino acid code). Cysteine residues forming intramolecular disulfide bridges are highlighted in red lettering.

⚲ Symptoms of tick paralysis usually appear after 5–7 days upon the attachment of the tick. Initially, there is a feeling of general weakness. Diarrhea, ataxia, paralysis of arms and legs, swelling of body parts, and breathing difficulties may follow. Fatalities have been reported. Sensitization can occur after repeated tick stings. In sensitized humans repeated exposure to these toxins may result in life-threatening anaphylaxis [54].

Proteins that may function as anesthetics, antiinflammatory agents, suppressors of the host's adaptive immune response, platelet inhibitors, and anticoagulants have been identified in the sialomes (salivary protein mixtures) of several tick species [5]. Some examples are described here.

The cystatin OmC2 was described as a constituent of the saliva of the soft tick *Ornithodoros moubata*. OmC2 is an effective broad-specificity inhibitor of cysteine cathepsins and may suppress the host's adaptive immune response [48]. Ticks also secrete several salivary serine protease inhibitors of the Kunitz-type into the wound. Some of these directly inhibit thrombin and have to be considered as true anticoagulants. Examples of such structurally similar proteins (~14 kDa) are ornithodorin of *Ornithodoros moubata*, savignin of *Ornithodoros savignyi*, monobin of *Argas monolakensis*, amblin of *Amblyomma hebraeum*, boophilin of *Rhipicephalus microplus*, and hemalin of *Haemaphysalis longicornis* [11]. Other Kunitz-type protease inhibitors (~7 kDa) inhibit factor Xa, an upstream protease in the blood coagulation cascade of vertebrates. Examples are TAP or FXaI of *Ornithodoros moubata* or *Ornithodoros savignyi*, respectively.

Proteins containing RGD domains interacting with integrins needed for platelet aggregation have been identified in saliva of different tick species. Savignygrin (7 kDa) of *Ornithodoros savignyi* [32], the 5 kDa variabilin of *Dermacentor variabilis* [59], and the 7 kDa ixodegrin of *Ixodes pacificus* [21] are examples of such proteins that act as inhibitors of platelet aggregation in the host.

In some tick species the saliva contains binding proteins for serotonin and/or histamine. The ~16 kDa lipocalins, monomine and monotonin, of *Argas monolakensis* mop up serotonin and histamine [33] as does the 22 kDa protein SHBP of *Dermacentor reticularis* [49]. As both agents are usually mediating itching or pain sensations in the host, it is assumed that these proteins act as anesthetics in the host.

The saliva of *Ixodes scapularis* inhibits platelet activation triggered by ADP, collagen, or platelet-activating factor (PAF). A salivary apyrase degrades ATP and ADP to AMP and orthophosphate suppressing platelet aggregation. The presence of prostaglandin E2 (PGE2) in tick salva inhibited interleukin 2 production by T-cell hybridomas in vitro which may point to an antiinflammatory and immune modulatory mechanism of PGE2 in tick saliva in vivo [43].

Fibrinolytic enzymes have been detected in tick saliva. Among those in *Ixodes scapularis* is a 37 kDa metalloprotease of the reprolysin family that degrades gelatin, fibrin(ogen), and fibronectin [20]. This enzyme may prevent blood coagulation or help to dissolve small blood clots that have already been formed at the sting site. Moreover, it may assist in forming the tissue cavern in host skin from which the tick ingests body fluids as it loosens cell-cell- and cell-matrix attachments.

🐛 Humans can avoid infection and envenomation by ticks by checking the entire body surface upon returning from the field and collecting or removing any ticks before or shortly after they have attached themselves to the skin. The transmission of *Borrelia* sp. occurs only after 12–24 h after tick infestation because the bacteria have to multiply in the tick stomach using fresh host blood before they become infectious. If the tick is readily removed, there is only a small risk of infection with these microorganisms. However, tick-borne encephalitis (TBE) virus in the salivary glands gets into the wound immediately after the sting. Thus, it is advantageous to remove ticks from the skin even before they sting. When ticks have already attached themselves they should be removed by pulling them out vertically using steel tweezers with pointed tips. The forceps should be positioned very close to the animal's attachment site in the host skin. The animal should not be twisted or squeezed as this may cause regurgitation of stomach content which may spur infection or envenomation. The sting site should be inspected every day over 2 weeks after the sting. A physician should be contacted if spreading reddening around the sting site is observed which may indicate that a bacterial infection with *Borrelia* sp. (potentially causing Lyme disease) has occurred.

🐾 Pets, domestic and farm animals, such as dogs, cats, sheep, and cattle, and also wild animals are infested by ticks. In Australia, many dogs die from infestation with *Ixodes holocyclus* which have been transmitted from bandicoots, possums, or kangaroos. A serious plague for poultry in almost all parts of the world is the pigeon tick, *Argas persicus*. Ticks are also carriers and vectors of numerous pathogens and may transmit these to livestock. The castor bean tick, *Ixodes ricinus* (Ph. 3.38), can transmit TBE (encephalitis virus) and Lyme disease (*Borrelia* bacteria) between domestic animals and to humans in Europe. The cattle tick *Rhipicephalus annulatus* (Southern United States, Mexico, South Africa) and related species transmit Texas fever to cattle. *Ornithodorus moubata* (Africa), *O. tholozani* (Asia), and other members of this genus transmit relapsing fever (also to humans!). Thogotovirus transmitted by *Ixodes* species in Africa causes mass abortions in sheep.

Ph. 3.38: Engorged female of the castor bean tick, *Ixodes ricinus* (Source: Paul Knaupe/iStock/Getty Images Plus).

References

[1]   Abroug F et al. (2020) Intensive Care Med 46(3): 401
[2]   Aghabiklooei A et al. (2014) Hum Exp Toxicol 33(10): 1081
[3]   Ahmadi S et al. (2020) Biomedicines 8(5): 118
[4]   Alewood D et al. (2003) Biochemistry 42(44): 12933
[5]   Ali A et al. (2022) Front Cell Infect Microbiol 12 809052
[6]   Bartholomew C (1970) Br Med J 1(5697): 666
[7]   Brückner A et al. (2017) Proc Natl Acad Sci U S A 114(13): 3469
[8]   Cabezas-Cruz A, Valdés JJ (2014) Front Zool 11(1): 47
[9]   Chand KK et al. (2016) Sci Rep 6(1): 29446
[10] Chippaux JP, Goyffon M (2008) Acta Trop 107(2): 71
[11] Chmelar J et al. (2012) J Proteom 75(13): 3842
[12] de Albuquerque KBC., de Araujo Lira AF (2021) J Arachnol 49(3): 402
[13] de Lima ME et al. (2016) *Phoneutria nigriventer* venom and toxins: A review. In: Gopalakrishnakone P, Corzo G, de Lima M, Diego-García E (eds.) Spider Venoms. Toxinology. Springer, Dordrecht, p. 71
[14] Ding J et al. (2014) Exp Biol Med 239(4): 387
[15] Diniz MRV et al. (2018) PLoS ONE 13(8): e0200628
[16] Estrada-Pena A, Jongejan F (1999) Exp Appl Acarol 23(9): 685

[17] Fatehi M et al. (1997) Neuropharmacology 36(2): 185
[18] Fernandes FF et al. (2022) J Venom Anim Toxins Incl Trop Dis 28 e20210042
[19] Forde A et al. (2022) Toxins 14(3): 219
[20] Francischetti IMB et al. (2003) Biochem Biophys Res Commun 305(4): 869
[21] Francischetti IMB et al. (2005) Insect Biochem Mol Biol 35(10): 1142
[22] Fusto G et al. (2020) J Venom Anim Toxins Incl Trop Dis 26 e20190100
[23] Fuzita FJ et al. (2016) BMC Genom 17(1): 716
[24] Gao Y-D, Garcia ML (2003) Proteins 52(2): 146
[25] Gomes PC et al. (2011) Toxicon 57(2): 266
[26] Gordon D et al. (2007) Toxicon 49(4): 452
[27] Gurevitz M et al. (2007) Toxicon 49(4): 473
[28] Himi T et al. (1990) J Neural Transm 80(1): 79
[29] Lajoie DM et al. (2013) PLoS ONE 8(8): e72372
[30] Langenegger N et al. (2019) Toxins 11(10): 611
[31] Lucas S (1988) Toxicon 26(9): 759
[32] Mans BJ et al. (2002) J Biol Chem 277(24): 21371
[33] Mans BJ et al. (2008) J Biol Chem 283(27): 18721
[34] M'Barek S et al. (2004) J Pept Sci 10(11): 666
[35] Mejri T et al. (2003) Toxicon 41(2): 163
[36] Miller C (1995) Neuron 15(1): 5
[37] Ni L (2021) Front Mol Neurosci 13 638839
[38] Nunes KP et al. (2013) Toxicon 69 152
[39] Peterson ME (2006) Clin Tech Small Anim Pract 21(4): 187
[40] Piser TM (1994) Pflügers Archiv 426(3): 214
[41] Possani LD et al. (1981) Carlsberg Res Commun 46(4): 207
[42] Poulsen MH et al. (2013) J Med Chem 56(3): 1171
[43] Ribeiro JM t al (1985) J Exp Med 161(2): 332
[44] Ribeiro LA et al. (1990) Toxicon 28(6): 715
[45] Rochat H et al. (1970) Eur J Biochem 17(2): 262
[46] Rochlin I, Toledo A (2020) J Med Microbiol 69(6): 781
[47] Russell JF (2012) Med History 23(4): 404
[48] Salát J et al. (2010) Biochem J 429(1): 103
[49] Sangamnatdej S et al. (2002) Insect Mol Biol 11(1): 79
[50] Santibáñez-López CE et al. (2016) Toxins 8(1): 2
[51] Saporito RA et al. (2007) Proc Natl Acad Sci U S A 104(21): 8885
[52] Sherby SM et al. (1987) Compar Biochem Physiol C – Toxicol Pharmacol 87(1): 99
[53] Sitges M et al. (1986) J Neurosci 6(6): 1570
[54] Stone BF et al. (1989) Exp Appl Acarol 7(1): 59
[55] Südhof TC (2001) Annu Rev Neurosci 24 933
[56] Swanson DL, Vetter RS (2006) Clin Dermatol 24(3): 213
[57] Tobassum S et al. (2020) Toxin Rev 39(3): 214
[58] Vink S et al. (2014) Toxicon 90 308
[59] Wang X et al. (1996) J Biol Chem 271(30): 17785
[60] Watt DD, Simard, JM (1984) J Toxicol – Toxin Rev 3(2–3): 181
[61] Welsh JH, Batty CS (1963) Toxicon 1(4): 165
[62] Wigger E et al. (2002) Toxicon 40(6): 749
[63] Wullschleger B et al. (2005) J Exp Biol 208(11): 2115
[64] Yan S, Wang X (2015) Toxins 7(12): 5055
[65] Zhao Y et al. (2019) Molecules 24(11): 2045

3.2.8 Myriapoda

The taxon Myriapoda is a subphylum of arthropods containing the Diplopoda (milli-pedes, two pairs of legs per segment) and the Chilopoda (centipedes, one pair of legs per segment). Their mouthparts lie on the underside of the head, with epistome and labrum forming the upper lip, and a pair of maxillae forming the lower lip. A pair of mandibles lies inside the mouth. While diplopods are saprophageous and need their venom glands and their toxins only for defensive purposes [15], the chilopods are ter-restrial predators. The first pair of walking legs (maxillipedes) is transformed into venom claws, so-called forcipules (Ph. 3.39) [6], with sclerotic tips that serve as hollow injection needles for venom produced in venom glands located at the bases of these extremities [10]. Prey is grabbed using these claws and injected with potent venom before it is shred by the mouthparts and ingested. The claws may also be used by the animals to defend themselves against attackers. Some chilopods, e.g., members of the genus *Scolopendra*, may even be dangerous to humans.

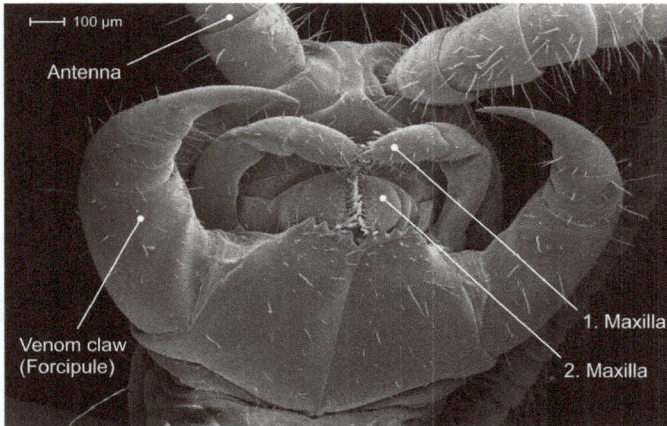

Ph. 3.39: Scanning electron microscopic image of the anterior bottom side of a centipede (*Lithobius forficatus*) showing the massive venom claws (forcipules) and the mouthparts (maxillae) (Source: Andy Sombke, with permission).

3.2.8.1 Diplopoda

Diplopods may rely on passive mechanisms to deter or repel potential predators. Being only very small (2 mm in length) and having a soft cuticle and no chemical de-fense, *Polyxenus fasciculatus* (Polyxenidae) makes use of its detachable bristles which entangle and disable predatory ants [7]. Most diplopods, however, carry pairs of de-fense glands (Fig. 3.16) in each of their segments that develop by invagination of the ectodermal cell layer that usually secretes the cuticula. The reservoir of the venom gland is filled with 2-hydroxy-2-phenylacetonitrile (mandelonitrile) which is synthesized by hypodermal cells from phenylalanine [15]. When threatened by potential predators, xystodesmid millipedes carrying type 3 defense glands transfer a certain amount of mandelonitrile from the gland reservoir into a reaction chamber. The epithelial cells of this chamber secrete an enzyme, α-hydroxynitrile lyase [4, 5] that mediates the rapid degradation of mandelonitrile to benzaldehyde and hydrogen cyanide which are released from the opening of the chamber (ozopore) in a more or less rapid manner. The toxic reaction products may harm or even kill predators like birds or rodents.

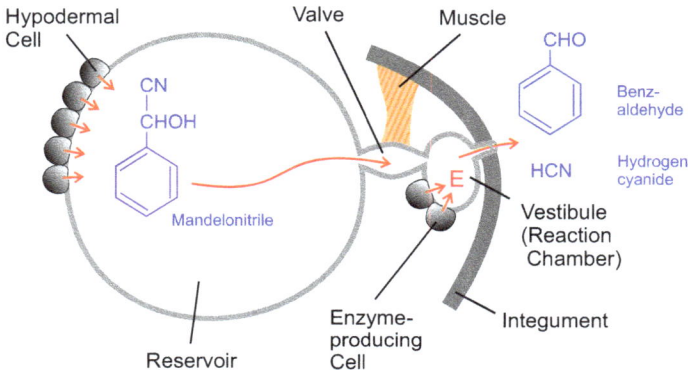

Fig. 3.16: Scheme of a type 3 defensive gland of a xystodesmid millipede (E - α-hydroxynitrile lyase).

The defensive secretions of *Pachyiulus hungaricus* which occurs in the southeast of Europe contain 2-methyl-1,4,-benzoquinone and 2-methoxy-3-methyl-1,4-benzoquinone which have, in addition to repellent functions against predators, potent antibacterial and antifungal functions [16].

In some groups of diplopods (e.g., in the Chordeumatida), droplets of sticky fluid are released from the sockets of setae that cover the body surface. The droplets trap small intruders which glide to the millipede's dorsum and are disposed off by getting stuck to environmental objects (stones, leaves, sticks, etc.) [15]. These droplets may also cover the mouthparts of potential predators and inhibit their movements. This may help to repel these predators.

3.2.8.2 Chilopoda

Most species of centipedes are not dangerous for humans. However, some tropical forms, e.g., the East Asian species *Scolopendra subspinipes* with a body length of up to 20 cm, preys primarily on arachnids (spiders, scorpions), but is large enough to overpower small vertebrates, such as mice or reptiles, or sting humans when feeling threatened. The LD_{50}s of these toxins are rather low (Tab. 3.5). Even a human fatality has been reported.

Tab. 3.5: LD$_{50}$ values in mice of chilopod venoms.

Species (distribution)	Body length (cm)	LD$_{50}$ (i.v.) (mg/kg BW)	LD$_{50}$ (i.m.) (mg/kg BW)
Scolopendra viridicornis (Brazil)	16–19	1.5	12.5
Scolopendra subspinipes (East Asia)	11–18	2.35	60
Otostigmus scabricauda (Colombia, Venezuela)	6–7	0.6	3.5
Crytops iheringi (Brazil)	6–9	7.5	17
Scolopocryptos ferrugineus (Colombia, Venezuela)	5–7	8	19.5

Proteomic and transcriptomic analyses of *Scolopendra subspinipes* venom samples revealed the presence of diverse neurotoxins which contain two to four intramolecular disulfide bridges. However, the disulfide framework is different from that found in neurotoxins of other venomous animals indicating that the centipede venoms have evolved independently and underwent convergent evolution [12, 17]. The neurotoxins act on excitable cells in insects as well as in vertebrates and affect voltage-gated cation channels (Na$^+$, K$^+$, Ca^{2+}, and TRPV channels) [3]. While most of the toxins inhibit these channels and are strongly insecticidal (LD$_{50}$ values in cockroaches are approximately 5 nmol/kg BW) [14], others, e.g., the toxin ω-SLPTXSsm1a, actually activate voltage-gated calcium channels in rat dorsal root ganglion (DRG) neurons while being only weakly insecticidal [17]. This may indicate that the toxin mixtures have bimodal functions. Some ingredients are directed against invertebrate prey, while others may be optimized to repel vertebrate predators.

Venoms of scolopendromorph centipedes are rich in these neurotoxins containing multiple cysteine residues, while the venoms of scutigeromorph centipedes are less complex, but rich in high-molecular-mass cytolysins, especially pore-forming toxins [12]. These toxins form multihomomeric complexes on the surface of cells in sting victims which then unfold their β-barrel regions which penetrate the plasma membrane and form transmembrane pores. The cell membrane becomes permeable to ions and small organic molecules which ultimately kills the affected cells [11].

Centipede venom contains serotonin, histamine, hemolytic phospholipases, cardiotoxins, and cytolytic substances [13]. Due to the presence of biogenic amines in the venoms, most centipede stings are painful, but otherwise more or less harmless. However, humans being stung by individuals of the large *Scolopendra* species may experience intense and long-lasting pain and swelling of the affected body parts. Erythema, induration, and/or tissue necrosis may occur as well [2, 8]. The long-lasting pain sensation associated with *Scolopendra subspinipes* stings are probably due to the presence of a toxin, RhTX, that activates transient receptor potential vanilloid 1 (TRPV1) channels with a similar potency as capsaicin [9, 19]. However, the venom of *Scolopendra subspinipes mutilans* contains a toxin named µ-SLPTX-Ssm6a, a unique 46-residue peptide that potently inhibits human voltage-gated sodium channels of the NaV1.7 subtype with an IC_{50} of ~25 nmol/L. Inhibition of such channels in dorsal root ganglia and sympathetic neurons results in efficient suppression of pain perception. The toxin may be suitable as a powerful analgesic [18].

Besides using the venom claws chilopods may use defense glands located at different positions of the body surface to defend themselves against potential predators. Such glands produce low–molecular-mass substances like benzoquinone (see Section 2.7.2), quinoline alkaloids (see Section 2.9.10), or cyanogenic substances (see Section 2.10).

☠ Stings of tropical scolopendids lead to severe, burning pain at the site of the sting, which subsides after about 2–3 h. Erythema, swelling, and redness of the sting site are also common symptoms. Systemic reactions (nausea, dizziness, headache, restlessness, tachycardia, or fever) have been described upon stings applied by *Scolopendra* or *Otostigmus* species, but seem to be rare.

💊 Unless there are systemic symptoms occurring upon chilopod stings there is no need for medical treatment. Local anesthetics may provide some relief from acute pain. Immersion of the affected limb in hot water may shorten the period of pain sensation [1].

References

[1]   Balit CR et al. (2004) J Toxicol Clin Toxicol 42(1): 41
[2]   Bush SP et al. (2001) Wilderness Environ Med 12(2): 93
[3]   Chu YY et al. (2020) Toxins 12(4): 230
[4]   Dadashipour M et al. (2015) Proc Natl Acad Sci U S A 112(34): 10605
[5]   Duffey SS, Towers GHN (1978) Can J Zool 56(1): 7
[6]   Dugon MM, Arthur W (2012) Evol Dev 14(1): 128
[7]   Eisner T et al. (1996) Proc Natl Acad Sci U S A 93(20): 10848
[8]   Fung HT et al. (2011) Hong Kong Med J 17(5): 381
[9]   Geron M et al. (2017) Toxins 9(10): 326
[10] Haug JT et al. (2014) Arthropod Struct Dev 43(1): 5
[11] Iacovache I et al. (2008) Biochim Biophys (BBA) – Acta Biomembr 1778(7–8): 1611
[12] Jenner RA et al. (2019) Mol Biol Evol 36(12): 2748
[13] Malta MB et al. (2008) Toxicon 52(2): 255
[14] Rates B et al. (2007) Toxicon 49(6): 810
[15] Shear WA (2015) Biochem Syst Ecol 61 78
[16] Stanković S et al. (2016) PLoS ONE 11(12): e0167249
[17] Yang S et al. (2012) Mol Cell Proteomics 11(9): 640
[18] Yang S et al. (2013) Proc Natl Acad Sci U S A 110(43): 17534
[19] Yang S et al. (2015) Nat Commun 6(1): 8297

3.2.9 Insects (Hexapoda)

The taxon Hexapoda comprises all arthropods with six walking legs. This group of animals, also called 'insects', is the class of animals that contains more different species (at least 1.5 million) than any of the other classes of animal. Many insects have biting or stinging mouthparts associated with venom glands, e.g., the true bugs (Heteroptera), such as the European common backswimmer, *Notonecta glauca*, or the dipterous insects (Diptera) comprising midges and horse flies. In the Hymenoptera (bees, wasps, and ants) the venomous stinger at the tip of the abdomen evolved from the ovipositor. This explains why only the females of these species are able to apply stings.

Beetles (Coleoptera) excrete their venoms from pygidial and prothoracic glands. Many of them also possess the ability to defend themselves by 'reflex bleeding', i.e., by releasing an oily fluid containing toxins from glands at the joints of the legs. Examples are the oil beetles (Meloidae) and the ladybirds (Coccinellidae). Sequestration of toxins (e.g., batrachotoxin) within certain tissues may render some beetle species poisonous for predators.

The caterpillars of butterflies (Lepidoptera) sequester different kinds of toxic molecules in body fluids and tissues and become poisonous for potential predators. Defense glands may also secrete such toxins when the animals are threatened. Caterpillars of many butterfly species carry venomous hairs that contain enzymes and toxic proteins. These hairs break easily or are actively shed when an animal is under attack. Their tips may penetrate the skin of the attacker and induce local inflammation of even allergic reactions.

The venoms of stinging insects containing biogenic amines (histamine, serotonin), peptides (mast cell degranulating peptide, MCP), and proteotoxins usually trigger local reactions in vertebrates, as the amounts of toxins reaching the central circulation are highly diluted in the body fluids of the victim. The local reactions (pain, swelling, and redness) are generally sufficient to scare off larger animals and humans. Only in cases of mass attacks systemic effects may occur even in humans.

One problem, however, may be that a human sting victim affected by venoms containing peptides and proteotoxins is sensitized against the venom ingredients. In case of repeated envenomation the immune system may overreact and generate anaphylactic reactions which can lead to death.

Insect peptide and proteotoxins often have antimicrobial effects and also serve to protect the animals themselves or their nests from bacterial and fungal growth.

3.2.9.1 Insects with Stinging or Biting Mouthparts

Assassin bugs (Reduviidae) are venomous insects that use their specialized mouthparts to transmit venom to their victim's body fluids. Most species prey on invertebrates, but some have developed a hematophagous lifestyle and feed on blood of mammals or humans. Insectivorous assassin bugs inject paralyzing (e.g., disulfide-rich peptide neurotoxins) or even lethal substances into their prey organisms to suppress flight and fight reactions. In addition, they inject digestive enzymes that liquefy the internal body material of the prey. The digested material is subsequently sucked into the digestive tract of the bug. The Australian assassin bug, *Pristhesancus plagipennis* (Ph. 3.40), feeds on bees and other insects. Transcriptomic analyses of its venom gland and proteomic analyses of its venom have shown that they contain more than a hundred proteins comprising disulfide-rich peptides, cystatins, putative cytolytic toxins, a triabin-like protein, proteases, catabolic enzymes, putative nutrient-binding proteins, and others [104, 105]. The enzymes and cytolytic venom components are likely used to mediate extraintestinal predigestion of prey organisms. The presence of triabin-like molecules points to the early origin of the triabins present in hematophagous bugs (e.g., *Triatoma pallidipennis*) where they function as potent inhibitors of thrombin [66]. The hematophagous South American triatomine bug *Triatoma infestans* ('kissing bug'; Ph. 3.41) is one of the common vectors of the protist *Trypanosoma cruzi*, the infectious agent of Chagas disease, also known as American trypanosomiasis [72].

Ph. 3.40: The common assassin bug, *Pristhesancus plagipennis*, from Queensland, Australia (Source: Arthur Clapham, Wikimedia, CC-BY).

Plant-feeding bugs like the greater milkweed bug *Oncopeltus fasciatus* from the southwest of the United States store large amounts of cardenolides (see Section 2.5.2), which they acquire from their host plants, in their epidermis right underneath the cuticula of the exoskeleton. When a predator handles such a bug, certain areas of the cuticula, which are thin and brittle, rupture and stored cardenolides are released from the epithelial cells as oily droplets. Exposure of mucous membranes of the predator to these cytotoxins (inhibition of the Na^+/K^+-ATPase) repels predators. The milkweed bug has developed mutations in its own Na^+/K^+-ATPase that render this enzyme insensitive to cardenolides [4, 17, 57].

Ph. 3.41: The 'kissing bug' *Triatoma infestans* from South America (Source: Bärbel Stock, Zoologische Staatssammlung München, Wikimedia, CC-BY-SA).

Aphids are another taxon of Hemiptera feeding on plant sap. Using members of the plant families Apocynaceae or Asclepiadaceae as host plants they take up cardenolides like strospeside or digoxin. An example for such an aphid is the tropical and subtropical milkweed aphid, *Aphis nerii* (Ph. 3.42). Certain amounts of these ingested toxins are sequestered within the body fluids and used as defense substances, but most of the material is rapidly excreted from the body with the liquid feces. The 'honeydew' excreted by *A. nerii* is composed of 46% (w/w) cardenolides [51].

Black flies (Simuliidae), biting midges (Ceratopogonidae), and mosquitos (Culicidae) are hematophagous animals that feed on body fluids of invertebates or vertebrates, respectively. The compositions of the salivary secretomes (sialomes) of members of these families are similar but have certain species-specific components as well [82]. The sialome of the biting midge, *Culicoides sonorensis* (Ph. 3.43), comprises 45 proteins, among them Kunitz-like serine protease inhibitors (anticoagulants), but also enzymes like trypsin or maltase [46]. Intradermal injection of saliva collected from female *Culicoides nubeculosus* into human skin produced edema, vasodilatation, and pruritus [45]. Bites of

Ph. 3.42: *Aphis nerii* on oleander flower buds (Source: Luis Fernández García L. Fdez., Wikimedia, CC-BY-SA).

certain midges may result in sensitization of humans and allergic reactions upon repeated exposure. An example is the Taiwanese midge *Forcipomyia taiwana* whose saliva contains three major allergens, Fort1 (24 kDa), Fort2 (35 kDa), and Fort3 (65 kDa) [16]. Some midges, such as many Phlebotominae (sand flies) and Simuliidae (black flies), are vectors of pathogens causing various diseases (e.g., bluetongue disease in ruminants).

Ph. 3.43: A female biting midge, *Culicoides sonorensis*, drinking blood through an artificial membrane (Source: Scott Bauer, US Department of Agriculture, public domain).

Tsetse flies (*Glossina* sp.) belong to the familiy of Glossinidae. *Glossina morsitans* (Ph. 3.44), a fully hematophagous species, is distributed throughout East Africa and equatorial Africa. Males and females of these flies transmit the infectious agents of the African trypanosomiasis ('sleeping disease'), *Trypanosoma brucei gambiense* or *Trypanosoma brucei rhodesiense*. The reservoirs of these flagellates are cattle, but the flies transmit the protists also to humans during blood meals. The saliva of tsetse flies contains several antihemostatic, antiinflammatory, and immunomodulatory proteins that may facilitate parasite transmission [49, 81]. The salivary proteins induce specific antibody responses in mammalian hosts that can be used as diagnostic tools. One ingredient of tsetse fly saliva, the antigen-5-related allergen, may induce anaphylactic reactions in humans repeatedly stung by tsetse flies [10].

Ph. 3.44: An adult tsetse fly, *Glossina morsitans* (Source: Alan R. Walker, Wikimedia, CC-BY-SA).

Horseflies are members of the Tabanidae family of Diptera. Different species of horseflies (genera *Tabanus*, *Chrysops*, and *Haematopota*) are distributed all over the world. The females sting animals including humans using their specialized mouthparts to obtain blood (Fig. 3.17). The mouthparts consist of a bundle of six chitinous stylets. Together with a fold of the fleshy labium they form the proboscis. Upon landing on a host animal, the fly thrusts its head downwards so that the stylets can slice into the skin. The stylets are moved up and down in a saw-like manner and enlarge the wound. Saliva is injected into the wound through the salivary duct in the hypopharynx. It contains anticoagulants to prevent blood clotting [41]. Blood and lymph collecting in the pool-like wound are lapped up by the spongeous labella [44]. Horsefly bites are itchy and painful for a few days. Allergic reactions in humans have been reported. Horseflies may be pathogen vectors. The filarial nematode worm *Loa loa* is transmitted by *Chrysops* sp. from human to human in Africa, anthrax bacteria may be spread among cattle and sheep, and the tularemia (rabbit fever) pathogen *Francisella tularensis* may be transmitted from animals to humans.

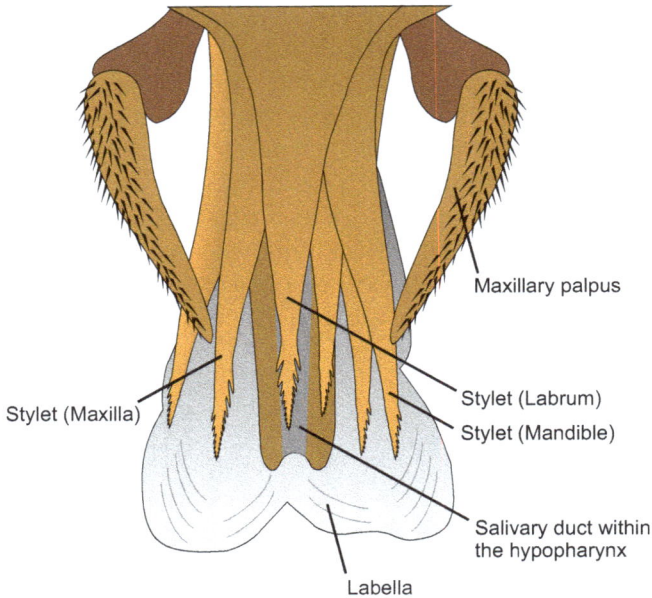

Fig. 3.17: Scheme of the mouthparts of a horsefly. Frontal view on the mouthparts of the horsefly, *Hybomitra affinis*, after removal of the labial folds (total width of the salivary duct at the tip of the hypopharynx is ~30 µm). Drawing after a microscopic image published in the April 2012 issue of *Micscape Magazine* (www.micscape.org) by Anthony Thomas, Canada.

Robber flies (Asilidae) feed mainly on other insects. They have very hard bristles, formed by the hypopharynx and the galeae, with which they penetrate the integument of prey insects. Murder flies (subfamily Laphriinae) can even pierce the hard exoskeletons of beetles. Asilid saliva is composed of insecticidal compounds and enzymes which predigest the prey [60].

Female mosquitos (Culicidae) have very specialized mouthparts (Fig. 3.18) which enable them to penetrate the multilayered vertebrate skin and insert the tips into small blood vessels. Injection of saliva containing anticoagulants and other biologically active components (e.g., antiinflammatory agents) and ingestion of host blood by the peristaltic pumping action of the pharynx musculature are synchronous actions which are possible because salivary duct and the food channel are perfectly isolated from each other over the entire length of the mouth parts.

Mosquito saliva is a complex mixture of small molecules, peptides, and proteins [11, 80]. The salivary glands of female *Aedes* mosquitos (e.g., the Asian tiger mosquito and *Aedes albopictus*; Ph. 3.45) express more than 50 salivary proteins, among them members of the D7 protein family and those of the antigen 5 family, a serpin (a putative anticoagulant), a secreted calreticulin, and a C-type lectin (may moderate host immune reactions), two proteins similar to mammalian angiopoietins (may stabilize or stimulate

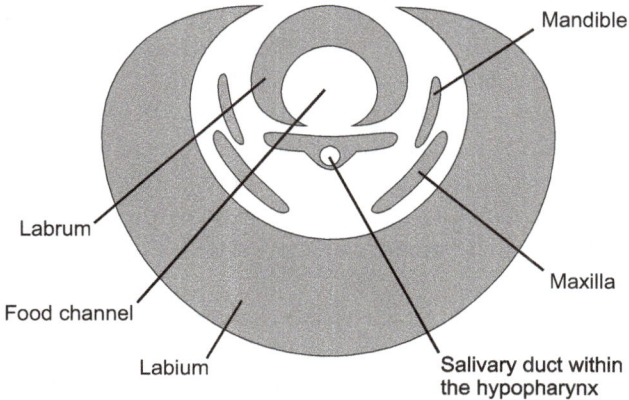

Fig. 3.18: Cross section of the elongated proboscis formed by the mouthparts of a mosquito. The diameters of the probosces range from 40 to 100 μm in different mosquito species. Note the separate ducts for saliva (formed by the hypopharynx) and host blood (food channel formed by the labrum).

the formation of new capillaries), a lysozyme (antimicrobial agent), as well as several serine proteases [101]. The D7 salivary proteins are odorant-binding proteins and bind host-produced eicosanoid and biogenic amine compounds. These molecules are released from cells in the host skin in response to the mosquito sting and activate vascular smooth muscle, platelet activation, and local inflammation. D7 proteins, however, mop up these mediators and suppress the respective host responses [39]. Apyrases in mosquito saliva destroy local ADP and avoid platelet activation in host blood vessels

Ph. 3.45: A female Asian tiger mosquito, *Aedes albopictus*, starting to ingest blood through an artificial membrane (Source: James Gathany, Centers for Disease Control and Prevention, US Department of Health and Human Services, Wikimedia, public domain).

[15]. *Aedes* mosquitoes have vasodilatory tachykinin decapeptides named sialokinins in their saliva [14], while saliva of *Anopheles* mosquitoes contains a 65 kDa peroxidase which destroys skin vasoconstricting norepinephrine and serotonin [83].

Different species of mosquitos may transmit the infectious agents of several diseases in humans and domestic animals. Among these pathogens are dengue, West Nile, Japanese encephalitis, Zika, chikungunya, and Rift Valley fever viruses. The vector's saliva interacts with local immune cells at the sting site. Some ingredients of mosquito saliva attenuate the response of the host's immune system which may enhance virus transmission and disease progression [29].

Termites or white ants (Isoptera) are more closely related to cockroaches than to ants and comprise about 3,000 species. They are adapted to live in subtropical and tropical regions of the world, especially in Africa, Australia, South America, and the Indomalayan area. Animals introduced to North America and Central Europe, e.g., the eastern subterranean termite (*Reticulitermes flavipes*) survive as long as the temperature of their colonies stays above freezing point and can cause damage by destroying structural wood in buildings. Termites live in social communities which may comprise up to millions of individuals. In addition to the winged sex animals, there are workers and soldiers. The latter are capable of mechanical defense through powerful mandibles or through the presence of a long, nose-shaped appendage on the forehead that is connected to a poison gland. The secretions are used for chemical defense. These animals may produce toxins in the salivary glands and lip glands as well. Defensive secretions of termites contain monoterpenes, sesquiterpenes, diterpenes (see Section 2.4), and polyketides (see Section 2.3) [77, 78].

3.2.9.2 Hymenoptera

The insect order Hymenoptera comprises the sawflies (Symphyta) and the stinging wasps, bees, and ants (Aculeata). Of toxicological relevance are many, but not all of the Aculeata because not all species are able to sting and of those that can, only the females have a stinger at the end of their abdomen. This is due to the fact that the stinger is evolutionarily derived from an ovipositor.

Although sawflies do not sting and are therefore considered harmless, there is one exception. The larvae of the Australian sawfly *Lophyrotoma interrupta* feed on leaves of *Eucalyptus* species. Feeding on fallen leaves with attached larva may be fatal for grazing animals, as the larvae accumulate lophorotoxin (Fig. 3.19), a benzoylated octapeptide containing four D-amino acid residues. Histopathology revealed petechial bleeding and tissue necroses in livers and kidneys of affected sheep. Similar substances have been found in larvae of European sawfly species [67].

Sawfly larvae may also use other substances as means of passive chemical defense. Larvae of the turnip sawfly, *Athalia rosae* (Ph. 3.46), feed on plants of the genus *Brassica*. These plants defend themselves against herbivores by storing mustard seed oil glucosides (glucosinolates; see Section 2.11) (Fig. 3.19) and the enzyme myrosinase in

A

C$_6$H$_5$-CO-D-Ala-D-Phe-Val-Ile-D-Asp-Asp-D-Glu-Gln

Lophorotoxin

B

Mustard oil-glucosinolate

Myrosinase

S=C=N—
=CH$_2$
Isothiocyanate

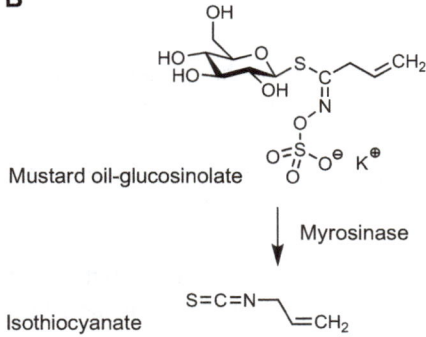

Fig. 3.19: Sawfly toxins. (A) Sequence of lophorotoxin of larvae of the Australian sawfly *Lophyrotoma interrupta*. (B) Degradation of mustard oil glucosinolate to isothiocyanate by *Athalia rosae* larvae.

different subcellular compartments in cells of their leaves. When herbivores feed on these leaves, enzyme and substrate get in contact with each other and the glucosinolates are broken down to yield isothiocyanates which induce stinging pain and usually deter herbivores. When larvae of *A. rosae* feed on such plants, they manage to sequester myrosinase and glucosinolates separately from each other in their own body. When attacked by predators, however, they secrete oily fluid from dorsal glands that

Ph. 3.46: Larva of the turnip sawfly, *Athalia rosae* (Source: Donald Hobern, Wikimedia, CC-BY).

contain myrosinase and glucosinolates which react to produce isothiocyanates. The production of droplets of such defense secretions ('easy bleeding' or 'reflex bleeding') efficiently deters predators [59].

Predatory hymenopteran species use their stingers for paralyzing or killing prey (mostly other insects). Bees and wasps use their stingers also for defense against attackers. Stingers of many bee species are too short to penetrate human skin. However, stings by workers of the honeybee, *Apis mellifera,* those from *Bombus* species (bumblebees), by leafcutter bees (*Megachile*), by Mason bees (*Osmia*), or by furrow bees (*Halictus*) can be effective also in humans. The stingers of many wasp species as well as those of fire ants penetrate human skin as well.

The compositions of venoms of honeybees and other bee species are very similar in quality and quantity, at least with respect to the major ingredients. Because the venom of the honeybee (*Apis mellifera*; Ph. 3.47) is the most studied of the bee venoms, it is presented here in greater detail. Bee venom is synthesized in paired venom glands located in the abdomen (Fig. 3.20). Venom is stored in the venom sac which contains approximately 300 µL of fresh venom (0.1 mg DW), of which 50–100 µL are injected into the victim during a sting. The stinging apparatus consists of venom sac, an alkaline gland that mixes a basic secretion to the venom transferred from the venom sac into the bulb and a stylet-like stinger that is flanked by two bristles or lancets. The lancets assist the stinger in penetrating the skin [108]. In bees the stinger is barbed and gets stuck in the elastic skin of humans or mammals when the bee tries to fly away or is passively torn away. The attached venom sac is usually torn out of the bee's abdomen and continues pumping venom into the tissue of the victim. The bee will die from this injury. If the bee stings another insect for defense, it can usually

Ph. 3.47: A worker of the honeybee, *Apis mellifera* (Source: ElementalImaging/iStock/Getty Images Plus).

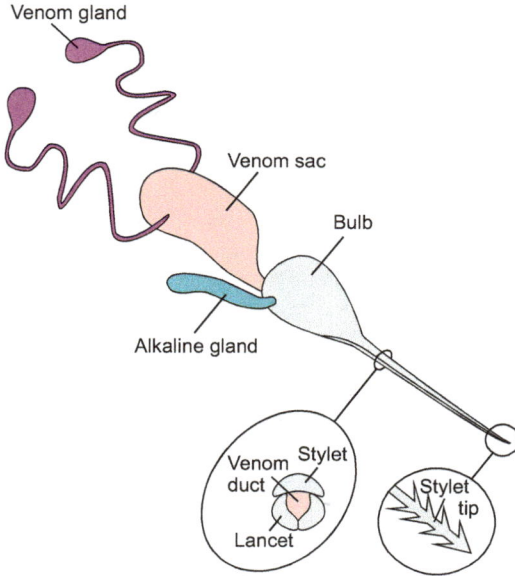

Fig. 3.20: Venom apparatus of the honeybee, *Apis mellifera* (further explanations in the text).

free the stinger from the brittle chitin shell of the victim without damaging the stinging apparatus.

The dangerous 'killer bees' (also called 'Africanized honeybees') have been cre-
ated in South America in 1957 when the African honeybee, *Apis mellifera ligustica*,
was introduced to Brazil and started to interbreed with already established European
bees. The amount of venom a 'killer bee' injects is actually no greater than that of
other bee species. The toxicities of these venoms are fairly similar as well. The killer
bee is more dangerous than other bees since agitated bees expel a very effective
alarm pheromone that calls other bees to the site of an attack even in distances of
1 km. In consequence, the victim is usually stung by several hundreds of bees. The
lethal dose for humans is between 500 and 1,400 stings per person [94].

The toxicity of bee venoms for humans and animals is based on the complex interaction of all their ingredients. Because of the small amounts of venom applied by a
bee during the sting, only local effects occur under normal circumstances. Systemic
reactions are only triggered upon receiving large numbers of stings. The LD_{50} of bee
venom for mice (i.v.) is about 6 mg/kg BW. However, when a person's immune system
has already been sensitized by previous stings he or she may show intense, potentially
life-threatening allergic reactions [18].

✱ Major allergens in bee venom are a phospholipase A_2 (10–12% of venom DW), a hyaluronidase
(1–3% of venom DW), an acid phosphatase (1–2% of venom DW), and melittin (~50% of venom DW)
[9, 71, 95]. Due to the high allergenicity of these proteins, an insect bite can trigger severe allergic

reactions in the entire organism. This includes anaphylactic shock. About 3.5% of the human population develops an IgE-mediated immediate-type allergy with potentially life-threatening symptoms. Anaphylaxis may be more severe in patients with mastocytosis (a buildup of mast cells in the skin or in internal organs) or in those patients being treated with ACE inhibitors or β-blockers. Symptoms usually become noticeable seconds or minutes after the sting. Anaphylactic reactions are often preceded by a prodromal stage with typical sensations such as tingling or burning in the palms or soles, a furry sensation on the tongue, or itching in the throat. Subsequent allergic reactions vary from person to person and range from skin irritation to gastrointestinal symptoms and cardiovascular problems and airway obstruction. Due to the increased vascular permeability and pronounced vasodilatation triggered by histamine and leukotrienes, there can be a severe drop in blood pressure and reduced perfusion of important organs, and in extreme cases anaphylactic shock.

A specific diagnosis of immediate-type allergic reactions to bee or wasp stings (skin test) and immunotherapy (hyposensitization) are possible using solutions of lyophilized bee venom. Hyposensitization should be carried out with caution by a specialist, especially in adults, because of possible allergic complications.

Persons known to suffer from insect allergy should always carry an emergency kit with them. It contains a fast-acting antihistamine, an orally ingestible glucocorticoid, and an autoinjector with epinephrine (adrenaline).

The main constituents of honeybee (*Apis mellifera*) venom are enzymes such as phospholipase A_2, hyaluronidase, and acid phosphatase as well as peptides and nonenzymatic proteins including melittin, MCD peptide, and apamin [37, 52]. Besides these peptides and proteins, vasoactive amines are enriched in bee venom: histamine, which causes pain and lowers blood pressure, dopamine, and catecholamines which may unfold local cytotoxicity in stung animals or humans [30].

Melittin is a cationic peptide comprising 26 amino acid residues (2.85 kDa) (Fig. 3.21). Despite its mostly hydrophobic character, melittin is soluble in aqueous solutions by forming tetramers exposing only the hydrophilic amino acid side chains [98]. Due to its cationic character (pI = 10) it associates with the negatively charged surface moieties of membrane lipids in cell membranes of skin cells in the affected organism. Changing their relative position to each other the melittin tetramers are able to integrate themselves into the plasma membrane and form transmembrane channels that are permeable for small ions. This results in sodium influx into the cell followed by release of

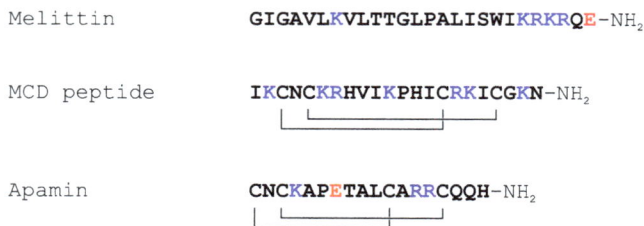

Melittin   **GIGAVLKVLTTGLPALISWIKRKRQE**-NH₂

MCD peptide   **IKCNCKRHVIKPHICRKICGKN**-NH₂

Apamin   **CNCKAPETALCARRCQQH**-NH₂

Fig. 3.21: Primary sequences (one-letter amino acid code) of biologically active peptide toxins in honeybee venoms. Basic amino acid residues are shown in deep blue color, acidic ones in red. The brackets indicate the patterns of disulfide bridges within these molecules.

potassium ions from the cytosol and ultimately in cell death [96, 110]. The LD_{50} of melittin in mice (i.v.) is 3.5 mg/kg BW.

When mast cells are affected by melittin, they release histamine which induces pain sensations, vasodilation, and edematous swelling of the affected limb. These nonspecific actions of melittin are aggravated by the activity of MCD peptide (cationic peptide with 22 amino acid residues and two disulfide bridges) (Fig. 3.21), another agent in bee venom that specifically activates mast cells [7, 30]. As mast cell activation in vivo is a strong signal for antibody production by immune cells [54], the effects of MCD peptide may enhance the allergic potential of bee venom.

Apamin is a cationic peptide composed of 18 amino acid residues (2.03 kDa) containing two disulfide bridges [31] (Fig. 3.21). It is able to cross the blood-brain barrier in vertebrates. It blocks Ca^{2+}-dependent potassium channels (SK channels) in neuronal cells of the central nervous system [102] which characterizes apamin as a potent neurotoxin. When injected with 0.5 mg/kg BW mice displayed incoordinated movements of the skeletal musculature followed by spasms, jerks, and convulsions of apparently spinal origin. The LD_{50} in mice (i.v.) is 4 mg/kg BW.

Peptides containing helix-loop-helix motifs were isolated from the venom of the bumblebee *Bombus pensylvanicus* (Ph. 3.48) [87]. The bombilitins are hemolytic peptides and activate phospholipase A_2 isoforms from different sources.

Ph. 3.48: A bumblebee (*Bombus pensylvanicus*) approaching a flower (Source: ElementalImaging/iStock/Getty Images Plus).

Wasps use their stingers to immobilize or kill prey or as defensive weapons to repel predators. Other than those in bees, wasp stingers have smooth surfaces without any barbs. Thus, wasps may sting not only other insects but also vertebrate animals repeatedly without harming themselves. The composition of wasp venom is largely similar to that of bee venom. However, the amounts of individual ingredients may differ substantially. Moreover, certain ingredients are missing (apamin, MCD peptide, and melittin) and replaced by others (kinins and mastoparans) (Tab. 3.6). The LD_{50}s of wasp venoms (mouse, i.p.), however, are in the same range as those of bee venoms [92].

Tab. 3.6: Ingredients of bee and wasp venoms (data from [24, 26, 34, 70]).

	Bees	Wasps and hornets
Transfer of venom protein per sting (μg)	50–147	1.4–5.0
Biogenic amines	Histamine Dopamin Catecholamines	Histamine Serotonin Catecholamines Acetylcholine
Peptides	Apamin MCD peptide Melittin	Kinins Mastoparans
Enzymes	Phospholipase A Hyaluronidase	Phospholipase A Phospholipase B Hyaluronidase

Wasp species that may get dangerous for humans (only multiple stings) are the Asian giant hornet (*Vespa mandarinia*), the greater banded hornet (*Vespa tropica*), the European hornet (*Vespa crabro*), and some of the European wasps (*Vespula vulgaris*, *Paravespula germanica*, and *Dolichovespula saxonia*).

🗣 Wasp stings induce sharp and very intense pain that may persist for several hours. Local edema and erythema are regularly appearing symptoms. Stitches in the mouth or in the throat are life-threatening due to the risk of swelling and obstruction of the airways. When a person has received several stitches systemic effects such as a drop in blood pressure, kidney damage, and circulatory failure may occur. As with bee venom, anaphylactic reactions are also possible.

Wasp venoms are rich in short peptides (9–18 amino acid residues) that are considered to be 'bradykinin-like peptides' or shorter 'kinins' because they contain sequences of amino acids that are typical for the vertebrate tissue hormone bradykinin (Fig. 3.22). The kinin system in vertebrates consists of endogenous tissue factors and blood proteins that play roles in inflammation, blood pressure control, blood coagulation, and in the generation of pain sensations [58]. It seems likely that these are the physiological systems targeted by kinins in wasp venoms as well. It is assumed that the sharp decrease in blood pressure and the intense pain sensations experienced by humans stung by wasps are, at least in part, due to the action of these kinins. In insects that are stung by wasps the kinins cause depletion of presynaptic transmitter stores in neurons of the central nervous system. This slowly induces an irreversible block of neuronal information processing [75].

Kinins

```
Polistes chinensis          SKRPPGFSPFR
Vespa mandarinia             GRPPGFSPFRID
Vespa simillima              ARPPGFSPFRIV
Paravespula maculifrons  TATTRRRGRPPGFSPFR
                                     | |
                                     XZ
```
bold - Bradykinin consensus sequ.; **P** - Hydroxyproline; X, Z - Glycosylations

Mastoparans

```
Vespa crabro         INLKALLAVALLIL-NH₂ Mastoparan C
Vespa mandarinia     INLKAIAALAKKLL-NH₂ Mastoparan-M
Vespula vulgaris     INWKKIKSIIKAAMN
```

Fig. 3.22: Primary sequences (one-letter amino acid code) of biologically active peptides (kinins and mastoparans) in venoms of different wasp species.

Kinins are used by females of the mammoth wasp (*Megascolia maculata*; Ph. 3.49) (up to 4 cm body length) to immobilize larva of the rhinoceros beetle (*Oryctes nasicornis*) in the last larval stage. The female wasp flies in loops about 15 cm above the ground. It tracks down the beetle larva using its sense of smell and burrows toward it. As soon as the wasp reaches the larva, it starts to bite and sting the larva without injecting any venom. This way, the wasp tries to force the larvae to change position so that the wasp can reach the bottom side of the larva with its stinger. When the larva begins to tire, the wasp stings it into ganglia of the ventral central nerve cord. The venom of scoliid wasps contains two variants of bradykinin which irreversibly block synaptic transmission to nicotinic acetylcholine receptors in the postsynaptic membrane of neurons in the insect nervous system [43]. The sting leads to immobility of the larva within

Ph. 3.49: A giant wasp, *Megascolia maculata*, from Europe and the Near East (Source: imv/iStock/Getty Images Plus).

three to five minutes. The mammoth wasp female then places an egg on the surface of the larva. When the wasp larva hatches, it bores its head into the paralyzed beetle larva and consumes the internal resources for its own growth and development.

Other peptide toxins in wasp venoms are the mastoparans (Fig. 3.22) which are basic peptides with 14 or 15 amino acid residues [33, 62]. Mastoparans are lyobipolar molecules whose basic amino acid residues interact with membrane lipids in target cells. Subsequent conformational changes are associated with membrane integration of these peptides which alters the transmembrane permeability. In vertebrates, some of the mastoparans have hemolytic effects, others activate phospholipase A_2, and some induce mast cell degranulation or induce the release of catecholamines from chromaffin cells of the adrenal medulla. Others induce serotonin release from platelets and platelet aggregation, and others may have antihypertensive or cardiotoxic effects [62]. Mastoparans may act through their ability to bind to calmodulin and prevent the activation of calcium-calmodulin-dependent enzymes, e.g., phosphodiesterases [62]. Some effects of mastoparans may be mediated via the receptor-independent activation of heterotrimeric GTP binding proteins [70].

Other peptides isolated from the venom of the European hornet *Vespa crabro* are referred to as crabrolins [2]. They release histamine from rat peritoneal mast cells and activate phospholipase A_2 isoforms from different sources. They belong to a group of so-called 'chemotactic peptides' of wasp venoms which have, besides mastoparan-like actions, chemotactic effects on vertebrate lymphocytes [62].

In addition to the peptides, proteotoxins with different functions in target organisms have also been isolated from wasp venoms. An example is mandarintoxin from the Asian hornet *Vespa mandarinia*. It is a basic protein (21 kDa) that blocks the influx of Na^+ ions into nerve cells through voltage-gated sodium channels and suppresses the generation of action potentials [90]. A protein called VOLF (*Vespa orientalis* lethal factor) was isolated from the venom of *Vespa orientalis*, a wasp species distributed from the Mediterranean region to East Asia. VOLF has strong toxic effects due to its ability to act as an acetylcholine esterase. It destroys the transmitter (ACh) in the synaptic cleft of the motor endplate and other acetylcholinergic synapses (LD_{50} (mouse, i.v.) = 80 µg/kg BW [85]).

✱ Although clearly less than bee venoms, wasp venoms have a certain allergenic potential. The venom of the North American wasp *Vespula maculifrons* contains five major allergenic proteins: Vmacl (97 kDa), a hyaluronidase (46 kDa), Vmac3 (39 kDa), the phospholipases A and B (34 kDa, each), and antigen 5 (22 kDa). Among these proteins the hyaluronidase demonstrated the most IgE binding with serum samples from allergic patients [35].

The venoms of wasp species, which paralyze other insects with stings in order to bring them in as prey or to lay their eggs in them, are remarkable with respect to their mechanism of action. An interesting example is a digger wasp (Sphecidae), the beewolf *Philanthus triangulum*. Females of this species prey exclusively on honeybees which they paralyze with a sting in order to be able to extract their honey sacs or to carry them to their nests for food for the brood. In addition to acetylcholine and L-glutamic acid, the venom contains three polyamines, the philanthotoxins (Fig. 3.23). Philanthotoxins inhibit the reuptake of the motor neuron transmitter glutamate into the presynaptic nerve terminal and block the ion channels at the glutamate receptor–ionophore complex in the postsynaptic cell membrane [5]. These channels are used for neuromuscular transmission in insects so that the stung bees are quickly paralyzed by the wasp philanthotoxins.

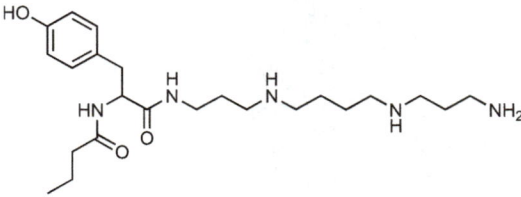

Fig. 3.23: Philanthotoxin 343, a component of beewolf (*Philanthus* sp.) venom.

Toxins produced by ants include aliphatic acids (see Section 2.1), biogenic amines (see Section 2.8), terpenes (see Section 2.4), pyrrole and pyridine alkaloids (see Sections 2.9.4 and 2.9.8), and their hydroderivatives, indolicidine and decahydroquinoline alkaloids, and pyrazine derivatives. In stinging ants, the venoms also contain peptide and proteotoxins [99, 103].

A hemolytic polypeptide (barbatolysin) has been isolated from the venom of harvester ants of the genus *Pogonomyrmex* [3, 91]. The LD_{50} (mice, s.c.) of crude venom isolated from *Pogonomyrmex maricopa* is approximately 0.2 mg/kg BW. Stings of the Australian hopper ant, *Myrmecia pilosula* (Myrmeciinae) (Ph. 3.50), are known for their allergenic potential. Analyses of venom composition of *M. pilosula* revealed the presence of histamine and several enzymes (e.g., a hyaluronidase and a phospholipase A_2) along with other pharmacologically active constituents. The latter encompass a variety of highly basic peptides called pilosulins which have only little sequence similarity to other Hymenoptera venom peptides. Five predominant peptides, which make up approximately 90% of the venom dry weight, have cytotoxic, hypotensive, histamine-releasing, and antimicrobial properties. Pilosulin 3 has been identified as one of the major allergens in the venom of this species [106]. Pilosulin-like peptides have also been found in investigations of the venom of the Asian stinging ant *Odontomachus monticola* [42], although this species belongs to a different ant family, the Ponerinae.

The members of the Ponerinae family of ants also produce peptide and proteotoxins. Venom of the Central American bullet ant, *Paraponera clavata*, blocks neurotransmission in target animals. Poneratoxin (PoTX, 25 amino acid residues) paralyzes insects [76, 97]. It also acts on vertebrates by inhibiting the signal transmission at the neuromuscular junction by pre- and postsynaptic mechanisms [32]. Fifteen peptides,

A jack jumper ant, *Myrmecia pilosula*, from Tasmania (Source: en:User:Ways, Wikimedia, CC-BY-SA).

the ponericins, which have antibiotic, insecticidal, and hemolytic activity, were iso-lated from the venom of *Neoponera goeldii* (Ponerinae) [69]. Stings of bullet ants are extremely painful.

The venom of fire ants, e.g., that of *Solenopsis invicta* (Ph. 3.51), contains oily alka-loids (solenopsins) derived from piperidine (see Section 2.9.6) [109] in combination with toxic proteins [22]. The alkaloids in the venom may inhibit enzymes (e.g., eNOS)

0.5 mm

A fire ant, *Solenopsis invicta* (Source: April Nobile, AntWeb.org, Wikimedia, CC-BY).

that are involved in signal transduction cascades in target organisms, while proteo-toxins may inhibit the activation of muscle cells (myotoxins) or function as cytolytic enzymes (phospholipases, metalloproteases). Several other proteins in *Solenopsis invicta* venom have allergenic potentials.

An agitated fire ant usually holds on to the skin of a human victim using its strong mandibles and applies several stings with a stinger located at the end of the abdomen. Stings cause burning sensations at the sting site followed by urticaria. The sting site swells substantially within a few hours, but swelling will spontaneously disappear within a few days if not infected. Already sensitized people may experience anaphylaxis following fire ant stings, especially when having received stings from many ants, a condition that requires emergency treatment.

First aid for fire ant stings includes disinfection of the sting site, topical application of anesthetics like benzocaine, and taking antihistamines. Severe allergic reactions to fire ant stings may be fatal if not immediately treated.

Many ants use low–molecular-mass molecules for defense and communication. Among these substances are monoterpenes called iridoids (see Section 2.4.2). The Argentine ant *Linepithema humile* (Dolichoderinae) produces iridomyrmecin (Fig. 3.24). This substance is an antibiotic and seems to disable individuals of competing ant species at least transiently but is allegedly not harmful for mammals [107]. The Australian cocktail ant, *Iridomyrmex nitidiceps*, generates iridodial and isovaleric acid as major defense substances [13].

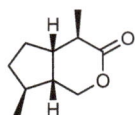

Iridomyrmecin

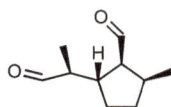

Iridodial

Fig. 3.24: Iridoids in ant venoms.

3.2.9.3 Beetles (Coleoptera)

Within the different life stages of the beetles (egg, larva, pupa, imago) compounds of numerous chemical characteristics occur as defense poisons. These substances can be conditionally secreted when animals are under stress or sequestered in tissues to render these animals unpalatable to potential predators. The substances used as defensive weapons are manifold: aliphatic and aromatic alcohols, aldehydes, and acids, cardioactive glycosides, polyketides, terpenes, pregnane derivatives, saponins, benzoquinone derivatives, and to a lesser extent, also peptide- and proteotoxins.

Some beetle species protect their egg clutches by impregnating them with toxins to repel animals that would usually feed on them. When adults of the predatory pirate bug *Orius laevigatus* (Anthocoridae) were offered eggs of ladybirds (*Harmonia axyridis* or *Adalia bipunctata*; Ph. 3.52) they rejected that kind of food, while adults of the common green lacewing, *Chrysoperla carnea* (Neuroptera; Chrysopidae), accepted these eggs. However, when *C. carnea* larvae were exclusively fed on ladybird eggs their development was slowed down or completely suppressed indicating that this kind of food, while acceptable for adult lacewings, is not appropriate for lacewing larvae. Female ladybirds impregnate their egg clutches with azaphenalene alkaloids (see Section 2.9.18) [1] or with adaline [100] (Fig. 3.25), which cause bad smell or taste of the egg clutches or are downright toxic for predatory animals [86]. These alkaloids have the ability to block nicotinic acetylcholine receptors (nAChRs) in nerve and muscle cells of affected animals [47].

Ph. 3.52: A two-spotted ladybird, *Adalia bipunctata* (Source: © Entomart.be, Wikimedia).

Larvae or pupae of beetles are also well protected from attacks by predators as is illustrated by an example of leaf beetles (Chrysomelidae) of the genus *Diamphidia* (Ph. 3.53) in Namibia. These beetles lay their eggs on the stems of shrubs and cover them with their feces which provide protection against water loss and predators. After hatching the larvae feed on the plant material and may diapause in the ground for up to several years during unfavorable periods. Bushmen of the San people dig beside *Commiphora* shrubs in search of larvae or pupae of the Bushman arrow-poison beetles *Diamphidia nigroornata* or *D. vittatipennis*. Squeezing the hemolymph of several larvae onto the shafts of their arrows generate poison arrows that they use for hunting. Animals that are hurt by these arrows get slowly paralyzed and can be

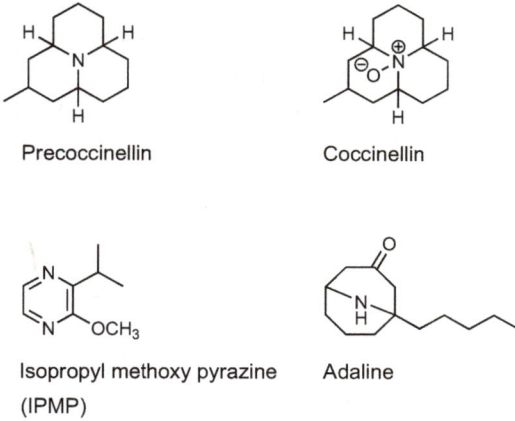

Precoccinellin

Coccinellin

Isopropyl methoxy pyrazine
(IPMP)

Adaline

Fig. 3.25: Defense substances of ladybirds.

easily hunted down. The poison of one arrow is sufficient to kill an adult giraffe. The hemolymph of *Diamphidia nigroornata* larvae and pupae, but not that of adult beetles, contains a basic protein of approximately 60 kDa, diamphotoxin (dimamphidiatoxin). It increases the permeability of cell membranes of animal cells for ions and small organic molecules. It damages nerve cells (neurotoxin) and blood cells (hemolysin). The lethal dose in mice (i.p.) is approximately 25 pg [19].

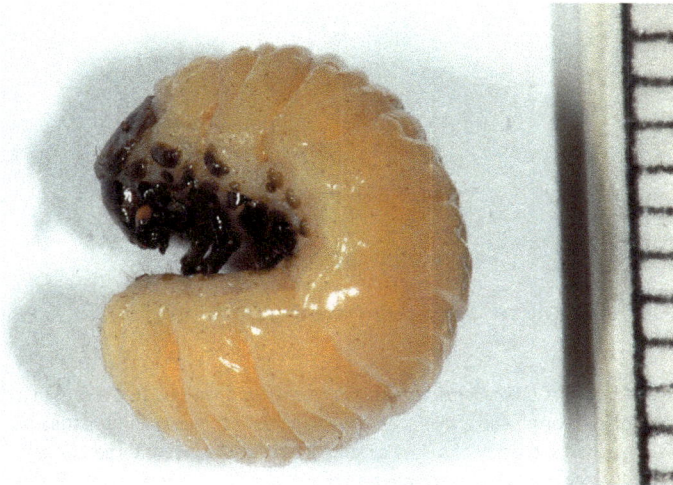

Ph. 3.53: A larva of the poison beetle, *Diamphidia* sp. (Source: Fritz Geller-Grimm, Wikimedia, CC-BY-SA).

There are many examples of chemical defenses in adult beetles as well. Some toxins are synthesized in the beetle's own metabolism, and others are products of symbiontic microorganisms living in the beetles. An example for the latter is pederin (Fig. 3.26) (see Section 2.3.8) which was isolated from hemolymph samples of rove beetles of the genus *Paederus* (Staphylinidae) (Ph. 3.54). The toxin is probably synthesized by *Pseudomonas* endosymbionts occurring in the females of these beetles and accumulates to high levels in the body fluids (up to 0.25% of an insect's wet weight). Males store only moderate amounts of pederin which are mainly acquired from their mothers via the eggs. Pederin is a cytotoxin (LD_{50} (mouse) = 0.14 mg/kg BW) due to its ability to suppress DNA replication and protein synthesis [27].

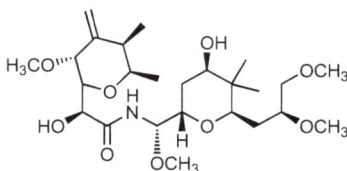

Fig. 3.26: Pederin, a defensive polyketide toxin in staphylinid beetles of the genus *Paederus*.

Ph. 3.54: A whiplash beetle, *Paederus* sp. (Source: Alvesgaspar, Wikimedia, CC-BY-SA).

☠ When humans get in close skin contact to *Paederus* beetles, which may happen due to the fact that these beetles sometimes show mass reproduction, this may cause skin rashes (*Paederus* dermatitis or Dermatitis linearis) which are slow in onset (delay of a days or so) but may last for up to three weeks.

Even small doses (1 ng/mL) of pederin have cytotoxic effects on animal cells. Pederin is a blocker of mitosis [27]. The LD_{50} of pederin in the mouse is 0.14 mg/kg body weight. Low amounts of pederin applied to human skin cause acute irritating contact dermatitis. At higher concentrations, pederin induces skin inflammation and vesicular rashes. Side effects are stinging, itching, and painful burning sensations [93]. These symptoms usually subside within 2 weeks upon exposure.

The only treatments known are topical application of antiinflammatory ointments (corticoids) or taking antihistamine drugs.

Water beetles of the genus *Dytiscus*, e.g., the great diving beetle (*Dytiscus marginalis*; Ph. 3.55), have defense glands in the prothorax as well as pygidial glands next to the anus. Both types of glands produce and secrete milky fluids enriched in 11-deoxy-corticosterone [89]. An amount of 0.4 mg of this corticoid could be extracted from one beetle (which is the same amount that can be extracted from the adrenal glands of 1,000 cows). Exposure of predators like fish or amphibian to such concentrations of steroids has anesthetizing and emetic effects which make this substance an effective deterrent [55].

Ph. 3.55: A diving beetle, *Dytiscus marginalis* (Source: B. Kimmel, Wikimedia, GNU-CC-BY).

Paired pygidial glands in the abdomen are also used as defense weapons in ground beetles of the Brachininae subfamily. *Brachinus crepitans* occurs in southern Europe, Northern Africa, the Middle East, and Asia. This beetle is able to directionally spray its toxin mixture toward an attacker ('bombardier beetle') by turning its body standing face to face with the opponent, lifting its own body from the ground and bending the abdomen forward (Fig. 3.27). The beetle then transfers some hydroquinone (Fig. 2.15), toluhydroquinone, and hydrogen peroxide from the toxin gland reservoir into a chitinous reaction chamber through a muscular valve. Annex glands lining the reaction chamber secrete catalase which results in the explosive degradation of hydrogen peroxide. The gas pressure in the chamber is released through the anteriorly directed openings of the

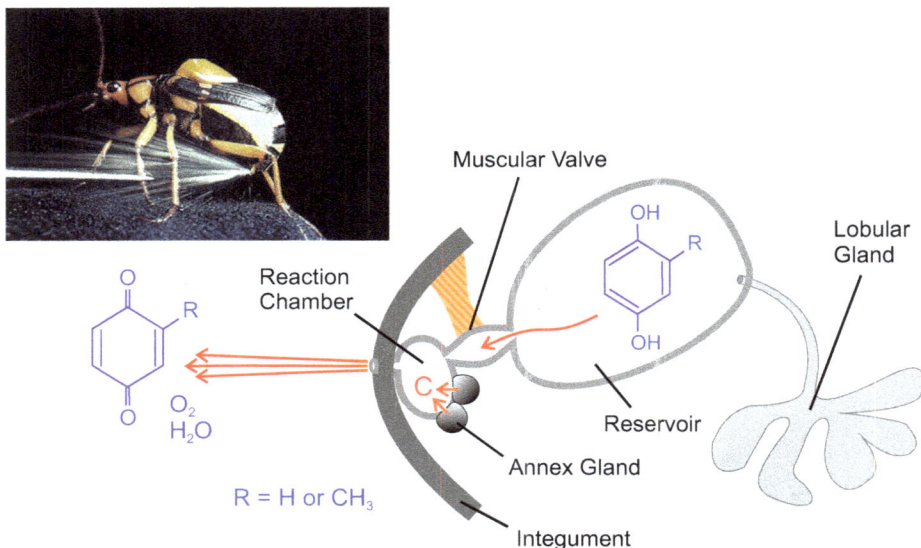

Fig. 3.27: Pygidial gland of the ground beetle, *Brachinus crepitans*. Pygidial gland cells secrete hydroquinone, toluhydroquinone, and hydrogen peroxide. This material is stored in a reservoir. When threatened, the beetle transfers a small portion through a muscular valve into a chitinous reaction chamber. Catalase (C) is secreted by annex glands into this chamber and mediates rapid degradation of hydrogen peroxide to oxygen and water. The developing gas pressure in the chamber is explosively released through the pygidial gland pores which allow the beetle to spray the quinones aimedly at the attacker. When getting in contact with air the material immediately gets sticky due to the high protein content. Together with the skin-irritating quinones, the sticky material effectively repels predators. The insert shows a spraying bombardier beetle in the lab upon mechanical stimulation using a steel needle (reproduced from [25] with permission granted by *Proceedings of the National Academy of Sciences*, USA).

pygidial glands so that the quinones are sprayed into the attacker's face up to a distance of several centimeters [25, 88].

Other beetles defend themselves by releasing an oily liquid from the intersegmental membranes of their walking legs when threatened by predators. This reaction has been called 'reflex bleeding' although the fluid is not hemolymph as initially thought but a glandular fluid that is usually enriched in substances that irritate the attacker's skin or have inflammatory properties. An example for such a defensive substance that acts as a predator deterrent [12] is cantharidin (Fig. 3.28) that occurs in the defense secretions of Eurasian blister beetles (Meloidae) like *Lytta vesicatoria* ('Spanish fly'; Ph. 3.56) or *Meloe violaceus* (the violet oil beetle, Ph. 3.57). Cantharidin (see Section 2.4.3) is a terpenoid and derived from farnesol which is partially degraded (loss of 5 C atoms) and transformed to a multicyclic derivative during biosynthesis [73]. It is synthesized in the gonads of male blister beetles who transfer almost the entire

Ph. 3.56: A 'Spanish fly' (*Lytta vesicatoria*) on a plant (Source: Siga, Wikimedia, GNU-CC-BY-SA).

Ph. 3.57: Male of the blister beetle, *Meloe violaceus* (Source: Jan-Peter Hildebrandt).

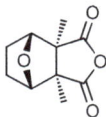

Fig. 3.28: Structure of cantharidin, the defensive substance of blister beetles.

amount of stored cantharidin to the females during mating. The females sequester portions of the material for their own defense in defense glands but use the remaining material to impregnate the eggs [65].

At the molecular level, cantharidin activates extracellular proteases which cleave cell-cell contacts in epidermal cells. This induces loosening of the skin tissue context and blister formation. Other molecular targets of cantharidin are protein phosphatases [48, 64]. Inhibition of these enzymes results in erratic protein phosphorylation and induces inflammation and apoptosis of epithelial cells. Cantharidin is used in medicine to remove skin warts [56]. Extracts of blister beetles have been used as aphrodisiacs ('Spanish fly') due to their ability to induce priapism upon oral ingestion [79]. Symptoms of cantharidin poisoning include burning of the mouth, dysphagia, nausea, hematemesis, hematuria, and dysuria [40]. There are several reports of fatal outcomes of cantharidin abuse (LD_{50} (human, p.o.) ~ 0.5 mg/kg BW). There is no known antidote.

Larvae of leaf beetles (Chrysomelidae) have dorsal defense glands (Ph. 3.58) which secrete a sticky fluid that contains plant-borne β-amyrin, curcurbitacins, pyrrolizidine alkaloids (see Section 2.9.4), phenolglucosides, or iridoids (see Section 2.4.2) as defense molecules against different kinds of generalistic predators [28]. The precursors of these substances are taken up from food plants, but the insect larvae generate the final products in their own metabolism [8, 50].

3.2.9.4 Stick Insects (Phasmatodea)

Anisomorpha buprestoides (Ph. 3.59), the two-striped walkingstick, is a stick insect (Phasmatodea) from the southeastern United States. Females are larger (70 mm) than males (42 mm). When threatened the animals spray substances that irritate and repel other insects, birds, or mammals. The secretions are enriched in monoterpene dialdehydes (see Section 2.4.2) in different stereoisomeric forms, especially anisomorphal (Fig. 2.8), dolichodial, and peruphasmal. The relative composition of these forms in the defense secretions changes with age of the animal [23].

3.2.9.5 Caterpillars of Butterflies (Lepidoptera)

Butterflies (Lepidoptera) are represented worldwide with around 165,000 species. The bodies and wings of the imagines are covered by scales that give them color and markings. The adult butterflies are mobile and may be able to escape an approaching predator. The larvae (caterpillars) or pupae of butterflies, however, are quite stationary and

Ph. 3.58: Activation of dorsal defense glands in a leaf beetle larva (Source: Antje Burse and Wilhelm Boland, with permission).

Ph. 3.59: A stick insect, *Anisomorpha buprestoides* (Source: Bugenstein, Wikimedia, CC-BY-SA).

not able to do so. Thus, caterpillars have developed several defense strategies against potential predators. Some caterpillars weave protective nets around their feeding sites to prevent predators from approaching them. Other caterpillars make themselves unpalatable to potential predators by accumulating toxins within their tissues or fend off predators by secreting toxic fluids. The caterpillars of some species combine different defensive strategies, e.g., by filling hollow cuticular hairs carrying barbs with toxic substances. These structures break easily off the cuticula when the caterpillar is attacked. The bristles penetrate the skin of the attacker, the barbs anchor the hairs in the skin, and the toxins are released. This generally causes inflammatory or even allergic reactions in the target animals.

These cuticular hairs or spines that often look like a pelt can affect humans as well, particularly in cases of mass infestations of orchards or forests. Loose hairs may induce harm upon being inhaled, or the urticating bristles may stick in the skin upon touching the caterpillars. The caterpillars of tropical butterflies can cause very serious poisoning. In Brazil, the fatality rate after contact with caterpillars of the giant silkworm moth, *Lonomia obliqua* (Saturniidae) (Ph. 3.60) is higher than that of snake bites. The LD_{50} (mouse, i.v.) of the *Lonomia* venom is 9.5 mg/kg BW. Other tropical and subtropical butterfly species with potentially harmful caterpillars are *Megalopyge opercularis* (the southern flanell moth; Megalopygidae) occurring in the southern states of the United States or the Australian billygoat plum stinging caterpillar of *Thosea penthima* (Limacodidae). The latter has numerous spikes equipped with venom sacs with which they deliver painful stings. One of the major components of the toxin mixture is histamine.

Ph. 3.60: Caterpillar of the giant silkworm moth, *Lonomia obliqua* (Source: Centro de Informações Toxicológicas de Santa Catarina, Wikimedia, public domain).

⚑ A case of human envenomation by a *Thosea* caterpillar sting on the forearm has been reported [38]. The sting caused immediate burning pain and local wheal formation. Pain radiated up the arm and persisted for 10 h. In addition severe chest pain was experienced by the patient that lasted for several hours.

In Europe, human poisoning including local rash formation (dermatitis) or breathing problems is possible with caterpillars of the families of Noctuidae (including the brown-tail moth, *Euproctis chrysorrhoea*) or the Notodontidae (including the oak processionary moth, *Thaumetopoea processionea*; Ph. 3.61), or, in the Mediterranean, by caterpillars of the pine processionary, *T. pityocampa*.

Ph. 3.61: Caterpillar of the oak processionary moth, *Thaumetopoea processionea* (Source: R. Altenkamp, Wikimedia, GNU-CC-BY-SA).

> ♨ The first step after contact with caterpillar hairs is to remove the hairs stuck to the skin with a strong jet of water or with adhesive tape. Avoid rubbing! Topical application of glucocorticoids may accelerate the disappearance of local reactions. Symptomatic treatment is generally sufficient. Medical aid should be sought upon contact with *Lonomia* species. Depending on the intensity of contact, the coagulability of the blood must be restored by infusions of antifibrinolytics, e.g., of aprotinin. Human plasma infusions generally worsen the patient's condition and should not be applied.

🐕 Dogs which have sniffed on or eaten caterpillars of *Thaumetopoea wilkinsoni* showed lesions in nose and mouth and tongue necrosis [6].

The caterpillars of other butterfly species, their pupae and imagines are passively poisonous because they absorb toxic secondary substances from plants and store them in their bodies, e.g., various aromatic compounds, nitrogen-containing metabolites such as alkaloids (see Section 2.9), aristolochic acids, cyanogenic glycosides (see Section 2.10), glucosinolates (see Section 2.11), and other sulfur-containing metabolites, isoprenoids such as cardiac glycosides, cucurbitacins, iridoid glycosides, and others. Some species are able to derivatize the toxin precursors taken up from their host plants or synthesize toxic substances all by themselves, e.g., the cyanogenic glycosides linamarin and lotaustralin in larvae of the European five-spot burnet moth *Zygaena trifolii* [61]. Larvae of *Zerynthia polyxena*, the southern festoon (Papilionidae), acquire aristolochic acid from members of the plant family Aristolochiaceae [84]. Aristolochic acids are effective deterrents of bird predators. Pyrrolizidine alkaloids (see Section 2.9.4) are the defensive substances accumulated in caterpillars of the cinnabar moth, *Tyria jacobaeae* (Erebidae) (Ph. 3.62). They are acquired from host plants belonging to the genera *Senecio* or *Jacobaea* (ragwort). To avoid self-poisoning these butterflies efficiently N-oxidize the alkaloids in the hemolymph using a senecionine *N*-oxygenase (SNO), a flavin-dependent monooxygenase with high substrate specificity for pyrrolizidine alkaloids. The animals store the nontoxic *N*-oxides which are efficient deterrents of potential bird predators [63].

Ph. 3.62: Adult (A) and caterpillar (B) of the cinnabar moth, *Tyria jacobaeae* (Source A: Thomas Hundtke, Wikimedia, CC-BY-SA; Source B: Quartl, Wikimedia, CC-BY-SA).

In many cases of caterpillars feeding on toxic host plants the toxins accumulating in the caterpillars are retained throughout the entire life span of that individual and may even being transferred to the eggs by the female butterflies to protect them from predation [68]. To cope with these toxic plant compounds butterflies have evolutionarily developed sophisticated pathways for sequestering such substances in tissues that are not sensitive to the toxins [74] or have developed mutations in the endogenous target molecules that render them resistant against these toxins [20]. An example is the monarch butterfly (*Danaus plexippus*; Ph. 3.63). Individuals of this species accumulate cardenolides (see Section 2.5.2) that would normally inhibit the Na^+/K^+-ATPase (sodium pump) needed for maintaining the ion gradients of these ions across the plasma membrane of its cells. A mutation in the α-subunit of this transport molecule, the exchange of an asparagine residue against histidine at amino acid position 122 (N122H), however, renders the pump insensitive against cardenolides [36, 53].

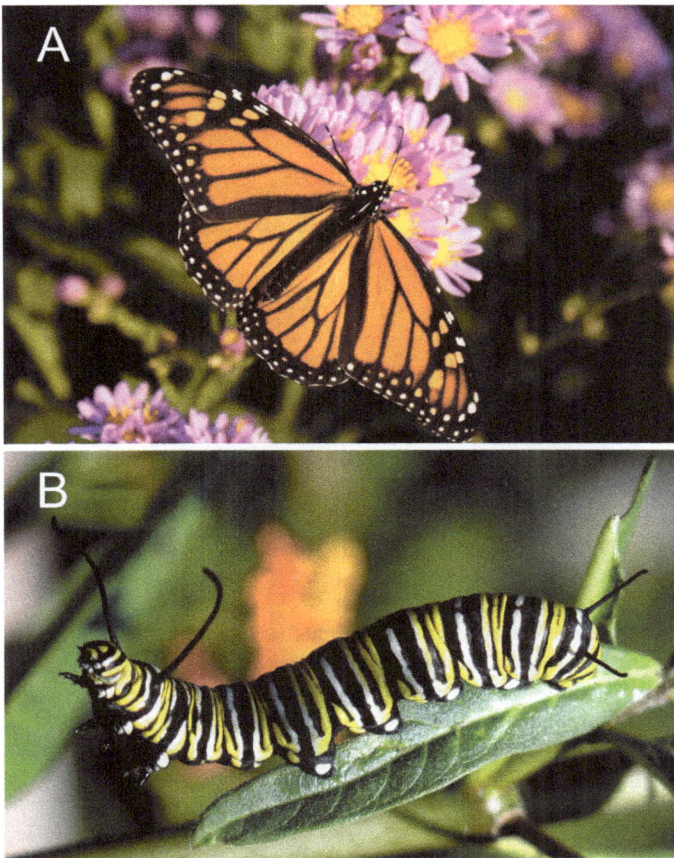

Ph. 3.63: Adult (A) and caterpillar (B) of the monarch butterfly, *Danaus plexippus* (Source image A: LyWashu, Wikimedia, CC-BY-SA; Source image B: Burkhard Mücke, Wikimedia, CC-BY-SA).

Other insects feeding on toxic host plants have developed similar genetic adaptations to avoid self-poisoning [21].

References

[1]   Alujas-Burgos S et al. (2018) Org Biomol Chem 16(37): 8218
[2]   Argiolas A, Pisano JJ (1984) J Biol Chem 259(16): 10106
[3]   Bernheimer AW et al. (1980) Toxicon 18(3): 271
[4]   Bramer C et al. (2015) Proc Royal Soc B – Biol Sci 282 1805 20142346
[5]   Brier TJ et al. (2003) Mol Pharmacol 64(4): 954
[6]   Bruchim Y et al. (2005) Toxicon 45(4): 443
[7]   Buku A (1999) Peptides 20(3): 415
[8]   Burse A, Boland W (2017) Z Naturforsch C, J Biosci 72(9–10): 417
[9]   Burzyńska M, Piasecka-Kwiatkowska D (2021) Int J Mol Sci 22(16): 8371
[10]   Caljon G et al. (2009) Insect Biochem Mol Biol 39(5): 332
[11]   Calvo E et al. (2009) BMC Genom 10 57
[12]   Carrel JE, Eisner T (1974) Science 183(4126): 755
[13]   Cavill GWK et al. (1982) Tetrahedron 38(13): 1931
[14]   Champagne DE, Ribeiro JM (1994) Proc Natl Acad Sci U S A 91(1): 138
[15]   Champagne DE et al. (1995) Proc Natl Acad Sci U S A 92(3): 694
[16]   Chen YH et al. (2005) Allergy 60(12): 1518
[17]   Dalla S et al. (2017) Insect Biochem Mol Biol 89 43
[18]   de Graaf DC et al. (2009) J Proteom 72(2): 145
[19]   de la Harpe J et al. (1983) J Biol Chem 258(19): 11924
[20]   Dobler S et al. (2011) Phytochemistry 72(13): 1593
[21]   Dobler S et al. (2015) Entomol Exp Appl 157(1): 30
[22   dos Santos Pinto JRA et al. (2012) J Proteome Res 11(9): 4643
[23]   Dossey AT et al. (2008) J Chem Ecol 34(5): 584
[24]   Edery H et al. (1978) Venoms of Vespidae. In: Bettini S (ed.) Arthropod Venoms. Handbook of Experimental Pharmacology/Handbuch der experimentellen Pharmakologie, Vol. 48. Springer, Berlin, Heidelberg, Germany, p. 691
[25]   Eisner T, Aneshansley DJ (1999) Proc Natl Acad Sci U S A 96(17): 9705
[26]   Fitzgerald KT, Flood AA (2006) Clin Tech Small Anim Pract 21(4): 194
[27]   Frank JH, Kanamitsu K (1987) J Med Entomol 24(2): 155
[28]   Gross J, Schmidtberg H (2009) Glands of leaf beetle larvae – Protective structures against attacking predators and pathogens. In: Jolivet P, Santiago-Blay J, Schmitt M (eds.) Research on Chrysomelidae, Vol. 2. Koninklijke Brill N.V., Leiden, Netherlands, p. 177
[29]   Guerrero D et al. (2020) Front Cell Infect Microbiol 10 407
[30]   Habermann E (1972) Science 177(4046): 314
[31]   Habermann E (1984) Pharmacol Therap 25(2): 255
[32]   Hendrich AB et al. (2002) Cell Mol Biol Lett 7(2): 195
[33]   Hirai Y et al. (1979) Chem Pharm Bull 27(8): 1945
[34]   Hoffman DR, Jacobson RS (1984) Ann Allergy 52(4): 276
[35]   Hoffman DR, Wood CL (1984) J Allergy Clin Immunol 74(1): 93
[36]   Holzinger F et al. (1992) FEBS Lett 314(3): 477
[37]   Ionete RE et al. (2013) Prog Cryogen Isotopes Sep 16(2): 97
[38]   Isbister GK, Whelan PI (2000) Med J Aust 173(11–12): 654

[39]   Jablonka W et al. (2019) Sci Rep 9(1): 5340
[40]   Karras DJ et al. (1996) Am J Emerg Med 14(5): 478
[41]   Kazimírová M et al. (2001) Pathophysiol Haemost Thromb 31(3–6): 294
[42]   Kazuma K et al. (2017) Toxins 9(10): 323
[43]   Konno K et al. (2002) Toxicon 40(3): 309
[44]   Lall SB, Davies DM (1971) J Med Entomol 8(6): 700
[45]   Langner KFA et al. (2007) J Med Entomol 44(2): 238
[46]   Lehiy CJ, Drolet BS (2014) PeerJ 2 e426
[47]   Leong RL et al. (2015) Neurochem Res 40(10): 2078
[48]   Li YM, Casida JE (1992) Proc Natl Acad Sci U S A 89(24): 11867
[49]   Li S et al. (2001) Insect Mol Biol 10(1): 69
[50]   Lorenz M et al. (1993) Angew Chem – Int Ed Engl 32(6): 912
[51]   Malcolm SB (1990) Chemoecology 1(1): 12
[52]   Matysiak J et al. (2011) J Pharm Biomed Anal 54(2): 273
[53]   Mebs D et al. (2000) Chemoecology 10(4): 201
[54]   McLachlan JB et al. (2008) Nat Med 14(5): 536
[55]   Miller JR, Mumma RO (1976) J Chem Ecol 2(2): 115
[56]   Moed L et al. (2001) Arch Dermatol 137(10): 1357
[57]   Moore LV, Scudder GGE (1986) J Insect Physiol 32(1): 27
[58]   Moreau ME et al. (2005) J Pharmacol Sci 99(1): 6
[59]   Müller C (2009) Phytochem Rev 8(1): 121
[60]   Musso JJ et al. (1978) Ann Soc Entomol Fr 14(2): 177
[61]   Nahrstedt A, Davis RH (1986) Phytochemistry 25(10): 2299
[62]   Nakajima T et al. (1985) Peptides 6(Suppl. 3): 425
[63]   Naumann C et al. (2002) Proc Natl Acad Sci U S A 99(9): 6085
[64]   Neumann J et al. (1995) J Pharmacol Exp Ther 274(1): 530
[65]   Nikbakhtzadeh MR et al. (2007) J Insect Physiol 53(9): 890
[66]   Noeske-Jungblut C et al. (1995) J Biol Chem 270(48): 28629
[67]   Oelrichs PB et al. (1999) Toxicon 37(3): 537
[68]   Opitz SEW, Müller C (2009) Chemoecology 19(3): 117
[69]   Orivel J et al. (2001) J Biol Chem 276(21): 17823
[70]   Ozaki Y et al. (1990) Biochem Biophys Res Commun 170(2): 779
[71]   Paull BR et al. (1977) J Allergy Clin Immunol 59(4): 334
[72]   Pérez-Molina JA, Molina I (2018) Lancet 391(10115): 82
[73]   Peter MG et al. (1977) Helvetica Chim Acta 60(8): 2756
[74]   Petschenka G et al. (2013) Proc Royal Soc B – Biol Sci 280 1759 20123089
[75]   Piek T (1991) Toxicon 29(2): 139
[76]   Piek T et al. (1991) Comp Biochem Physiol Part – C: Toxicol Pharmacol 99(3): 481
[77]   Prestwich GD (1983) Annu Rev Ecol Syst 14 287
[78]   Prestwich GD et al. (1977) J Chem Ecol 3(5): 579
[79]   Prischmann DA, Sheppard CA (2002) Am Entomol 48(4): 208
[80]   Ribeiro JMC et al. (2007) BMC Genom 8 6
[81]   Ribeiro JMC, Francischetti IMB (2003) Annu Rev Entomol 48(1): 73
[82]   Ribeiro JMC et al. (2010) Insect Biochem Mol Biol 40(11): 767
[83]   Ribeiro JMC, Nussenzveig RH (1993) J Exp Biol 179 273
[84]   Rothschild M et al. (1972) Insect Biochem 2(7): 334
[85]   Russo AJ et al. (1983) Toxicon 21(1): 166
[86]   Santi F, Maini S (2006) Bull Insectology 59(1): 53
[87]   Schievano E et al. (2003) J Am Chem Soc 125(50): 15314

[88] Schildknecht H et al. (1968) Zeitschrift für Naturforschung 23b: 1213
[89] Schildknecht H et al. (1966) Angew Chem – Int Ed Engl 5(4): 421
[90] Schmidt JO (1982) Annu Rev Entomol 27 339
[91] Schmidt JO, Blum MS (1978) Science 200(4345): 1064
[92] Schmidt JO et al. (1986) Toxicon 24(9): 907
[93] Sendur N et al. (1999) Dermatology 199(4): 353
[94] Sherman RA (1995) West J Med 163(6): 541
[95] Sobotka AK et al. (1976) J Allergy Clin Immunol 57(1): 29
[96] Soliman C et al. (2019) PLoS ONE 14(10): e0224028
[97] Szolajska E et al. (2004) Eur J Biochem 271(11): 2127
[98] Terwilliger TC, Eisenberg D (1982) J Biol Chem 257(11): 6010
[99] Touchard A et al. (2016) Toxins 8(1): 30
[100] Tursch B et al. (1973) Tetrahedron Lett 14(3): 201
[101] Valenzuela JG et al. (2002) Insect Biochem Mol Biol 32(9): 1101
[102] Vincent JP et al. (1975) Biochemistry 14(11): 2521
[103] von Sicard NA et al. (1989) Toxicon 27(10): 1127
[104] Walker AA et al. (2017) Mol Cell Proteomics 16(4): 552
[105] Walker AA et al. (2016) Toxins 8 43
[106] Wanandy T et al. (2015) Toxicon 98 54
[107] Welzel KF et al. (2018) Sci Rep 8(1): 1477
[108] Wu J et al. (2014) PLoS ONE 9(8): e103823
[109] Yi GB et al. (2003) Int J Toxicol 22(2): 81
[110] Zorilă B et al. (2020) Toxins 12(11): 705

3.2.10 Crustacea

There are only a few Arthropoda of the subphylum Crustacea that have been identified as poisonous or venomous animals. Eating crabs, crayfish, or lobsters which had lived in stagnant and nutrient-rich water in which dinoflagellates and other microorganisms thrive may be harmful because these microorganisms may produce toxins that accumulate in crustaceans through the food web. Land-dwelling crabs may consume plant material that is rich in certain defensive toxins. Thus, saxitoxin (see Section 2.9.14) or cardenolides (see Section 2.5.2) may be present in such individuals which, when consumed by humans, may induce gastrointestinal or, in some cases, cardiological problems.

> ⚑ A 63-year-old female patient from New Caledonia with a previous history of cardiovascular and metabolic dysfunctions who had eaten of the coconut crab, *Birgus latro* (Ph. 3.64), was brought to the hospital because she had serious cardiac problems (first-degree atrioventricular block and atrial pauses). Analyses of serum and urine samples indicated that she suffered from cardenolide poisoning. The patient was given 760 mg of digoxin-specific Fab antibody fragments. Shortly after the infusion the electrocardiogram improved and the patient survived [2].

Such cases of passive toxicity in crustaceans may also occur through microbial toxins like saxitoxin. Consumption of contaminated crab meat results in 'diarrhetic shellfish poisoning'. Gastrointestinal problems and paralysis of cardiac and ventilatory muscles

Ph. 3.64: A giant coconut crab or 'robber crab', *Birgus latro* (Source: Olivier Lejade, Wikimedia, CC-BY-SA).

may occur as saxitoxin is a potent blocker of voltage-gated sodium channels and inhibits the generation of action potentials in excitable cells (neurons and muscle cells) [1].

Ectoparasitic crustaceans may be considered venomous as they release chemical mediators into the host skin wounds. Examples are fish lice (Argulidae). Investigations on the nutritive strategy of adults of *Argulus japonicus* have revealed that they are obligated blood feeders [6]. Although not studied yet, this feeding strategy requires the production and secretion of anticoagulants into the host skin wound.

The juvenile stages of the gnathiid isopod, *Paragnathia formica*, are hematophagous ectoparasites of fishes and secrete three subtypes of salivary trypsin inhibitors (18, 21, and 22 kDa) and a strong anticoagulant into their hosts [3].

Another class of Crustacea, the Remipedia, includes species which have been recognized as venomous animals as well. Remipedia are up to 5 cm long blind animals living in coastal aquifers containing saline groundwater. The multisegmented animals swim on their backs using multiple lateral appendages as propelling devices. The heads of the predatory animals carry fangs (maxillules) connected to secretory glands generating a combination of digestive enzymes and toxins [4]. When grabbing prey, the pointed fangs penetrate the integument and venoms are injected into the body cavity of the victim. The venom apparatus and its functions have been studied in detail in the remipede *Xibalbanus tulumensis* showing that remipedes can inject their venom into prey organisms in a controlled manner. Transcriptomic profiling of the venom glands revealed the presence of a diversity of enzymes, especially proteases as well as a paralytic neurotoxin [5].

References

[1] Catterall WA (1980) Annu Rev Pharmacol Toxicol 20 15
[2] Maillaud C et al. (2012) Toxicon 60(6): 1013
[3] Manship BM et al. (2012) Parasitology 139(6): 744
[4] van der Ham JL, Felgenhauer BE (2007) J Crustacean Biol 27(1): 1
[5] von Reumont BM et al. (2014) Mol Biol Evol 31(1): 48
[6] Walker PD et al. (2011) Crustaceana 84(3): 307

3.2.11 Fishes (Pisces)

There are probably more than 2,900 species of venomous fish [26, 27] and many species of poisonous fish worldwide. Venomous fish possess a venom apparatus that may be used for defense (most cases) or against potential prey (only a few cases) [12]. Poisonous fish impregnate their body surface (toxic mucus) or selected tissues within the body with toxins which they have produced themselves or taken up from the environment via the food web or from (endo)symbiontic microorganisms.

The trait of being venomous has been evolutionarily developed multiple times in different taxa of fish (Tab. 3.7). Molecular studies have revealed that venom glands and venom applicators have evolved four times in cartilaginous fish, once in eels (Anguilliformes), once in catfish (Siluriformes), and 12 times in the other taxa of Osteichthyes. Venom application occurs through dorsal spines (95% of all venomous fish species), venomous fangs (2%), cleithral spines (2%), or opercular or subopercular spines (1%) [26]. Besides being useful in defense against predators, venom systems in fish may have other functions as well. The venom of the fangblenny, *Meiacanthus grammistes*, which is transferred to target organisms through grooved canine teeth, causes hypotension, neurological, and proinflammatory effects. These may act to disorient attackers, but may also play important roles in intra- and interspecific competition. Blennies are known for being strongly territorial. They fight aggressively for their plots and try to hit competitors with their venomous spines. Inducing disorientation and drowsiness in a competing fish would increase the probability of the competitor becoming prey of larger fish which effectively removes the competitor from the area [12].

The spiny dogfish (*Squalus acanthias*; Ph. 3.65) carries two grooved dorsal spines which are derived from placoid scales and located right in front of the dorsal fins. When a dogfish is attacked it flicks its tail using the massive body muscles and rams these spines into the skin of its opponent. Venom is effectively transferred from the reservoir of the venom gland into the skin wound through the grooves alongside the spines. Stings are very painful and cause local erythema and edema [7]. When injected in mice, venom gland extracts induce initial hyperactivity followed by lethargic behavior and hindleg paralysis. The composition of the venom has not been studied in detail.

Tab. 3.7: Selected fish taxa including venomous fish species.

Systematic position			Example species	Common name
Chondrichthyes 'cartilaginous fish'	Myliobatiformes	Dasyatidae	*Dasyatis pastinaca*	Common stingray
		Urolophidae	*Urolophus fuscus*	Banded stingray
		Myliobatidae	*Myliobatis* sp.	Eagle ray
	Squaliformes	Squalidae	*Squalus acanthias*	Spiny dogfish
Osteichthyes 'bony fish'	Anguilliformes	Monognathidae	*Monognathus ahlstromi*	Paddletail onejaw
	Siluriformes	Plotosidae	*Plotosus lineatus*	Striped eel catfish
		Ariidae	*Netuma bilineata*	Bronze catfish
		Ictaluridae	*Noturus gyrinus*	Tadpole madtom
	Trachiniformes	Trachinidae	*Trachinus draco*	Greater weeverfish
			Echiichthys vipera	Lesser weeverfish
	Acanthuriformes	Acanthuridae	*Ctenochaetes strigosus*	Spotted sturgeonfish
			Acanthurus coeruleus	Atlantic blue tang
	Blenniformes	Blenniidae	*Meiacanthus grammistes*	Striped blenny
	Gobiesociformes	Gobiesocidae	*Acyrtus artius*	Papillate clingfish
			Arcos nudus	Padded clingfish
	Perciformes	Scorpaenidae	*Scorpaena scrofa*	Red scorpionfish
			Pterois volitans	Red lionfish
			Scorpaenopsis gibbosa	Humpback scorpionfish
			Dendrochirus zebra	Zebra turkeyfish
			Sebastes norvegicus	Atlantic redfish
		Synanceiidae	*Synanceia horrida*	Estuarine stonefish
			Apistus carinatus	Ocellated waspfish
			Inimicus japonicus	Devil stinger
	Trachiniformes	Uranoscopidae	*Astroscopus guttatus*	Northern stargazer
	Batrachoidiformes	Batrachoididae	*Thalassophryne maculosa*	Cano toadfish

Ph. 3.65: The spiny dogfish, *Squalus acanthias* (Source: US National Oceanic and Atmospheric Administration, Wikimedia, public domain).

The tail of a stingray, e.g., in the southern stingray, *Hypanus americanus* (Ph. 3.66), is decorated with an erectable and barbed spine that may well reach a length of 10 cm in some species. When disturbed, the stingray may flip its tail upward and hurts its attacker by inflicting deep skin wounds. Rupture of the epithelial sheeth covering the spine results in transfer of venom into the victim's tissue. These spines work so effectively in penetrating human skin that they have been used by the Mayas in bloodletting rituals [11].

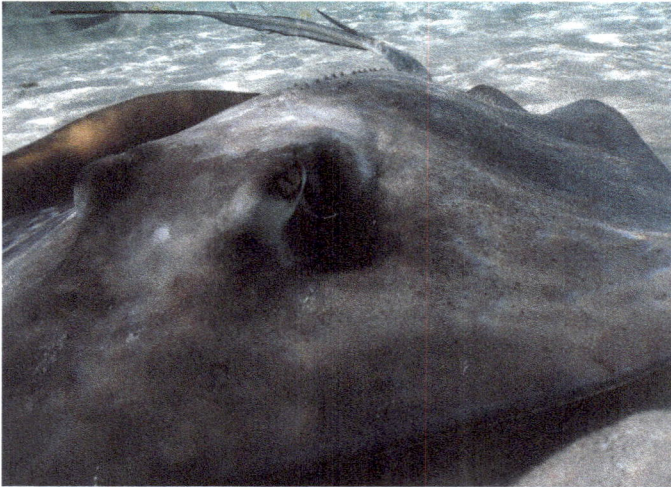

Ph. 3.66: The southern stingray, *Hypanus americanus*, from the Caribbean (Source: Rob Atherton/iStock/Getty Images Plus).

🪱 The composition of stingray venom is not well investigated due to the instability of the venom components. Stings are very painful as the venom contains serotonin. Cardiac arrhythmia and sudden hypotension may occur in victims. Stings may result in tissue necrosis due to the presence of proteases in the venom. Complications may occur by secondary bacterial infection. Human deaths have been reported after powerful stings occurred to the chest.

Venomous spines in bony fish are generally derived from skeletal elements [32]. The lateral spines associated with the operculum (e.g., in the Trachinidae) or the cleithrum as well as the dorsal spines which may or may not be integrated into the dorsal fin (e.g., in the Scorpaenidae) are defensive weapons. If predators try to catch these fishes, they erect or spread out the spines which have pointed ends and may penetrate the skin of the attacker during the fight. Venom is released from the reservoir of the venom gland at the base of the spines. Humans may get hurt if they step on such a fish or try to catch it. In many cases the brittle spines break upon impact so that parts remain in the wound. Secondary bacterial infections occur frequently in such cases.

Besides carrying dorsal spines, some venomous catfish species have spines with serrated anterior edges in their pectoral fins. These fins may be spread out in an attempt to fend off an attacker but can also be used to hurt an opponent when the fish points the spines forward and suddenly advances toward its victim. The venom reservoirs associated with the spines are covered by a thin layer of epithelium which ruptures upon impact so that venom is released into the skin wound of the opponent.

Only a few fish species use fangs to deliver venoms to other organisms, the jawed eels (Monognathidae) and lampreys (Petromyzonidae). An example is the paddletail one-jaw (*Monognathus ahlstromi*), a small (~10 cm body length) anguilliform deep sea fish living in the North Pacific. The lower jaw of this fish is well developed and carries several small teeth while the upper jaw is completely reduced in the adults. There is only one large tooth originating from the ethmoidal bone (Fig. 3.29) that is hollow and connected to a pair of ethmoidal glands that are supposed to produce venom [24]. It is unknown whether this apparatus is used for obtaining prey or rather serves defensive purposes.

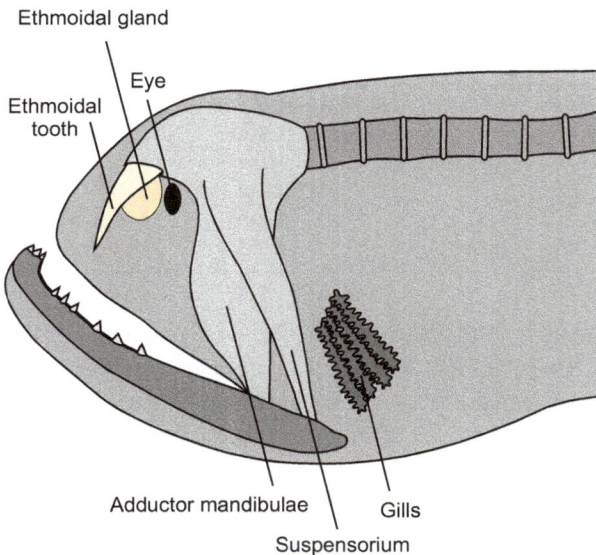

Fig. 3.29: Scheme of the head of *Monognathus ahlstromi* (redrawn after [24]).

Generally, fish venoms are proteins and have cardiovascular, neuromuscular, and inflammatory effects in target organisms. In addition, they contain cytotoxins and pain-inducing agents [15, 32]. Table 3.8 shows an overview of identified fish toxins and their effects in target organisms.

Tab. 3.8: Toxins of selected venomous fish and their effects in target organisms.

Fish family	Species	Toxin	Major effects in target organisms
Plotosidae (catfish)	Plotosus lineatus	Plototoxin	Hypotension
	Plotosus canius	Toxin-PC	Cytolysis
			Lethality
			Cardiac arrest
			Lethality
Synanceiidae (stonefish)	Synanceia horrida	Stonustoxin	Neurotoxicity
	Inimicus japonicus	Ij-Tx	Cardiac problems
			Hypotension
			Lethality
			Cytolysis
			Cytolysis
Scorpaenidae (scorpionfish, lionfish)	Dendrochirus zebra	Dz-Tx	Cardiac problems
	Scorpaena plumieri	Sp-CTx	Cytolysis
	Pterois volitans	Pv-Tx	Cardiac problems
	Paracentropogon rubripinnis	Pr-Tx	Hypotension
			Cytolysis
			Cytolysis
Trachinidae (weeverfish)	Trachinus draco	Dracotoxin	Cytolysis
	Echiichthys vipera	Trachinine	Lethality
			Hypotension
			Cardiac problems
			Cytolysis
			Lethality

Cardiovascular effects are observed in animals which have been envenomated by the zebra turkeyfish (*Dendrochirus zebra*; Ph. 3.67). Heart and ventilation rates are slowed down. The LD_{50} (mouse, i.v.) is 1.1 mg/kg BW. A fall in blood pressure has been observed in animals affected by toxins of scorpionfish as *Scorpaena scrofa* or *Apistus carinatus* (Ph. 3.68). Toxins of stonefishes, *Synanceia verrucosa* (Ph. 3.69) or *S. horrida*, have hypotensive effects (relaxation of vascular smooth muscle induced by nitric oxide released from endothelial cells [20]), but may also inhibit the electrical conductivity in atrioventricular heart muscle cells. Even paralysis of skeletal muscle has been observed. The LD_{50} (i.v., mice) of crude venom preparations of *S. horrida* is 0.4–0.6 µg/kg [25]. Thus, this species is considered one of the most toxic fish. Similar effects are induced by venom components of weeverfish (*Trachinus* sp.; Ph. 3.70). Stings by weeverfish may additionally induce necrotic effects around the sting site. The LD_{50} of weeverfish toxin in mice (i.v.) is 0.1 mg/kg BW. Catfish venoms are assumed to be homologs of defensive toxins of skin glands [31]. They have neurotoxic and hemolytic properties and induce severe pain, paralysis in skeletal muscles, or respiratory distress in affected animals and humans. An extract obtained from venom glands of the catfish *Plotosus canius* (Ph. 3.71) was toxic to mice with an LD_{50} of 0.2 mg/kg BW [1].

Ph. 3.67: The zebra lionfish, *Dendrochirus zebra* (Source: Jens Petersen, Wikimedia, GNU-CC-BY).

Ph. 3.68: The longfin waspfish, *Apistus carinatus* (Source: Rickard Zerpe, Wikimedia, CC-BY-SA).

Ph. 3.69: The stonefish, *Synanceia verrucosa* (Source: Rob Atherton/iStock/Getty Images Plus).

Ph. 3.70: The greater weeverfish, *Trachinus draco* (Source: wrangel/iStock/Getty Images Plus).

☠ Stonefish stings cause extreme pain, swelling that spreads over the affected extremity, nausea, vomiting, diarrhea, chills, sweating, and tachycardia. Skin necroses around the sting site do not heal easily. Pain can prevail for weeks. Fatalities due to stonefish stings occur frequently.

Cytolysis and tissue necrosis upon envenomation by fish toxins are generally associated with the actions of proteases or pore-forming/membrane damaging toxins [16] as

Ph. 3.71: The Gray eel-catfish, *Plotosus canius* (Source: Vaikoovery, Wikimedia, GNU-CC-BY).

phospholipases do not occur in fish venoms [4] with the exception of fangblennies whose venoms contain phospholipases A_2 [5].

Blenny venom contains unusual defense peptides. One of them blocks opioid receptors in target organisms which induces central anesthesia in the target organism. The resulting loss in pain sensitivity and a concomitant fall in blood pressure brought about by the actions of neuropeptide Y and enkephalins in blenny venom irritate the predator so that the blenny has a chance to escape [5]. The opioid peptide is currently under investigation for a potential use as a painkiller in medicine.

The toadfish, *Thalassophryne nattereri* (Batrachoididae), uses sharp hollow spines on the first dorsal fin and on the operculum for its defense against larger predators [19]. The spines are connected to venom glands that produce proteins of the natterin family of which members occur also in other animal taxa and may have completely different functions [17]. Natterin in venom of *Thalassophryne nattereri* induces inflammatory reactions at the sting site in target organisms [18] by acting as an aerolysin-like pore-forming toxin and destroying cells. This results in neutrophil activation, inflammation, and tissue necrosis. Consequently, toadfishes are avoided by predatory fish species.

The presence of natterins in venom glands associated with spines in *Thalassophryne nattereri* as well as in skin glands of several fish species provides a nice example for certain continuities between venoms and poisons. The oriental catfish *Plotosus lineatus* contain proteinaceous toxins (toxins I and II with 35 or 37 kDa, respectively) in its defensive skin secretions which are natterin-like molecules [29]. These toxins induce the formation of edema, inflammation, and tissue destruction when they get in contact with mucous membranes of target animals. The catfish uses these substances passively, i.e., they are only applied to other organisms when these attack the catfish. In this context, the natterins are clearly poisons.

Another impressive example of a poisonous fish is the Japanese pufferfish, *Takifugu rubripes* (Ph. 3.72). In its original habitat, the internal organs and the skin contain high concentrations of tetrodotoxin (see Section 2.9.7) [21], which is a highly efficient ($K_d = 10^{-19}$ mol/L) blocker of voltage-gated sodium channels in excitable cells in humans and animals [3, 14] and is highly toxic (LD_{50} (mouse, p.o.) = 10 µg/kg BW). Tetrodotoxin is synthesized by symbiontic bacteria of the strains *Vibrio, Bacillus, Micrococcus, Acetinobacter,* or *Alteromonas*) which the fish acquires from the environment. When pufferfishes are raised in the absence of these bacteria, they are free of tetrodotoxin. It is absolutely deadly for a predator (including humans) to ingest a pufferfish as a whole. However, skilled cooks are able to prepare thin slices of pufferfish muscle which contains only moderate amounts of tetrodotoxin for human consumption. The slices of uncooked pufferfish muscle ('fugu') are considered a delicacy by Japanese gourmets. Consumption induces tingling sensations in mouth and tongue.

Ph. 3.72: The Japanese puffer fish, *Takifugu rupripes* (Source: Taku_S/iStock/Getty Images Plus).

Other cases of secondary toxicity in fish can be observed in groupers (Serranidae) (Ph. 3.73). These large fish stand at the upper end of the marine food web in tropical and subtropical regions. Shortly after big storms or human construction activities have severely disturbed local coral reefs it occurs that these fish become poisonous due to the accumulation of toxins in their tissues [13]. The toxins are highly poisonous polyethers, brevetoxin (see Section 2.3.4), ciguatera toxin, or maitotoxin (see Section 2.3.10). They are produced by Dinoflagellata (*Gambierdiscus toxicus* or *Karenia brevis*) which show mass reproduction in disturbed ecosystems. The toxins delay the open times of voltage-activated cation channels in excitable cells of sensitive animals

Ph. 3.73: A grouper (Serranidae) (Source: Lingbeek/iStock/Getty Images Plus).

and humans resulting in overexcitation of neurons in the central nervous system, muscle cramps, and spastic paralysis. Maitotoxin is the most toxic nonprotein compound in nature: LD_{50} (mouse, p.o.) = 170 ng/kg BW [10].

Some fish species enrich the mucus of the body surface or the inner organs with self-made toxic substances. The secreted toxins that stay associated with the mucus at the body surface are named 'ichthyocrinotoxins', those that accumulate in internal organs like gonads or are used to impregnate the gametes are called 'ichthyotoxins'. Ichthyocrinotoxins have been isolated and characterized from the skin secretions of the smooth trunkfish, *Lactophrys triqueter* (Ph. 3.74), and other family members of the Ostraciidae.

Ph. 3.74: A smooth trunkfish, *Lactophrys triqueter* (Source: johnandersonphoto/iStock/Getty Images Plus).

The toxins are β-substituted choline chloride esters (acetoxy-, butyryloxy-, valeryloxy-, or caproyloxy-) of palmitic acid, so-called pahutoxins [8, 9]. Pahutoxins (see Section 2.8.3) interact with cell membranes in the target organisms and are potent cytolysins [2].

The Moses sole, *Pardachirus marmoratus* (Pleuronectiformes; Soleidae) (Ph. 3.75), lives in the Red Sea and repels predatory sharks by secreting a milky fluid from dorsal skin glands [6]. These secretions contain a 34 amino acid peptide called 'pardaxin' and several derivatives of steroidal glycosides named 'pavoninines' (Fig. 3.30). Using hound sharks (*Mustelus griseus*) as experimental animals, researchers obtained evidence that the peptide pardaxin is probably an antifeedant that acts on the gustatory sense of the shark, while the pavoninines act as repellents by stimulating the shark olfactory sense [28]. The pavoninines are saponins and may harm cell membranes in surface tissues of predators. Thus, they are also cytotoxic substances [30]. Perfusion of the buccal cavity and the gills of a dogfish shark (*Squalus acanthias*) with solutions containing pardaxin resulted in increases in epithelial urea permeability [23] indicating

Ph. 3.75: The Moses sole, *Pardachirus marmoratus* (Source: Vitalii Kalutskyi/iStock/Getty Images Plus).

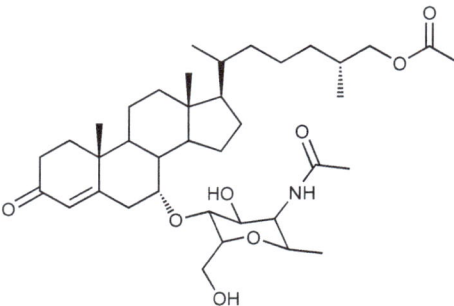

Fig. 3.30: Pavoninine-1 in skin secretions of the Moses sole, *Pardachirus marmoratus*.

that pardaxin damages epithelial cells as well. In addition to cytotoxic actions on vertebrate cells, pardaxin may also have antimicrobial potencies [22].

References

[1]   Auddy B et al. (1995) Nat Toxins 3(5): 363
[2]   Boylan DB, Scheuer PJ (1967) Science 155(3758): 52
[3]   Bucciarelli GM et al. (2021) Toxins 13(8): 517
[4]   Campos FV et al. (2021) Toxins 13(12): 877
[5]   Casewell NR et al (2017) Curr Biol 27(8): 1184
[6]   Clark E, George A (1979) Environ Biol Fish 4(2): 103
[7]   Evans HM (1920) Br Med J 1(3087): 287
[8]   Goldberg AS et al. (1988) Toxicon 26(7): 651
[9]   Goldberg AS et al. (1982) Toxicon 20(6): 1069
[10] Gusovsky F, Daly JW (1990) Biochem Pharmacol 39(11): 1633
[11] Haines HR et al. (2008) Lat Am Antiq 19(1): 83
[12] Harris RJ, Jenner RA (2019) Toxins 11(2): 60
[13] Holmes MJ et al. (2021) Toxins 13(8): 515
[14] Katikou P et al. (2022) Mar Drugs 20(1): 47
[15] Kiriake A et al. (2013) Toxicon 70 184
[16] Kreger AS (1991) Toxicon 29(6): 733
[17] Lima C et al. (2021) Toxins 13(8): 538
[18] Lima C et al. (2021) Int Immunopharmacol 91 107287
[19] Lopes-Ferreira M et al. (2014) J Venom Anim Toxins Incl Trop Dis 20 35
[20] Low KSY et al. (1993) Toxicon 31(11): 1471
[21] Noguchi T, Arakawa O (2008) Mar Drugs 6(2): 220
[22] Oren Z, Shai Y (1996) Eur J Biochem 237(1): 303
[23] Primor N (1985) Experientia 41(5): 693
[24] Raju SN (1974) Fish Bull 72(2): 547
[25] Saggiomo S et al. (2021) Mar Drugs 19 302
[26] Smith WL et al. (2016) Integr Comp Biol 56(5): 950
[27] Smith WL, Wheeler WC (2006) J Heredity 97(3): 206
[28] Tachibana K et al. (1984) Science 226(4675): 703
[29] Tamura S et al. (2011) Toxicon 58(5): 430
[30] Williams JR, Gong H (2004) Lipids 39(8): 795
[31] Wright JJ (2009) BMC Evol Biol 9(1): 282
[32] Ziegman R, Alewood P (2015) Toxins 7(5): 1497

3.2.12 Amphibia

Amphibia generally have soft integuments which provide only limited mechanical
protection against predators. As the integument is relatively thin and usually moist,
amphibians ought to protect themselves against being colonized by microorganisms.
Numerous skin glands produce secretions containing toxins or antibiotics to repel

predators or inhibit the growth of microorganisms, respectively. Skin secretions fulfilling these tasks have been detected in the orders Anura/Salientia (frogs and toads) and Urodela/Caudata (newts, salamanders). The biological active ingredients are generally biogenic amines, steroids, alkaloids as well as peptides. Antimicrobial peptides from amphibian skin glands are effective against microorganisms in concentrations as low as 10^{-5}–10^{-3} mol/L.

3.2.12.1 Frogs and Toads (Anura)

Large numbers of different toxic peptides from the skin secretions of frogs and toads have been isolated and at least partially characterized. Some examples are listed in Tab. 3.9. When these peptides are absorbed in the gastrointestinal tract, through other mucous epithelia or through skin wounds of predators they unfold diverse unwanted reactions like blood pressure imbalances, histamine release, pain sensations, or numbness [2, 7–9, 14, 18, 22, 25, 26, 29, 33, 35–37]. These effects protect these animals efficiently against attacks by predators or from being colonized by microorganisms.

Tab. 3.9: Selected toxic peptides in skin secretions of frogs and toads and their effects in target animals.

Peptide	Amino acids	Producing species	Species distribution	Properties
Protirelin	3	*Bombina orientalis*	Northeast of China	Identical with mammalian thyreoliberin (TRH), may induce large changes in blood pressure
Dermorphin, Deltorphin	7	*Phyllomedusa rhodei* *Phyllomedusa bicolor*	South America	Long-lasting peripheral and central opiate-like activities, high affinity for μ- or δ-opioid receptors
Bradykinin	9	*Rana temporaria* *Pelophylax esculentus*	Europe	Vasodilation, induction of pain sensations
Caerulein	10	*Litoria caerulea* *Xenopus laevis*	Northern Australia South Africa	Smooth muscle contraction, untimely release of digestive enzymes, attenuation of pain reception
Physalaemin	11	*Physalaemus* sp. *Phyllomedusa* sp.	South America	Increases salivation, potent vasodilator with hypotensive effects
Granuliberin R	12	*Glandirana rugosa*	Japan	Degranulation of mast cells

Tab. 3.9 (continued)

Peptide	Amino acids	Producing species	Species distribution	Properties
Temporin-10a	13	*Rana temporaria*	Europe	Antimicrobial peptide
Ranatensin R	17	*Glandirana rugosa*	Japan	Increase in blood pressure
Kassinakinin S	17	*Kassina senegalensis*	Africa	Histamine release from mast cells
Ranacyclin E	17	*Pelophylax esculentus*	Europe	Forming pores in bacterial cell membranes, cyclic antimicrobial peptide
Maximins S	18	*Rana temporaria*	Europe	Antimicrobial peptides
Kassinatuerin-1	21	*Kassina senegalensis*	Africa	Antimicrobial peptides, cytotoxicity
Brevinins	19–23	*Clinotarsus curtipes*	India	Antimicrobial peptides, cytotoxicity
Sauvagine	40	*Phyllomedusa sauvagei*	South America	Antimicrobial peptide, structural similarity to corticotropin releasing hormone (corticoliberin)
Esculentin-1	46	*Pelophylax esculentus*	Europe	Antimicrobial peptide

Neotropical frogs of the Dendrobatidae familiy carry defensive glands on their skin. When threatened they secrete droplets of fluid that contain high concentrations of batrachotoxin (BTX) (Fig. 2.42) as well as other alkaloids like pumiliotoxins, histrionicotoxins (see Section 2.9.8), gephyrotoxins, or decahydroquinolines (see Section 2.9.10) [12]. The precursors of these substances probably originate from food organisms of these frogs (ants, mites, millipedes, or beetles) [11, 30]. Thus, the composition of the poison varies somewhat according to the diet of the frogs. Batrachotoxin (see Section 2.9.16) blocks the inactivation of voltage-activated sodium channels in excitable cells of animals and humans. The resulting overexcitation in motor neurons results in spastic paralysis of the target animal [34]. It has been reported that natives in South America catch these frogs and smear some of their skin secretions onto the tips of their hunting arrows. Using these poison arrows for hunting is very efficient as only small skin wounds inflicted by these arrows in prey animals are sufficient to induce rapid paralysis.

The question how these frogs avoid self-intoxication is still a matter of debate. Some dendrobatid frogs have mutations in a subunit of their voltage-activated sodium channels that may render these channels batrachotoxin-insensitive. However, a recent study has revealed that 'sponge proteins' like saxiphilin may sequester toxin molecules in the body fluids of the frogs and prevent the frogs from intoxicating themselves [1].

The alkaloid content of skin secretions of dendrobatid frogs does not only deter predators but may also have antimicrobial functions as well. Different skin gland alkaloids have been shown to inhibit the growth of Gram-positive (*Bacillus subtilis*) and Gram-negative bacteria (*Escherichia coli*) as well as the growth of the fungus *Candida albicans* [20]. The results indicate that skin alkaloids of poison frogs may be protective against skin infections.

Epibatidine (Fig. 2.42) is a complex alkaloid in skin secretions of the dendrobatid frog *Epipedobates anthonyi* (Ph. 3.76) [10]. This frog skin alkaloid (see Section 2.9.8) has analgesic potency that is 200-fold greater than that of morphine. Thus, it has been investigated in an attempt to develop new analgesics for use in medicine. However, complications with access and benefit-sharing regulations of the Convention on Biological Diversity as well as species identification problems have slowed down any further studies [3].

Ph. 3.76: The white-striped poison dart frog, *Epipedobates anthonyi* (Source: sandergroffen/iStock/Getty Images Plus).

The skin secretions of two Brazilian hylid frogs (*Corythomantis greeningi* (Ph. 3.77) and *Nyctimantis brunoi*) are very toxic due to the presence of proteolytic and fibrinolytic agents. Unlike other poison dart frogs these species are equipped with skull spines capable of injecting secretions of skin glands into other organisms via headbutting when feeling threatened. These frogs are truly venomous animals [17]. *C. greeningi* has larger head spines and skin glands producing larger volumes of secretion, but *N. brunoi* has the more lethal venom. The LD_{50} (mouse, i.p.) is 0.16 mg/kg BW.

Skin secretions of many frog and toad species contain catecholamines (adrenaline, noradrenaline) or bioactive analogs of serotonin like bufotenin [21] (Fig. 3.31).

Ph. 3.77: Habitus of *Corythomantis greeningi* and an SEM image of its skull (images from [17] with permission from Current Biology).

The latter has been found in skin secretions of the European green toad, *Bufotes viridis*, the Brazilian toad *Rhinella rubescens*, and the Colorado River toad, *Incilius alvarius* (Ph. 3.78). Bufotenin (see Section 2.8.6), or 5-hydroxy-*N,N*-dimethyltryptamine, is structurally related to the neurotransmitter serotonin and binds to its receptors in the central nervous system in animals or humans upon ingestion of toad skin, secretions of toad skin glands, or toad eggs. Bufotenin is a psychoactive drug and induces hallucinations [16]. The LD_{50} of bufotenin in rodents (p.o.) is 200–300 mg/kg BW.

Ph. 3.78: The Sonoran Desert toad, *Incilius alvarius* (Source: wrangel/iStock/Getty Images Plus).

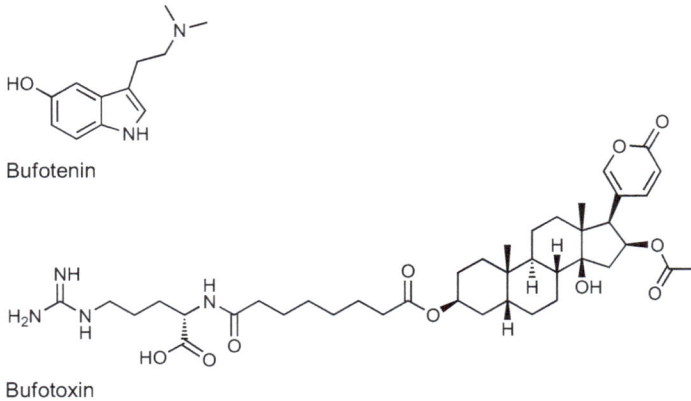

Bufotenin

Bufotoxin

Fig. 3.31: Skin gland toxins of toads.

Ph. 3.79: The European green toad, *Bufotes viridis* (Source: Zdenek Macat/iStock/Getty Images Plus).

The skin secretions of many toads (e.g., *Rhinella marina* or *Bufotes viridis*; Ph. 3.79) are rich in bufotoxin, a heat-stable steroidal compound resembling plant cardenolides (Fig. 3.31). These so-called bufadienolides inhibit the Na^+/K^+-ATPase of animal or human cells which results mainly in cardiac problems (bradycardia, cardiac arrest).

☠ Toad poisoning occurs upon ingestion of toad meat or eggs or use of allegedly aphrodisiac pills (such as 'love stone') or of traditional medicines such as 'Chan Su' [4]. Patients who had consumed toad meat develop gastrointestinal symptoms (nausea) within 2 h after ingestion. Electrocardiograms show sinus bradycardia. Shock and even cardiac arrest may happen [31]. The LD_{50} (mouse, p.o.) is 0.3–1.1 mg/kg BW. The overall mortality rate in humans who had eaten toad meat and/or toad eggs is approximately 8%.

📖 Patients intoxicated by toad poisons ought to seek immediate medical attention. Digoxin-specific antibody fragments (DsFab) may be applied to prevent cardiac arrest. Otherwise, supportive symptomatic care should be provided.

3.2.12.2 Salamanders (Urodela)

When threatened by predators, European salamanders defend themselves by secreting droplets of sticky fluid containing toxins from dorsal or lateral skin glands. In extreme cases the fluid may even be sprayed at the attacker. The secretions contain steroidal alkaloids (see Section 2.9.16). The two major alkaloids, samandarine and samandarone (Fig. 3.32), were identified in the skin secretions of the European fire salamander (*Salamandra salamandra terrestris*; Ph. 3.80) [24]. These substances are derived from cholesterol by modifications in the A and D rings (Figs. 2.23 and 2.50). These substances have predator-repellent as well as antimicrobial effects [6, 13, 19, 27]. When animals or humans ingest salamander tissues or skin secretions, these substances act on the central nervous system. Samandarine may cause seizures or partial paralysis of the skeletal muscles by as yet unkown mechanisms. The LD_{50} (mouse, s.c.) is 3.4 mg/kg BW.

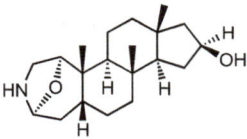

Samandarine

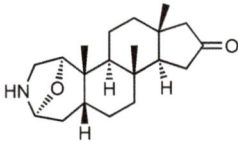

Samandarone

Fig. 3.32: Steroidal alkaloids in salamanders.

Besides cardiac glycosides, salamanders produce adhesive compounds in their skin glands. The skin secretions of four salamander species (*Plethodon shermani, Plethodon glutinosus, Ambystoma maculatum,* and *Ambystoma opacum*) have been investigated with respect to their biocompatibility using human and mouse cells as test systems. While the adhesives of *Plethodon shermani* turned out to be cytocompatible, the adhesives of the other species showed cytotoxic properties [32]. This indicates that the adhesives may result in irritation as well as avoidance reactions in predators trying to catch these amphibians.

Ph. 3.80: The European fire salamander, *Salamandra salamandra* (Source: o2beat/iStock/Getty Images Plus).

Some newt species endemic to the west of North America are particularly toxic due to the presence of tetrodotoxin (see Section 2.9.7) in the skin and in skin secretions. The toxin is probably produced by symbiontic microorganisms or is acquired from food organisms. An individual of the rough-skinned newt, *Taricha granulosa* (Ph. 3.81), contains enough tetrodotoxin to kill an adult human when the toxin is ingested or gets in contact with skin wounds [15]. Newts are resistant against the toxin but are protected from becoming prey of vertebrate predators except of garter snakes (genus *Thamnophis*, Colubridae) which are resistant against tetrodotoxin as well [5].

Ph. 3.81: The rough-skinned newt, *Taricha granulosa* (Source: Wirestock/iStock/Getty Images Plus).

3.2.12.3 Caecilians (Apoda)

Caecilians are snake-like amphibians (order Gymnophiona). As in Anura and Caudata, the caecilian skin is rich in mucous glands. Only recently, researchers investigated the skin glands of *Siphonops annulatus* (Ph. 3.82) more throroughly and found out that they resemble those of salamanders [23]. The toxins of these glands are not characterized in detail, but the bioactive fractions contain cytotoxic compounds that may also affect eukaryotic pathogens like *Leishmania* or the trypanosomes that trigger Chagas disease [28].

Ph. 3.82: The caecilian *Siphonops annulatus* (Source: I. R. Dias, Wikimedia, CC-BY).

References

[1]   Abderemane-Ali F et al. (2021) J Gen Physiol 153(9): e202112872
[2]   Abraham P et al. (2014) Biochimie 97 144
[3]   Angerer K (2011) Innovation: Eur J Soc Sci Res 24(3): 353
[4]   Bick RJ et al. (2002) Life Sci 72(6): 699
[5]   Brodie ED, Brodie ED (1990) Evolution 44(3): 651
[6]   Brodie ED et al. (1991) Biotropica 23(1): 58
[7]   Chen T et al. (2005) Biochem Biophys Res Commun 337(2): 474
[8]   Conlon JM et al. (2005) Peptides 26(11): 2104
[9]   Conlon JM et al. (2012) Gen Comp Endocrinol 176(3): 513
[10]   Daly JW et al. (2000) Nat Prod Rep 17(2): 131
[11]   Daly JW et al. (1994) J Chem Ecol 20(4): 943
[12]   Daly JW et al. (2005) J Nat Prod 68(10): 1556
[13]   Erjavec V et al. (2017) Medycyna Weterynaryjna – Veterinary Medicine – Sci Pract 73(3): 186
[14]   Erspamer V et al. (1989) Proc Natl Acad Sci U S A 86(13): 5188
[15]   Hague MT et al. (2016) Ecol Evol 6(9): 2714
[16]   Hitt M, Ettinger DD (1986) N Engl J Med 314(23): 1517
[17]   Jared C et al. (2015) Curr Biol 25(16): 2166
[18]   Kim JB et al. (2001) J Pept Res 58(5): 349
[19]   Lüddecke T et al. (2018) Sci Nat 105(9): 56
[20]   Macfoy C et al. (2005) Zeitschrift für Naturforschung C 60(11–12): 932

[21] Maciel NM et al. (2003) Comp Biochem Physiol B Biochem Mol Biol 134(4): 641
[22] Mangoni ML et al. (2003) Biochemistry 42(47): 14023
[23] Mauricio B et al. (2021) Toxins 13(11): 779
[24] Mebs D, Pogoda W (2005) Toxicon 45(5): 603
[25] Montecucchi PC et al. (1981) Int J Pept Protein Res 17(3): 316
[26] Montecucchi PC, Gozzini L (1982) Int J Pept Protein Res 20(2): 139
[27] Pereira KE et al. (2018) J Exp Biol 221(14): 183707
[28] Pinto EG et al. (2014) J Venom Anim Toxins Incl Trop Dis 20(1): 50
[29] Ponti D et al. (1999) Eur J Biochem 263(3): 921
[30] Saporito RA et al. (2007) Proc Natl Acad Sci U S A 104(21): 8885
[31] Trakulsrichai S et al. (2020) Ther Clin Risk Manage 16 1235
[32] von Byern J et al. (2017) Toxicon 135 24
[33] Wabnitz PA et al. (1999) Rapid Commun Mass Spectrom 13(24): 2498
[34] Wang SY, Wang GK (2003) Cell Signal 15(2): 151
[35] Wang T et al. (2005) Biochem Biophys Res Commun 327(3): 945
[36] Yasuhara T et al. (1979) Chem Pharm Bull 27(2): 492
[37] Yasuhara T, Nakajima T (1975) Chem Pharm Bull 23(12): 3301

3.2.13 Reptiles and Birds (Sauropsida)

Sauropsida are a monophyletic subgroup of the Tetrapoda which includes the traditional classes of reptiles (Reptilia) and birds (Aves). The reptiles are no longer accepted as a systematic group due to the fact that several lineages summarized under the term 'reptiles' have different evolutionary orgins. Moreover, the crocodiles are much closer related to birds than to other modern groups of reptiles. In the context of this book, we subdivide the Sauropsida into the Lepidosauria, which encompass lizards (Lacertilia) and snakes (Serpentes) on one hand, and on the other hand, the Archosauria with the birds (Aves) as the relevant group.

3.2.13.1 Lizards (Squamata)

Of toxicological significance among the lizards are members of the families Varanidae (monitor lizards) and Helodermatidae (beaded lizards). Monitor lizards are large lizards in the genus *Varanus*. The most prominent member of this genus is the Komodo dragon, *Varanus komodoensis* (Ph. 3.83). These predatory and scavenging animals produce saliva that has toxic ingredients which may assist them in hunting. Among the salivary proteins are enzymes, e.g., kallikrein and phospholipase A_2 (PLA$_2$), and proteins that are subdivided in natriuretic peptides, the AVIT peptides (cysteine-rich peptides with approx. 80 amino acid residues), and the CRISP proteins.

Kallikreins are serine proteases that cleave kininogens, thereby producing peptides like bradykinin which induce pain sensations in affected organisms [86] and may stun the lizard's prey. PLA$_2$ inhibits platelet aggregation and provokes long-lasting bleeding. In addition, PLA$_2$ assists in destruction and digestion of animal tissues [2]. Natriuretic peptides induce hypotension, shock, and loss of consciousness in bitten

Ph. 3.83: The Komodo dragon, *Varanus komodoensis* (Source: ANDREYGUDKOV/iStock/Getty Images Plus).

animals [38]. The AVIT peptides carry a typical N-terminal amino acid sequence (Ala-Val-Ile-Thr). They bind to selected G protein-coupled receptors in excitable cells of the victim and induce hyperalgesia and contractions in smooth muscle (painful cramping) [38, 59]. The cysteine-rich secretory glycoproteins (CRISP proteins) in the lizard's saliva induce hypothermia in bitten animals and interfere with signal transduction processes in neuronal cells by inhibiting cyclic nucleotide-gated calcium channels [38].

Only a few lizard species are venomous. The Gila monster (*Heloderma suspectum*) and the Mexican beaded lizard (*Heloderma horridum*) are the only known venomous lizards. These animals live a secret life in the southwest of the US or in Mexico, respectively, sitting in dens or under stones during the day and foraging at night. The preferred food is bird or reptile eggs that they search using their sensitive senses of smell and taste. Occasionally, they prey on small birds, mammals, frogs, other lizards, or insects. The lizards are usually not aggressive but may use their strong jaws to bite intruders disturbing them in their resting places. The venom glands of *Heloderma* sp. are modified salivary glands located on either side of the reptile's lower jaw. Each gland has separate ducts leading to the bases of grooved teeth on each side of the lower jaw (Fig. 3.33). When biting, the lizards hang on to the victim and chewing movements of their teeth massage the venomous saliva into the wounds.

The venom contains enzymes and small molecule mediators [106]. A hyaluronidase loosens the victim's tissues at the bite site so that the other venom components can easily diffuse away from the bite site and into the systemic circulation. Serotonin and kallikrein (gilatoxin of *H. suspectum*, 30 kDa [135] are mediators of intense pain sensations. An L-amino acid oxidase and a phospholipase A may be instrumental for digestive processes.

Fig. 3.33: Head of *Heloderma suspectum* with scheme of the venom gland and grooved teeth in the lower left jaw (image of *H.s.*: MonsterDoc, Wikimedia, CC-BY-SA).

Several small peptides have been detected in the venoms of *Heloderma* species: helospectin (38 amino acid residues [137]), helodermin (35 amino acid residues [109]), and exendin-3 and -4 (39 amino acid residues, each) which contain amino acid sequences similar to the mammalian peptide hormones vasoactive intestinal peptide (VIP), secretin, or glucagon-like peptide-1 (GLP-1), respectively (Fig. 3.34). *Heloderma* venom peptides have the potential for being used as pharmaceuticals. Exendin-4 is an agonist for the glucagon-like peptide-1 receptor and exhibits antidiabetic and antitumor effects [22]. It is used as an antidiabetic drug [47].

Alignment of human vasoactive intestinal peptide (VIP) with *Heloderma* helospectin:

VIP, *Homo*	HSDAVFTDNYTRLRKQMAVKKYLNSILN
Helospectin, *Heloderma*	HSDATFTAEYSKLLAKLALQKYLESILGSSTSPRPPSS

Alignment of human secretin with *Heloderma* helodermin:

Secretin, *Homo*	HSDGTFTSELSRLREGARLQRLLQGLV
Helodermin, *Heloderma*	HSDAIFTQQYSKLLAKLALQKYLASILGSRTSPPP

Alignment of human GLP-1 with *Heloderma* exendin-4:

GLP-1, *Homo*	HAEGTFTSDVSSYLEGQAAKEFIAWLVKGR
Exendin-4, *Heloderma*	HGEGTFTSDLSKQMEEEAVRLFIEWLKNGGPSSGAPPPS

Fig. 3.34: Sequence alignments of *Heloderma* venom peptides with human peptide hormones (one-letter amino acid code). Identical amino acids in the respective positions of the peptides are indicated by red letters.

Helothermin in the venom of *Heloderma horridum* is a 25.4 kDa protein containing eight S-S bridges. It is a blocker of ryanodine receptor channels in heart and skeletal muscle and thus inhibits the contractility of these organs [87]. The LD_{50} (mouse, i. v.) is 0.135 mg/kg BW.

Gila monsters and beaded lizards are immune to the effects of their own venoms.

> 🦎 Envenomation by a Gila monster in vertebrates including humans results in excruciating pain that may persist for 24 h, a rapid fall in blood pressure, respiratory irregularities, tachycardia, hypothermia, edema, and hemorrhage in the gastrointestinal tract, the lungs, eyes, liver, and the kidneys. The amount of venom which is transferred during a Gila monster bite is generally not fatal to healthy adult humans, although a bite can be life-threatening for people with preexiting heart problems. Symptomatic treatment of the patient is usually sufficient. An antiserum (antidote) is not available.

3.2.13.2 Snakes (Serpentes)
Of the 3,000 species of snakes (suborder Serpentes) approximately 450 are venomous and potentially dangerous for humans. The venomous snakes belong to the following four families:
− Colubridae (vipers), living mainly on the African continent;
− Elapidae, distributed in Asia, Australia, The Americas, and Africa;
− Viperidae (vipers), living in Europe, Asia, and Africa;
− Lamprophiidae, found mostly in Africa, but also in parts of Europe and Asia.

Most colubrids are not venomous or have venom that is not harmful to humans. However, species of the genus *Boiga* (cat snakes; Ph. 3.84) may cause medically significant injuries. In addition, the African boomslang (*Dispholidus typus*; Ph. 3.85), the African twig snakes (genus *Thelotornis*), and snakes belonging to the Asian genus *Rhabdophis* (keelback snakes) have been reported to cause human casualties. Most colubrids have an opisthoglyphous ('rear-fanged') dentition (Fig. 3.35), meaning that they have elongated teeth (one to three pairs) located in the backs of their upper jaws. These teeth are grooved at their frontal planes and the ducts of the venom glands (Duvernoy's glands) terminate at the bases of these teeth. When biting a piece of prey, these snakes hold on to their victims and massage the venom into its tissues by chewing movements of the jaws. This type of dentition and function may have evolved several times in the evolution of snakes [14, 138]. Some colubrids like those of the genus *Rhabdophis*, however, have only short teeth all along their upper jaws. This homodont dentition is called 'aglyphous dentition'. This indicates that the dentition type is not strictly associated with the phylogenetic position of a given snake species.

The Elapidae family encompasses venomous snake species which are characterized by their permanently erect frontal fangs at each side of the upper jaw (proteoglyphous dentition). Exceptions are the death adders (genus *Acanthophis*) of Australia (Ph. 3.86) and Indonesia which have erectable fangs. Typical elapid fangs are hollow

Ph. 3.84: A *Boiga* snake (Source: dwi septiyana/iStock/Getty Images Plus).

Ph. 3.85: A boomslang (*Dispholidus typus*) from South Africa (Source: Willem Van Zyl/iStock/Getty Images Plus).

and function as hypodermic needles. Secretions of the paired venom glands are actively ejected from reservoirs and transported to the base of the fangs. However, the venom is portioned and the amount applied during a bite is exactly adjusted to the size of the prey. 'Dry bites' without any venom transfer to the victim may be applied for defensive purposes. In some species (spitting cobras) the openings of the fangs are frontaly directed and the venom is sprayed toward an attacker or toward a prey animal over a distance of up to 2 m. When the venom of spitting cobras, e.g., of *Naja pallida* (Ph. 1.1), hits mucous surfaces of other animals the components induce local damage and may diffuse into the systemic circulation of the victim. As these actions do not require injection of the toxin mixture, some researchers prefer to label these toxins as 'toxungens' instead of 'venoms' [89]. Many members of the elapid

Ph. 3.86: A central Australian death adder, *Acanthophis pyrrhus* (Source: Ken Griffiths/iStock/Getty Images Plus).

family are also known for their threat display of rearing upward while spreading the neck-flap. This is typically seen in Asian cobras.

The sea snakes (subfamily Hydrophiinae) have many features in common with the Elapidae. The sequences of the short neurotoxic peptides in members of the two families are very similar. Thus, some researchers consider the sea snakes being members of the Elapidae (the sea snakes forming the subfamily Hydrophiinae) while other reseachers think that the differences between members of the Elapidae (e.g., the presence of long and short neurotoxic peptides in the venoms) and the sea snakes (only short neurotoxic peptides in the venoms) justify the separation of the two groups in different families.

Viperidae (vipers) have long frontal fangs at both sides of their upper jaws. The fangs have hinges at their bases formed by short maxillary bones that can rotate back and forth. Using this apparatus, the snake folds the venom teeth backward when the mouth is shut. As soon as the snake opens its mouth, the fangs are unfolded and permit deep penetration and efficient injection of venom when biting prey. The hollow fangs (solenoglyphous dentition; Fig. 3.35) are regularly replaced by new ones. The venom teeth are connected to venom glands with reservoirs from which fine-tuned doses of venom are injected into prey.

The pitvipers (Crotalidae) (Ph. 3.87) are considered a family on their own by some researchers while most others prefer to include them as the subfamiliy Crotalinae into the Viperidae. Pitvipers possess a unique sensory organ, the facial pit, or fossa between the eye and the nostril on either side of the head (Fig. 3.36). These organs enable them to sense subtle fluctuations in thermal radiation from distant objects, to precisely localize warm-blooded prey even in total darkness, and to identify resting sites with optimal conditions for thermoregulation [68].

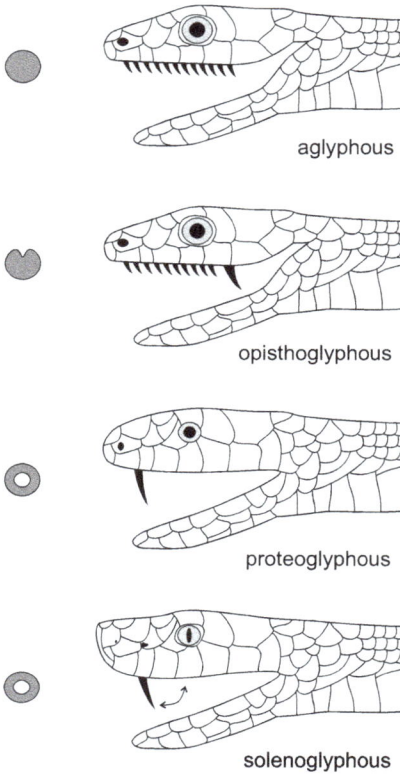

Fig. 3.35: Dentition types in snakes.

Ph. 3.87: A rattlesnake (*Crotalus catalinensis*) from the Island of Catalina (Source: NNehring/iStock/Getty Images Plus).

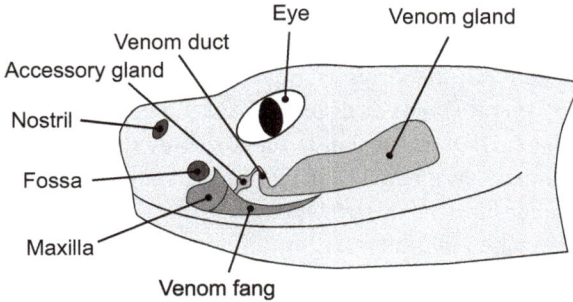

Fig. 3.36: Localization of the venom gland and accessory structures of the venom apparatus of the head of the Northern Pacific rattlesnake, *Crotalus oreganus.*

The family Lamprophiidae encompasses very diverse groups of snakes with different life styles and morphological traits. Tooth morphology and dentition within the Lamprophiidae is more variable than within any other snake family. While most members of this family are not or only moderately venomous, some are highly venomous and dangerous to humans (e.g., burrowing vipers of the genus *Atractaspis*). Burrowing vipers carry enormously developed fixed venom fangs on each side of their upper jaws that slide downward along grooves at the outside of the lower jaw when the snake shuts its mouth (Fig. 3.37). These fangs are hollow and connected to venom glands in the upper jaw.

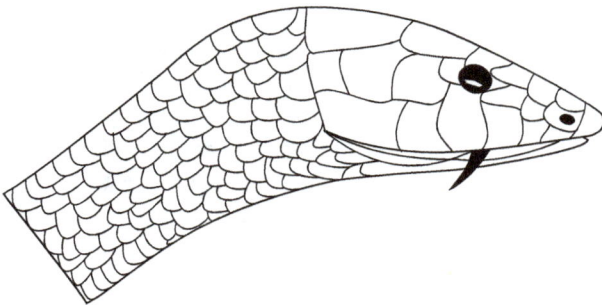

Fig. 3.37: Head of *Atractaspis* sp.

The venom glands of snakes, whose reservoirs are surrounded by muscle tissue that allows exact dosing of the amount of venom that is applied during a bite, are located below or behind the eyes (Fig. 3.36) and may extent into the body beyond the head region in some species, e.g., in the blue coral snake *Calliophis bivirgata*. Recent investigations of very different animal species with oral venom glands have revealed that there is a highly conserved network of approximately 3,000 genes encoding nonsecreted housekeeping proteins strongly coexpressed with the toxins, a 'metavenom network' [6]. The similarities of the protein patterns in a large variety of investigated species [143] allow the conclusions that all the oral venom glands originated from salivary glands of nontoxic ancestors and that the terms that have been invented to describe the different phenotypes of venom glands in different snake species may not be justified in an evolutionary context. This conclusion is supported by the observation that the same developmental genes are active during the formation of snake venom glands [102] and salivary glands of nontoxic species [111].

Different components of snake venom are synthesized by different cell types in the venom gland [102] and may be stored over extended periods without being degraded. To avoid enzymatic damage of venom gland tissues and venom components, proteolytic enzymes seem to be kept in inactive states by the addition of inhibitors or by keeping them in a denatured state under acidic conditions [77]. In addition, snakes are resistant against potential effects of their own venoms by having evolved mutations or posttranslational modifications in potential endogenous molecular targets rendering these targets insensitive to the toxins [131, 147] or by raising antibodies against their own venom components that capture and neutralize them as soon as they get into the snakes body fluids [125].

Snakes are ambush hunters which are not able to hunt down or follow injured prey over long distances. Thus, the main purpose of snake venoms is the induction of rapid paralysis in prey organisms. Front-fanged snakes like vipers (Viperidae) feed on various vertebrate and invertebrate prey species and have a broad spectrum of venom components of which each is optimized against a specifc target mechanism or molecule in one of these prey species. Rear-fanged snakes (many species of colubroid snakes like the brown tree snake, *Boiga irregularis*), however, typically produce venoms with lower complexity than front-fanged snakes, but there are exceptions to this rule. In some snake species, venoms are taxon-specific with respect to the target animals, i.e., lizards and birds may be highly susceptible to the venom while mammals are largely unaffected. It has been concluded from such observations that primarily prey preferences may have evolutionarily shaped the toxin mixtures in rear-fanged snakes [78]. However, snake venom composition is not only genetically defined in a given snake species but depends on the age of the snake, the period of resynthesis after a bite, and certain environmental factors like the external temperature or prey availability [43, 114, 146].

Snake venom is an aqueous (50–90% water), viscous (mucus), colorless, or slightly yellow liquid. When freeze-dried and cooled, it can be stored for long periods without loss of biological activity. Dry material consists of 90% (w/w) polypeptides and proteins. The remaining 10% consists of trace amounts of nucleotides (adenosine, flavonucleotides), metal ions (calcium, zink, aluminium, etc.), amino acids, and carbohydrates. With only a few exceptions, the protein components of snake venoms are not able to cross the blood-brain barrier in vertebrates [94]. This means, on one hand, that virtually all initial effects and symptoms of snake bite envenomation are caused by functional changes in the victim's peripheral tissues. On the other hand, this means that a bitten person experiences everything that is happening to him or her with full consciousness. Despite the inability of snake venom components to cross the blood-brain barrier, they may induce secondary complications in the brain by inducing infarction or hemorrhage [51].

The peptide and protein toxins in snake venoms can be classified in major groups according to their structural and functional characteristics [64]:

- Phospholipases A_2 (PLA$_2$)
- Serine proteases
- Metalloproteases
- Three-finger toxins (3FTxs)
- Protease inhibitors
- Lectins

The proteins within each of these groups share remarkable similarities in their primary, secondary, and tertiary structures, but may have entirely different functions in the target organisms.

Enzymes in Snake Venoms

Enzymes in snake venom support rapid distribution of venom compounds in the victim's tissues and serve to start predigestion. Hyaluronidases loosen the elastic tissue matrix by destroying hyaluronic acid. Mobilized water molecules then allow rapid diffusion of toxins away from the bite site [41].

Secreted phospholipase A_2 (sPLA$_2$) hydrolyzes cellular plasma membrane lipids and contributes to cytolysis and tissue necrosis in target organisms by mechanisms dependent on or independent of its catalytic activity [62]. Several isoforms of neurotoxic PLA$_2$ are present in venoms of different snake species (α- and β-forms). These proteins, on one hand, have enzymatic activities and may eventually destroy the presynaptic endings of motor neurons or, on the other hand, have the ability of acutely blocking transmitter release from the synaptic terminal of motor neurons by as yet unknown mechanisms [110]. It has been postulated that an α-sPLA$_2$-mediated disbalance of the rate of synaptic vesicle fusion with the presynaptic plasma membrane and the rate of membrane retrieval may result in the sessation of transmitter release [108]. Alternatively, β-isoforms of neurotoxic sPLA$_2$s may interact with cytochrome C oxidase in the inner mitochondrial membrane of neuronal cells and suppress ATP production. Such a mechanism has been described for the secreted PLA$_2$ of the European horned viper, *Vipera ammodytes* (Ph. 3.88) [124]. Notexin of the Australian tiger snake, *Notechis scutatus* (Ph. 3.89), is another type of synapse-destroying enzyme [144]. It is a basic protein of 119 amino acid residues (13.6 kDa) that is internally crosslinked by seven disulfide bridges. In experimental mice, the toxin causes paralysis of the ventilatory muscles (dyspnea) and paralysis of the hind legs. The victims die two to three days after the administration of a lethal dose [57]. Taipoxin of the Australian taipan (*Oxyuranus scutellatus*; Ph. 3.90) is another basic protein with all characteristics of a neurotoxic PLA$_2$ [33]. In this case, the toxin is a complex of three or four homologous

Ph. 3.88: An European horned viper, *Vipera ammodytes* (Source: AL-Travelpicture/iStock/Getty Images Plus).

Ph. 3.89: An Australian tiger snake, *Notechis scutatus* (Source: Ken Griffiths/iStock/Getty Images Plus).

subunits [75]. Some types of PLA$_2$, however, form complexes with other proteins potentiating their toxicity. Crotoxin of the South American rattlesnake *Crotalus durissus* is an example of cooperation of two nontoxic proteins, a basic 13 kDa PLA$_2$ homolog and an acidic 8.4 kDa protein, forming a neurotoxic complex [35, 48].

The metalloproteases in snake venom have probably evolved from nontoxic ancestor molecules, which were beneficial in remodeling the extracellular matrix, the ADAMs (a disintegrin and metalloprotease) [15, 40]. Like the housekeeping proteins, metalloproteases in snake venoms digest extracellular matrix molecules, however, in a completely unregulated fashion thereby damaging connective tissue or blood vessels in the victim

Ph. 3.90: A coastal taipan (*Oxyuranus scutellatus*) from Western Australia (Source: JohnCarnemolla/iStock/ Getty Images Plus).

[16]. Some snake venom metalloproteases, e.g., in the venom of the monocled cobra (*Naja kaouthia*; Ph. 3.91), specifically destroy von Willebrand factor, a protein present in vertebrate blood plasma that mediates platelet aggregation in injured blood vessels [53, 81]. These actions induce bleeding and the formation of blood blisters in the skin [84]. Thus, such toxins have been named 'hemorrhagins'. They act either by disruption of small blood vessels (rhexis) or by loosening the cell-cell adhesions between endothelial cells (diapedesis) [10, 50, 55, 90] (Tab. 3.10). Furthermore, they induce hemolysis and block hemostasis [44, 65, 112].

Ph. 3.91: A monocled cobra, *Naja kaouthia* (Source: Tom Brakefield/iStock/Getty Images Plus).

Tab. 3.10: Examples of hemorrhagins inducing extravasation.

Toxin	Snake species	Mode of action
Bilitoxin-1	*Agkistrodon bilineatus*	Rhexis
Atrolysin a	*Crotalus atrox*	Rhexis
HR-1, HR-2a, HR-2b	*Protobothrops flavoviridis* *Craspedocephalus gramineus*	Diapedesis
BaH 1	*Bothrops asper*	Diapedesis

L-Amino acid oxidases are present in the venoms of many snake species. The functional significance of these enzymes is not entirely clear. They may contribute to tissue destruction by the production of hydrogen peroxide (H_2O_2) [76] that, in turn, induces apoptosis in cells of affected tissues [34, 149].

The joint actions of these enzymes in the venoms of snakes [7] may be responsible for the generation of large areas of severe bruises and blood blisters in snake bite victims or the large-scale tissue necroses.

The venoms of almost every species of elapid snakes (Elapidae) contain acetylcholine esterase (AChE), an enzyme that degrades the neurotransmitter acetylcholine (ACh). In vertebrates, ACh carries the neuronal signal of motor neurons to the effector muscle cells in skeletal muscles [36]. Destruction of ACh in the synaptic cleft of the neuromuscular junction (Fig. 3.38) results in the inability of animals to induce voluntary muscle contractions and in a rapid onset of flaccid paralysis. Deaths of target animals or humans occur due to inactivity of the ventilatory muscles.

Kallikrein-like enzymes in snake venoms [37, 92] hydrolyze endogenous kininogens in the victims which results in the production of biologically active kinins. These mediators activate G protein-coupled receptors in different cell types resulting in vasodilation and hypotension (shock), in an increase in vascular permeability (edema) [60], and in the activation of pain receptors in the victims [97].

Three-Finger Toxins in Snake Venoms

A large and diverse group of nonenzymatic proteins and peptides in snake venoms are the three-finger toxins (3FTxs). Functionally, they may be classified as neurotoxins, myotoxins, and hemotoxins because they interfere with the respective organ systems (peripheral neurons, skeletal/cardiac muscle, or blood cells) in target animals or humans. Despite the diversity in structure and function of these toxins they seem to be evolutionarily related to each other [19, 39, 95, 126]. More than 30 subgroups of these snake toxins have been defined based on structure-function relationsships. These subgroups include acetylcholine esterase inhibitors, antiplatelet toxins, L-type calcium channel blockers, type IA and IB cytotoxins, type I, II, and III α-neurotoxins,

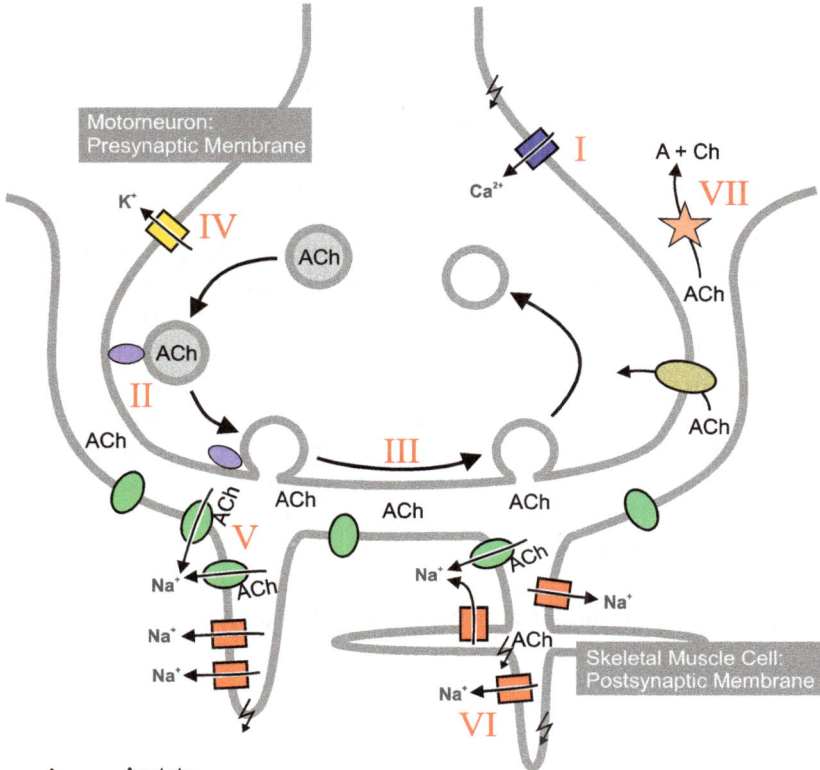

Fig. 3.38: Neuromuscular junction and snake venoms. Roman numbers correspond to partial processes of neuromuscular transmission that are affected by certain snake toxins as described in the text.

nonconventional (neuro)toxins, type A, B, and C muscarinic receptor antagonists, and synergistic toxins with mixed functions.

These toxins in snake venoms usually have the shape of a three-fingered hand (Fig. 3.39). The palm area, in which the peptide chain is cross-linked by disulfide bridges, forms a flat disk. Three hairpin-shaped fingers hang down like tentacles from this disk, each consisting of two antiparallel chains consisting of β-sheet structures that contain many hydrophobic amino acid residues [49]. The structure-function relationships of these proteins have been thoroughly studied using site-directed mutagenesis and functional testing of the variants [25]. It is assumed that the fingers have cytotoxic functions by inserting themselves into the plasma membranes of target cells while the palm stays on the cell surface (Fig. 3.39). The resulting disturbance of the plasma membrane lipid configuration allows ions to pass through the plasma membrane which results in changes in the electrical properties of nerve and muscle cells (overexcitation, cramps, etc.) [72]. In addition, the structural rearrangements of membrane lipids induced by the 3FTxs allow phospholipases in the snake venoms to access and degrade the membrane lipids and ultimately destroy the affected cell. This is another example of different toxin components working in conjunction to elicit toxic effects in the target organisms. Elapid venoms like those of *Naja* sp. (Ph. 3.91) and *Hemachatus* sp. have such cytolytic properties mediated by three-finger toxins [56, 96, 119].

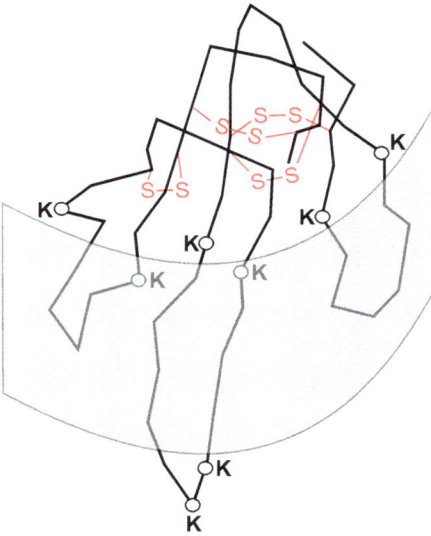

Fig. 3.39: Proposed conformation of cardiotoxin V^{II}4 of *Naja mossambica*. The 3D structure of the toxin is stabilized by four intramolecular disulfide bridges (red). The gray arc indicates the plasma membrane of the target cell in which the three fingers of the toxin (three-finger toxin) get inserted. Lysine residues (K) in the toxin molecule associate with the interfaces between the lipid environment of the membrane and the aqueous spaces of extracellular space and cytosol and assist in proper membrane insertion of the toxin.

The neurotoxins of the three-finger toxin family may act as monomers or as multimeric protein complexes [24]. They generally impair neuromuscular transmission [91] (Fig. 3.38). Depending on their amino acid sequences and tertiary structures, these α-neurotoxins can be subdivided into short-chain neurotoxins (60–62 amino acid residues and four disulfide bridges), long-chain neurotoxins (66–75 amino acid residues stabilized by five disulfide bridges), atypical long-chain neurotoxins, and nonconventional three finger neurotoxins (Tab. 3.11). These variants have evolved from a common ancestor, they are homologs [19, 39, 126]. As their fingers contain less hydrophobic amino residues in comparison with the cytolytic toxins, they are not cytolytic but interfere with the postsynaptic mechanisms of neuromuscular transmission. Although they are generally called 'neurotoxins', they are actually myotoxins blocking the nicotinic acetylcholine receptors (nAChRs) in the postsynaptic membrane (muscle cell membrane) of the motor end plate [17, 118]. The short toxins generally function in a reversible fashion while many of the long toxins block nAChRs irreversibly. Some of the long toxins have the additional ability to block the α7-subtype of the nAChR [85, 117] which is the subtype mediating the postsynaptic effects of signal transmission from neuron to neuron in the central and the peripheral nervous systems in vertebrates [101]. In these cases, the toxins function as true neurotoxins. An example for this subgroup of three-finger neurotoxins is the α-bungarotoxin (Tab. 3.11) of the Chinese krait, *Bungarus multicinctus* (Ph. 3.92).

Ph. 3.92: The Chinese krait, *Bungarus multicinctus* (Source: Briston, Wikimedia, CC-BY-SA).

Individual toxins in snake venoms are highly specific for binding to and altering the function of target molecules in prey animals like ion channels or plasma membrane receptors. This makes snake venoms a highly valuable toolbox for researchers investigating signal transduction processes or electrical processes in animal and human cells [83].

There are other 3FTx proteins that target a variety of additional protein targets to exert their toxic effects. For example, L-type calcium channels are targeted by calciseptine, a toxin isolated from the venom of the African black mamba, *Dendroaspis polylepis*

Tab. 3.11: Three-finger toxins in snake venom blocking nicotinic acetylcholine receptors in vertebrate muscle cells (data from [91]).

Toxin subtype	Structure	Primary molecular target	Secondary molecular target	Examples of snake species and toxins
Short-chain toxin	60–62 Amino acid residues, 4 disulfide bridges	High affinity for nAChRs in skeletal muscle (K_d = 10^{-9}–10^{-11} mol/L)		Elapidae, e.g., erabutoxin a of *Laticauda semifasciata* or toxin-α of *Naja nigricollis*
Long-chain toxins	66–74 Amino acid residues, 4 disulfide bridges plus an additional one at the tip of loop II	High affinity for nAChRs in skeletal muscle (K_d = 10^{-9}–10^{-11} mol/L)	High affinity for the neuronal nAChR (α7) (K_d = 10^{-8}–10^{-9} mol/L)	Elapidae, e.g., α-bungarotoxin of *Bungarus multicinctus* or α-cobratoxin of *Naja kaouthia*
Atypical long-chain toxins	69 amino acid residues, 4 disulfide bridges	High affinity for nAChRs in skeletal muscle (K_d = 10^{-11} mol/L)		Toxins Lc-a and Lc-b of the hydrophid snake *Laticauda colubrina*
Nonconventional toxins	65–67 Amino acid residues, 4 disulfide bridges plus an additional one at the tip of loop I	Moderate-to-high affinity for nAChRs in skeletal muscle (K_d = 10^{-6}–10^{-8} mol/L)	Candoxin of *Bungarus candidus* is a high affinity antagonist of the neuronal nAChR (α7) (K_d = 5 × 10^{-8} mol/L)	Elapidae, e.g., candoxin of *Bungarus candidus*, WTX of *Naja kaouthia*, or Wntx-5 of *Naja sputatrix*

(Ph. 3.93) [21]. Calciseptine is a smooth muscle relaxant and an inhibitor of cardiac contractions in target animals and humans and causes hypotensive effects or even cardiac arrest [139, 145]. Other 3FTx proteins in black mamba venom, the mambalgins, act on acid-sensing ion channels (ASICs). As ASICs are usually involved in the generation of pain sensations in vertebrates [20], the ASIC-inhibitory actions of mambalgins have analgesic effects [23]. Another 3FTx homolog in venoms of mamba species, e.g., that of the green mamba, *Dendroaspis viridis* (Ph. 3.94), is dendroaspin [129, 142]. Unlike other 3FTxs, dendroaspin contains an Arg-Gly-Asp-(RGD)-motif and functions as an inhibitor of platelet aggregation and platelet adhesion by binding to adhesion molecules of the GPIb and GPIIb/IIIa subtypes (GP stands for 'glycoprotein') (Fig. 3.40) [136, 140]. This interaction inhibits the process of primary hemostasis (first step in the physiological wound closure reaction in animals and man [30]) and induces sustained bleeding.

Several small proteins called muscarinic toxins (MTs) have been isolated from venom of the eastern green mamba (*Dendroaspis angusticeps*) and other mamba species [58]. They have high selectivities and binding affinities for individual subtypes of

Ph. 3.93: A black mamba, *Dendroaspis polylepis*, occurring in Eastern and Southern Africa (Source: Mark Kostich/iStock/Getty Images Plus).

Ph. 3.94: The African green mamba, *Dendroaspis viridis* (Source: Wirestock/iStock/Getty Images Plus).

the metabotropic muscarinic acetylcholine receptor (mAChR) in mammals. MT2 activates the mAChRs of the subtypes m1, m3, and m5 which are coupled to the calcium signaling system. MT7, however, is a potent noncompetitive antagonist at the m1 subtype [12]. Moreover, other aminergic toxins belonging to the family of 3FTxs have been identified in venoms of different mamba species which interact with different types of metabotropic (G protein coupled-)receptors, e.g., adrenergic or dopaminergic receptors [8].

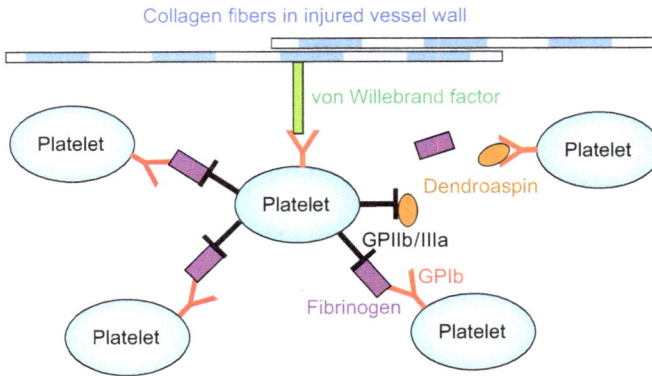

Fig. 3.40: Scheme of platelet aggregation at a site of vascular injury in mammals. When mammalian blood vessels are damaged, collagen fibers get exposed to the blood stream. A 227 kDa protein, the 'von Willebrand factor', binds to exposed collagen fibers and provides a docking site for glycoprotein Ib (GPIb), an adhesion protein associated with the plasma membranes of platelets. Platelets bound to the site of injury expose glycoprotein IIb/IIIa (GPIIb/IIIa) on their cell surfaces. These integrins bind fibrinogen molecules from the circulating blood which results in attachment and activation of additional platelets at the site. The resulting platelet aggregate forms a plug that prevents blood from leaving the vessel. Thus, platelet aggregation is the first phase of hemostasis. Snake venom proteins like dendroaspin of the green mamba (*Dendroaspis viridis*) block the glycoproteins and inhibit platelet aggregation, thereby inducing prolonged bleeding in bite victims.

The neuromuscular junction and the functionally disrupting effects of snake toxins

Skeletal muscles in vertebrates are generally innervated by motor neurons using acetylcholine (ACh) as the relevant transmitter. When an action potential (AP) reaches the presynaptic nerve terminal it activates L-type voltage-gated calcium channels (Fig. 3.38). Calcium influx depolarizes the plasma membrane. Such a membrane depolarization in conjunction with elevated concentrations of free calcium ions in the cytosol induces membrane attachment and fusion of acetylcholine-filled vesicles with the help of auxiliary proteins (docking and fusion agents). Release of ACh from the vesicle lumen into the synaptic cleft rapidly increases the ACh concentration in this narrow extracellular space. Binding of ACh to nicotinic acetylcholine receptors (nAChRs) in the postsynaptic membrane (muscle cell plasma membrane with its T-tubulus system) results in activation of these ionotropic receptors and allows limited amounts of sodium ions to enter the cytosol of the skeletal muscle cell. The resulting partial depolarization of the muscle cell, in turn, activates muscle-specific voltage-activated sodium channels, which allows massive influx of sodium ions into the muscle cell cytosol and the generation of action potentials (APs). These induce rapid calcium release from intracellular stores (sarcoplasmic reticulum) into the cytosol and induce muscle contraction. Depolarization of the presynaptic membrane activates voltage-gated potassium channels. The resulting efflux of K^+ ions repolarizes the neuronal cell and terminates transmitter release and signal transmission at the neuromuscular junction. Residual ACh in the synaptic cleft is either taken up through specific carriers into the synaptic ending (ACh reuptake) or enzymatically destroyed to yield acetate and choline by the acetylcholine esterase (AChE).

Different snake toxins interfere with these physiological processes to disrupt voluntary muscle movements in prey organisms. Some snakes utilize mixtures of substances of which different components block different partial processes of normal neuromuscular signal transmission (synergistic action).

I.   Calciseptine in mamba (*Dendroaspis* sp.) venom blocks L-type voltage-gated calcium channels in presynaptic endings of motor neurons. β-Bungarotoxin of the krait *Bungarus multicinctus* has the same ability.

II. β-Bungarotoxin of the krait *Bungarus multicinctus* disrupts trafficking of ACh-filled vesicles to the inner surface of the presynaptic membrane by interfering with docking and fusion proteins.

III. Neurotoxic phospholipases A_2 of different snake venoms (see above) interfere with vesicle fusion and transmitter release into the synaptic cleft and finally destroy the synaptic ending.

IV. Dendrotoxins of mamba (*Dendroaspis* sp.) venoms block voltage-gated potassium channels in the presynaptic membrane. This results in prolonged membrane depolarization and transmitter release which transiently overactivates the muscle cell but rapidly exhausts the transmitter stores in the neuronal ending.

V.   α-Bungarotoxin of the krait *Bungarus multicinctus* and many other snake toxins of the three-finger toxin family (3FTx) block the nAChRs in the postsynaptic membrane. This cuts the muscle cells off the signal transmission by the motor neuron. Similar actions (postsynaptic blockage of neuromuscular transmission) have been described for venom components of the Australian death adder (*Acanthophis antarcticus*), the acanthophins a–d [122].

VI. Crotamine of rattlesnake (*Crotalus* sp.) venoms inhibits the activation of voltage-gated sodium channels in the muscle cell membrane. This prevents the formation of action potentials and results in flaccid paralysis of the skeletal muscle.

VII. Cobra (*Naja* sp.) venom contains acetylcholine esterases which destroy the transmitter in the synaptic cleft and suppress signal transmission to nAChRs in the muscle cell membrane. In contrast, fasciculins in mamba (*Dendroaspis* sp.) venoms inhibit endogenous AChEs in prey animals which results in prolongation of nAChR activation by endogenously released acetylcholine. This results in cramps of skeletal muscles or even spastic paralysis.

Other Nonenzymatic Proteotoxins in Snake Venoms

Mamba (*Dendroaspis* sp.) venom contains peptide homologs of the mammalian basic pancreatic trypsin inhibitor, a Kunitz-type protease inhibitor [116], called dendrotoxins [120]. The dendrotoxins (~7 kDa, 57–60 amino acid residues) do not have any protease inhibitory functions [45] but are powerful blockers of voltage-gated potassium channels of the subtype Kv1. These channels usually mediate the exit of K^+ ions from the cytosol to the extracellular space during the repolarization phase of an action potential in a neuronal cell. Inhibition of these channels results in a prolongation of the depolarization phase of the action potential and a marked increase in transmitter release (acetylcholine in case of vertebrate motor neurons) [46]. Finally this induces exhaustion of the transmitter release system and flaccid paralysis of the skeletal muscle as the relevant effector system. The dendrotoxins are a good example for neurotoxins with presynaptic effects.

Sarafotoxins are small (21 amino acid residues and 2 disulfide bridges) cardiotoxic peptides in venoms of *Atractaspis* species [67, 132] with high structural similarity to mammalian endothelins. The toxins are able to bind to G protein-coupled endothelin receptors on the surface of target cells (cardiac muscle, vascular smooth muscle, bronchial muscle) and activate calcium signaling in these cells. Sarafotoxins may cause bronchial spasms, lung edema, cardiac arrest, and death in mice or rats [69] within minutes of intravenous administration (LD_{50} = 15 µg/kg BW).

Snake Toxins Affecting Blood Coagulation

Venoms of many snake species contain components that interfere with the blood coagulation system of their vertebrate victims. Procoagulants, e.g., the prothrombin-activating factor in *Echis carinatus* (Ph. 3.95) venom [113], induce rapid blood coagulation within the circulatory system which results in thrombus formation and microcirculatory problems in capillary beds of vital organs in bite victims. Anticoagulants in other snake venoms inhibit the blood coagulation cascade ('secondary hemostasis'; Fig. 3.41) and induce sustained bleeding that weakens the bite victim.

Ph. 3.95: The common saw-scaled viper, *Echis carinatus*, inhabits many parts of Asia, Arabia, and Northern Africa (Source: Lakeview_Images/iStock/Getty Images Plus).

Because venom components of different snake species interact with different partial processes in the multifactorial blood coagulation cascade in vertebrates, we briefly review the normal function of this system.

Blood coagulation may be initiated by injuries of the vascular system (intrinsic pathway) or by tissue destruction in close vicinity of blood vessels (extrinsic pathway). When blood vessels are injured, acid mucopolysaccharides in the extracellular matrix of the vessel walls come in contact with blood cells. A serine protease, coagulation factor XII (Fig. 3.41) which is associated with the plasma membrane of platelets, binds to these mucopolysaccharides and gets activated. The active protease (factor XIIa) has two substrates available in blood plasma. Kininogens are proteolytically processed to biologically active kinins (generation of pain sensation), and coagulation factor XI is cleaved to yield the active factor XIa. The proteolytic activity of factor XI cleaves another pro-protein in blood plasma, factor IX, which activates this factor (factor IXa). Factor XI may also be proteolytically processed and activated by the protease thrombin and factor IX may alternatively be cleaved by factor VIIa which illustrates

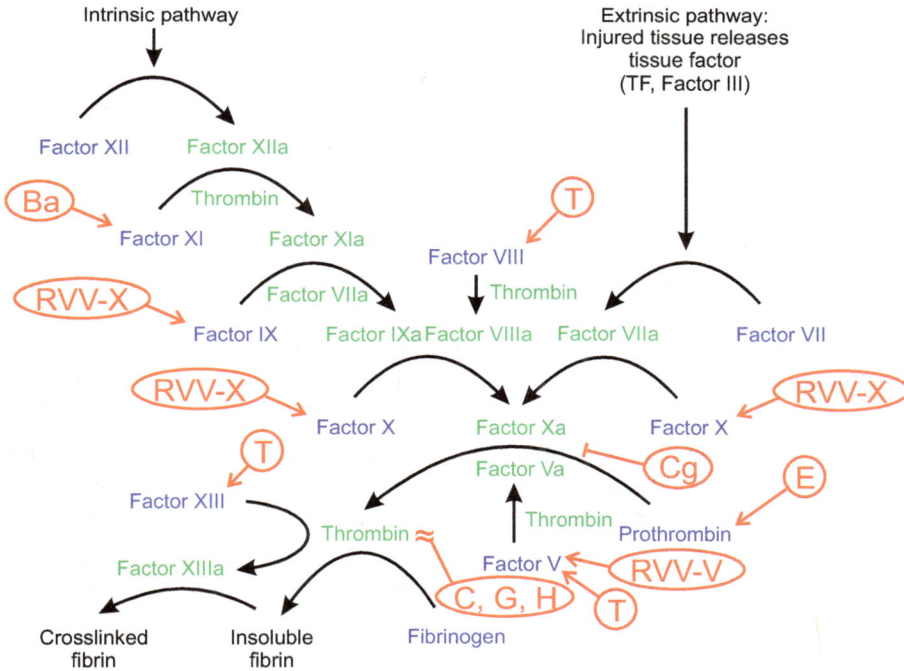

Fig. 3.41: Blood coagulation cascade and molecular targets of snake venom components. ←, Activation of the target; ⊢, inhibition of the target; ≈, acting like the target; Ba, toxin of *Bitis arietans*; C, crotalase of *Crotalus adamanteus*; Cg, prothrombinase inhibitor of *Craspedocephalus gramineus*; E, ecarin of *Echis carinatus*; G, gabonase of *Bitis gabonica*; H, habutobin/flavoxobin of *Protobothrops flavoviridis*; RVV-V and RVV-X, toxins of *Daboia russelii*; T, thrombocytin of *Bothrops atrox*.

the fact that the intrinsic pathway of the coagulation cascade has various starting points. The active form of factor IX, factor IXa, is again an active protease that cleaves and activates the plasma factor X using factor VIIIa as a cofactor (activation complex).

At the level of factor Xa the intrinsic pathway of the blood coagulation cascade fuses with the extrinsic pathway because factor X may also be cleaved and activated by factor VIIa, an active protease whose activation is mediated by tissue factor (TF, factor III) which is located at the cell surface of subendothelial cells around blood vessels. Tissue destruction involving blood vessels brings these two components together and initiates the extrinsic pathway of blood coagulation.

Factor Xa cleaves the proprotein prothrombin which circulates in blood plasma. Alternatively, prothrombin may be cleaved and activated by factor Va. This generates the active serine protease thrombin at the site of vascular or tissue injuries. Thrombin has the ability to proteolytically process plasma fibrinogen and form shorter fibrin fibers which are insoluble in aqueous media like blood. In parallel, thrombin activates factor XIII. Its activated form, factor XIIIa, induces crosslinking of the insoluble fibrin fibers. This results in the formation of a dense meshwork of fibrin fibers which traps blood cells and other proteins. The resulting scab stops further bleeding and forms a protective wound cover which supports subsequent wound healing that occurs underneath.

The proteolytic activities of factors VIIa, IXa, Xa, and of thrombin are dependent on the presence of high concentrations of calcium ions, i.e., blood coagulation occurs only at sites of calcium accumulation. Such calcium accumulation is induced by platelet aggregation and activation because activated platelets expose negatively charged head groups of plasma membrane phospholipids at the extracellular cell surface which provides binding sites for plasma calcium ions. This mechanism ensures that blood coagulation occurs only at sites of tissue or vascular injury and not in other places of the circulatory system.

There are several systems in vertebrate circulatory systems which inhibit or reverse spontaneous formation of blood clots. Protein C in plasma is an inhibitor of the coagulation factors Va and VIIIa with protein S serving as a cofactor. This process is dependent on the presence of vitamin K. Antithrombin inhibits the activities of factor Xa and thrombin. In addition, vertebrates have a fibrinolytic system that is able to dissolve blood clots as long as they are small, thereby avoiding the initiation of thrombosis. Tissue plasminogen activator (tPA) cleaves and activates plasminogen present in blood plasma. The active protease plasmin then destroys the fibrin meshwork and dissolves the blood clot.

Certain snake venom toxins prolong blood coagulation time in their victims. The respective proteins or glycoproteins have molecular masses ranging from 6 to 350 kDa [63]. The anticoagulants act by different mechanisms which are not fully understood.

Phospholipases A_2 in snake venom inhibit the activation of factor X and of prothrombin by enzymatic and nonenzymatic mechanisms. Examples are the three anticoagulatory proteins (CM-I, CM-II, and CM-IV) from *Naja nigricollis* (black-necked spitting cobra) [31] or the PLA_2 in the venom of the common European viper, *Vipera berus*

(Ph. 3.96), that acts by disturbing the association of coagulation factors [9]. Serine proteases in snake venom may destroy certain essential coagulation factors like factor V or
factor VIII or may directly activate protein C. Protein C stimulates fibrinolysis through
its interaction with plasminogen activator inhibitor [32]. Venoms of copperhead snake
species belonging to the genus *Agkistrodon* (e.g., the American moccasins) contain such
protein C activators [3, 66] which are glycoproteins with molecular masses of approximately 36–40 kDa and prolong blood clotting times substantially [82]. The three-finger
toxins hemextin A and hemextin B of the South African rinkhals, *Hemachatus haemachatus* (Ph. 3.97), form a complex that prevents blood clotting by inhibiting the factor III-mediated initiation of the extrinsic pathway of blood coagulation [4]. Another
group of snake venom anticoagulants belongs to the C-type lectin-related proteins.
Bothrojaracin has been isolated from venom of the South American jararaca, *Bothrops jararaca* (Ph. 3.98). It inhibits binding of fibrinogen to thrombin ($K_i = 15$ nmol/L)
which indicates that this toxin is a blocker of exosite I in the thrombin molecule and
prevents the substrate from binding to the enzyme [148].

The most common coagulopathy in snake bite victims is the 'venom-induced consumption coagulopathy' (VICC), a condition brought about by the activation of the coagulation pathway by certain components of snake venoms like thrombin-like enzymes,
prothrombin activators, or factor X activators [52]. This condition is often accompanied
by proteolysis of circulating fibrinogen by snake venom proteases or metalloproteases. A
fibrinogenolytic protease has been isolated from the venom of *Macrovipera lebetina* and
a metalloprotease, fibrolase, from the venom of the copperhead, *Agkistrodon contortrix*
(Ph. 3.99) [130]. In addition to fibrinogen destruction, thrombin-like or other proteases initially induce rapid blood coagulation which forms many small blood clots. This condition
activates the endogenous fibrinolytic plasmin system in the victim. Rapid degradation of

Ph. 3.96: The common European viper, *Vipera berus* (Source: taviphoto/iStock/Getty Images Plus).

Ph. 3.97: The South African rinkhals, *Hemachatus haemachatus* (Source: Willem Van Zyl/iStock/Getty Images Plus).

Ph. 3.98: The South American lancehead, *Bothrops jararaca* (Source: Sostenes Pelegrini/iStock/Getty Images Plus).

filamentous fibrin in conjunction with loss of circulating fibrinogen renders the blood incoagulable (paradoxical effect of procoagulants in snake venoms).

Thrombin-like enzymes in snake venoms are closely related to each other and are widely distributed in species of several pit viper genera (*Agkistrodon, Bothrops, Crotalus, Lachesis,* and *Trimeresurus*) in some vipers (e.g., *Bitis* sp.; Ph. 3.100 or *Cerastes* sp.; Ph. 3.101), and in the boomslang, *Dispholidus typus* (Ph. 3.85) [100]. Habutobin is such a thrombin-like enzyme in the venom of the Japanese habu snake, *Protobothrops flavoviridis* [18]. It cleaves the Arg16-Gly17 bond in the α chain of rabbit fibrinogen and, when applied in vivo to rabbits, rapidly consumes the circulating fibrinogen [88]. Habutobin has been shown to release plasminogen activators (tissue-type plasminogen activator (t-PA) and urokinase-type plasminogen activator (u-PA)) from cultured endothelial cells of the bovine pulmonary artery into the culture medium [127]. Batroxobin, or reptilase, is an enzyme isolated from the venoms of the snake species *Bothrops atrox* and *Bothrops jararaca*. It is a serine protease [54] which has only a moderate anticoagulant potency when applied in a living organism in contrast to its strong anticoagulatory effect in vitro. The reason for this difference is that reptilase, other than thrombin, does not activate coagulation factor XIII. The fibrin formed by reptilase is therefore more easily degradable than the crosslinked fibrin meshwork generated by thrombin. By rapid conversion of the full complement of circulating fibrinogen to fibrin and its subsequent enzymatic degradation (consumption coagulopathy), reptilase, leads to a fibrinogen deficiency in the bloodstream and in an inability of the blood to coagulate. Preparations of reptilase and of similar venom components of other snake species are used therapeutically in the dissolution of blood clots or as diagnostic tools [70, 80].

Synergism of Different Venom Components
Snake venoms generally contain more than one biologically active ingredient. The individual substances act at least additively and sometimes even synergistically on physiological processes in bite victims. This makes the envenomation very rapid in its onset and efficient for the snake as only minor amounts of venom have to be utilized to achieve the goal of paralyzing prey. A few examples may illustrate the material and functional complexity of snake venoms.

- Tiger snake (*Notechis scutatus*; Ph. 3.89) venoms contain potent presynaptic neurotoxins [57], procoagulants, hemolysins, hemorrhagins, and myotoxins [144].
- The venom of *Enhydrina schistosa* (Ph. 3.102), the common sea snake occurring in the tropical Indo-Pacific, is made up of highly potent neurotoxins and myotoxins. The average amount of venom per bite is approximately 8.5 mg, while the lethal human dose is estimated to be 1.5 mg. The LD_{50} value (mice, s.c.) is 0.164 mg/kg BW [13].
- The venom of the Chinese krait, *Bungarus multicinctus* (Ph. 3.92), contains two long-chain toxins, α- (10.3 kDa) and κ-bungarotoxins (9.6 kDa), and a short-chain toxin, β-bungarotoxin (2.9 kDa) [105]. The long- toxins inhibit the neuronal isoform (α7) of the nicotinic acetylcholine receptor in the postsynaptic membrane of neuron–neuron synapses [85], while β-bungarotoxin has enzymatic activity as a phospholipase A_2 and blocks synaptic transmission at the presynaptic side. The

Ph. 3.99: The broadband copperhead (eastern copperhead), *Agkistrodon contortrix*, inhabits the southeast of the USA (Source: David Kenny/iStock/Getty Images Plus).

Ph. 3.101: An Arabian horned viper, *Cerastes gasperettii* (Source: reptiles4all/iStock/Getty Images Plus).

joint actions of these toxins efficiently suppress the neuronal activity in the motoric nervous system of bite victims.

– The venom of the Indian cobra, *Naja naja* (Ph. 3.103), contains powerful myotoxins acting on mammalian skeletal muscle and on cardiac muscle cells, respectively [29, 79, 128]. Some of these toxins have enzymatic activities as well and hydrolyze the pyrophosphate linkages in ATP, NAD^+, and coenzyme A resulting in rapid metabolic failure in target organisms [1].

– Recent -omics studies have revealed that venom of the king cobra (*Ophiophagus hannah*; Ph. 3.104) contains a high diversity of three-finger toxins that include α-

Ph. 3.100: A Gaboon viper (*Bitis gabonica*) swallowing a rat (Source: Jan-Peter Hildebrandt).

neurotoxins, muscarinic toxin-like proteins, neurotoxins, cardiotoxins, and cytotoxins with some regional variation in toxin quality and quantity [133]. The sorting of long and short variants of neurotoxins is variable in these venoms, but they are highly abundant ingredients in any case [98, 134]. Besides the neurotoxins, king cobra venom contains ohanin, a 12 kDa protein that induces hypolocomotion and hyperalgesia in mice [103], a phospholipase A_2, and cysteine-rich secretory proteins (CRISPs) including ophanin, a protein blocking smooth muscle contraction by inhibiting voltage dependent calcium channels [93]. The β-cardiotoxin (β-CTX) in king cobra venom belongs to the three-finger toxin family and binds to adrenergic receptors in vertebrate cardiomyocytes mediating negative chronotropic effects [73, 104].

– The proteome of the eastern green mamba venom (*Dendroaspis angusticeps*) is composed of 42 distinct proteins and the nucleoside adenosine. The predominant proteins are three-finger toxins including aminergic toxins, which act on muscarinic and adrenergic receptors, and fasciculins, which are choline esterase inhibitors that cause spontaneous, involuntary muscle contraction, and relaxation (muscle fasciculation) [71]. Other ingredients of mamba toxins are the dendrotoxins which are structurally homologous to Kunitz-type protease inhibitors but do not have any inhibitory potencies on proteases [45]. The venom of the black mamba, *Dendroaspis polylepis* (Ph. 3.93), contains dendrotoxin 1 which inhibits voltage-gated K^+ channels in the presynaptic membrane in neurons. It also inhibits Ca^{2+}-sensitive K^+ channels in rat skeletal muscle [46]. Dendrotoxins 3 and 7 inhibit muscarinic acetylcholine receptors of the subtypes 4 and 1, respectively,

Ph. 3.102: The hook-nosed sea snake, *Enhydrina schistosa*, from the tropical Indo-Pacific regions (Source: Vikramonice, Wikimedia, CC-BY-SA).

Ph. 3.103: A juvenile of the Indian biocellate cobra, *Naja naja* (Source: Sindhu Ramchandran, Wikimedia, CC-BY).

Ph. 3.104: The king cobra (*Ophiophagus hannah*) inhabits jungles in Southern and Southeast Asia (Source: takeo1775/iStock/Getty Images Plus).

compromising the functions of heart and smooth muscles, glandular secretion, as well as the release of neurotransmitters in the victim [58]. Another Kunitz-like protein present in venom of the green mamba (Ph. 3.94) is calcicludine. It blocks neuronal voltage-activated calcium channels [115]. Individually, most of these components do not exhibit potent toxicity in vitro but are assumed to have synergistic effects in vivo.

Snake Bite Envenomation in Humans

Approximately 15% of the 3,000 snake species worldwide are considered to be dangerous to humans. Because many cases of snake bites in humans go unreported, there are no accurate incidence data for many regions of the world. It is estimated that approximately 2.5 million bites occur per year. Over 100,000 of these are fatal. Most venomous snake bite accidents occur in tropical or subtropical regions or during warm seasons when snakes are active and many people are outdoors.

To avoid snake bites, sturdy shoes, preferably long boots, and long trousers should be worn in risk areas. One should never climb or jump over obstacles, e.g., fallen logs. Do not walk in tall grass or thick vegetation. A flashlight must be used at dusk and at night. Hikes without a companion should be avoided.

The amount of venom released during a snake bite is dependent upon the snake species, the kind of bite (defensive bite, prey bite), the size, and age of the snake, and depends upon the time that has passed between the animal's last bite and the actual incident. In experiments with the rattlesnake *Crotalus rubber lucasensis*, only 15% of the amount of venom obtained from the initial venom extraction could be obtained from another bite 7 days later, 59% after 27 days, and 90% after 54 days [11]. This indicates that venom production in snakes is a costly and time-consuming process [99, 121].

The average amount of venom per bite is between 10 and 18 mg (DW) for *Vipera aspis* and *Vipera berus* and between 180 and 750 mg for the puff adder, *Bitis arietans* (Tab. 3.12). The maximum amount of venom is transferred during a bite of *Bothrops jararacussu* and amounts up to 1,530 mg DW [11]. The total amount of venom in an individual of the Australian inland taipan, *Oxyuranus microlepidotus*, which is considered the most venomous snake in the world (LD$_{50}$ (mouse, s.c.) = 2 μg/kg BW), is said to be sufficient to kill more than 200,000 mice. Similar toxicity levels have been reported for the Australian coastal taipan, *Oxyuranus scutellatus* [33]. A single bite usually uses only part of the stored venom, so multiple bites can occur in quick succession with the same intensity.

Tab. 3.12: Selected venomous snakes and their venom toxicities.

Species	Venom per bite (mg DW)	LD$_{50}$, mice (mg/kg BW)		Deadly dose in humans (mg/75 kg BW)
		i.p.	i.v.	
Agkistrodon piscivorus	90–145	5.1	4.0	
Bitis arietans	180–750	3.7		
Bitis gabonica	450–600	2.0	0.8	14.0
Boiga irregularis	2–11	31.0		
Bothrops atrox	70–160	3.8	4.3	
Bungarus caeruleus	10		0.1	6.0
Bungarus multicinctus	5–20	0.1	0.1	
Crotalus adamanteus	370–700	1.9	1.7	
Crotalus atrox	175–320	3.7	4.2	
Crotalus horridus	75–150	2.9	2.6	
Crotalus viridis	35–100	2.3	1.6	
Daboia russelii	130–250		0.1	42.0
Dendroaspis polylepis	100–120	1	0.3	10.0
Echis carinatus	12		2.3	5.0
Enhydrina schistosa	8–10		0.1	1.5
Lachesis muta	280–450	5.9		
Micrurus fulvius	2–6	1.0		

Tab. 3.12 (continued)

Species	Venom per bite (mg DW)	LD$_{50}$, mice (mg/kg BW)		Deadly dose in humans (mg/75 kg BW)
		i.p.	i.v.	
Naja naja	170–250	0.6	0.3	15.0
Notechis scutatus	35–65	0.1		3.0
Ophiophagus hannah		0.2	0.1	
Protobothrops flavoviridis		5.1	3.5	
Pseudechis australis	180			
Vipera berus	10–18	0.8	0.6	75.0

Snake bite envenomation is a complex medical emergency that affects not only the site of the bite but also involves multiple organ systems as well [105]. There are large individual differences in the courses of the envenomation syndromes indicating that close monitoring of each patient is required. Consultation with a physician who is experienced in diagnosis and treatment of snake bites is essential. Detailed guidelines for management and treatment of human envenomations by the most dangerous venomous snakes are provided by Gold et al. [42].

The World Health Organization (WHO) has released general instructions how to proceed when a human has been bitten by a snake:

- Immobilization: It is important to immobilize the affected body part. The affected limb should be positioned low (below the level of the heart).
- Pressure bandage: A pressure bandage at the bite site is recommended. Binding off the limb or stopping the blood flow to and off the limb is not recommended.
- Identification of the biting snake: For further treatment, it is important to identify the species of snake that caused the bite. Cell phone photos of the snake and the bite site are recommended. Active pursuit or attempting to capture the snake is discouraged because venomous snakes may have enough residual venom for a second dangerous bite. If the snake had escaped, the victim should be immediately asked what the snake looked like. This is important in case the bitten person loses consciousness.
- Rapid transport to a medical center: The chances of survival after being bitten by venomous tropical snakes depend largely on the time between the incident and the onset of therapy. Therefore, telephone contact should be made immediately with a treatment center and the fastest mode of transport should be chosen.
- Life support therapy during transportation of a patient: A venous access is necessary and should be placed by an emergency doctor. If required, shock treatment and artificial respiration should be provided.

3.2.13.3 Birds (Aves)

Until recently, toxicity was not considered a relevant trait in birds, but there are obviously interesting cases [74]. Birds did not develop primary toxicity during evolution, but

some species have acquired the ability of using toxins produced by other organisms to impregnate their tissues or their feathers without harming themselves. This may either deter predators or may be useful in fighting parasites or infectious microbes.

Examples of poisonous birds are the hooded pitohui (*Pitohui dichrous*; Ph. 3.105), which is a member of the Old World oriole family (Oriolidae), and the blue-capped ifrit (*Ifrita kowaldi*), a member of the Ifritidae family. Both species are native to Papua New Guinea. These insectivorous birds acquire steroidal alkaloids, mainly batrachotoxin and homobatrachotoxin, from food organisms like melyrid beetles of the genus *Choresine* [28]. The birds accumulate these toxins in internal tissues and skin and secrete them through skin glands onto their feathers [26, 27]. Up to 20 µg of these substances have been isolated from the skins of individual birds. Batrachotoxin and homobatracho-toxin (see Section 2.9.16) inhibit the intrinsic inactivation of voltage-gated sodium channels during an action potential which prolongs the action potential duration in neurons and muscle cells [61] and results in burning sensations, numbness, or even spastic paralysis in animals that try to prey on these birds. The LD_{50} in mice (p.o.) is approximately 2 µg/kg BW.

Other examples of birds which are at least temporarily poisonous upon taking up certain types of food are European quail (*Coturnix coturnix*) which feed on plant seeds toxic to humans during migration, and the spur-winged goose (*Plectropterus gambensis*) which accumulates cantharidin (see Section 2.4.3) from feeding on melyrid beetles [5]. There may be other cases of secondary toxicity in birds which have not been studied yet [74, 141].

Members of several bird species voluntarily expose themselves to insect toxins, mostly toxins of ants, which is a behavior called 'anting' or 'self-anointing' [123]. The

bird may pick up an ant using its bill and rub it on its feathers (active anting). Alternatively, the bird may sit on top of an anthill and perform dust-bathing movements (Ph. 3.106). This induces the ants to spray formic acid onto the bird (passive anting). Formic acid (see Section 2.1) has insecticidal, miticidal, fungicidal, as well as bactericidal properties [107].

References

[1]   Achyuthan KE, Ramachandran LK (1981) J Biosci 3(2): 149
[2]   Arbuckle K (2009) Biawak 3(2): 46
[3]   Bakker HM et al. (1993) Blood Coagul Fibrinolysis 4(4): 605
[4]   Banerjee Y et al. (2005) J Biol Chem 280(52): 42601
[5]   Bartram S, Boland W (2001) Chem Bio Chem 2(11): 809
[6]   Barua A, Mikheyev AS (2021) Proc Natl Acad Sci U S A 118(14): e2108106118
[7]   Bickler EP (2020) Toxins 12(2): 68
[8]   Blanchet G et al. (2014) Biochimie 103 109
[9]   Boffa M-C, Boffa GA (1976) Biochim Biophys Acta – Enzymol 429(3): 839
[10]   Borkow G et al. (1995) Toxicon 33(10): 1387
[11]   Boquet P (1964) Toxicon 15 5
[12]   Bradley KN et al. (2003) Toxicon 41(2): 207
[13]   Broad AJ et al. (1979) Toxicon 17(6): 661
[14]   Broeckhoven C, du Plessis A (2017) Biol Lett 13(8): 20170293
[15]   Casewell NR (2012) Toxicon 60(4): 449
[16]   Casewell NR et al. (2012) Toxicon 60(2): 119
[17]   Chandrasekara U et al. (2022) Toxins 14(8): 528
[18]   Damm M et al. (2018) Molecules 23(8): 1893

[19]   Dashevsky D et al. (2021) Toxins 13(2): 124
[20]   Deval E et al. (2010) Pharmacol Therap 128(3): 549
[21]   de Weille JR et al. (1991) Proc Natl Acad Sci U S A 88(6): 2437
[22]   Ding X et al. (2006) Hepatology 43(1): 173
[23]   Diochot S et al. (2012) Nature 490(7421): 552
[24]   Doley R, Kini RM (2009) Cell Mol Life Sci 66(17): 2851
[25]   Dufton MJ (1984) J Mol Evol 20(2): 128
[26]   Dumbacher JP et al. (1992) Science 258(5083): 799
[27]   Dumbacher JP et al. (2000) Proc Natl Acad Sci U S A 97(24): 12970
[28]   Dumbacher JP et al. (2004) Proc Natl Acad Sci U S A 101(45): 15857
[29]   Dutta S et al. (2017) J Proteom 156 29
[30]   Estevez B, Du X (2017) Physiology 32(2): 162
[31]   Evans HJ et al. (1980) J Biol Chem 255(8): 3793
[32]   Fay WP, Owen WG (1989) Biochemistry 28(14): 5773
[33]   Fohlman J et al. (1976) Eur J Biochem 68(2): 457
[34]   Fox JW (2013) Toxicon 62 75
[35]   Fraenkel-Conrat H (1982) J Toxicol – Toxin Rev 1(2): 205
[36]   Frobert Y et al. (1997) Biochim Biophys Acta (BBA) – Protein Struct Mol Enzymol 1339(2): 253
[37]   Fry BG (2005) Genome Res 15(3): 403
[38]   Fry BG et al. (2009) Proc Natl Acad Sci U S A 106(22): 8969
[39]   Fry BG et al. (2003) J Mol Evol 57(1): 110
[40]   Giorgianni MW et al. (2020) Proc Natl Acad Sci U S A 117(20): 10911
[41]   Girish KS et al. (2002) Mol Cell Biochem 240(1–2): 105
[42]   Gold BS et al. (2002) N Engl J Med 347(5): 347
[43]   Gren ECK et al. (2017) Geographic variation of venom composition and neurotoxicity in the rattlesnakes *Crotalus oreganus* and *C. helleri:* Assessing the potential roles of selection and neutral evolutionary processes in shaping venom variation. In: Dreslik MJ, Hayes WK, Beaupre SJ, Mackessy SP (eds.) The Biology of Rattlesnakes II. ECO Herpetological Publishing and Distribution, Rodeo, New Mexico, USA, p. 228
[44]   Gutiérrez JM et al. (2016) Toxins 8 93
[45]   Harvey AL (1997) Gen Pharmacol 28(1): 7
[46]   Harvey AL, Robertson B (2004) Curr Med Chem 11(23): 3065
[47]   He L et al. (2017) Sci Rep 7(1): 1791
[48]   Hendon RA, Fraenkel-Conrat H (1971) Proc Natl Acad Sci U S A 68(7): 1560
[49]   Hider RC, Khader F (1982) Toxicon 20(1): 175
[50]   Huang T-F et al. (1984) Toxicon 22(1): 45
[51]   Huang YK et al. (2022) Toxins 14(7): 436
[52]   Isbister GK (2010) Semin Thromb Hemost 36(4): 444
[53]   Ito M et al. (2001) Biochemistry 40(14): 4503
[54]   Itoh N et al. (1987) J Biol Chem 262(7): 3132
[55]   Jia L-G et al (1997) J Biol Chem 272(20): 13094
[56]   Jiang Y et al. (2011) BMC Genom 12 1
[57]   Karlsson E et al. (1972) Toxicon 10(4): 405
[58]   Karlsson E et al. (2000) Biochimie 82(9–10): 793
[59]   Kaser A et al. (2003) EMBO Rep 4(5): 469
[60]   Kashuba E et al. (2013) Biomarkers 18(4): 279
[61]   Khodorov BI (1985) Prog Biophys Mol Biol 45(2): 57
[62]   Kini RM (2003) Toxicon 42(8): 827
[63]   Kini RM (2006) Biochem J 397(3): 377

[64]	Kini RM, Doley R (2010) Toxicon 56(6): 855
[65]	Kini RM, Koh CY (2016) Toxins 8(10): 284
[66]	Klein JD, Walker FJ (1986) Biochemistry 25(15): 4175
[67]	Kloog Y et al. (1988) Science 242(4876): 268
[68]	Krochmal AR, Bakken GS (2003) J Exp Biol 206(15): 2539
[69]	Lal H et al. (1995) Br J Pharmacol 115(4): 653
[70]	Lan D et al. (2021) Front Neurol 12 716778
[71]	Lauridsen LP et al. (2016) J Proteom 136 248
[72]	Lauterwein J, Wüthrich K (1978) FEBS Lett 93(2): 181
[73]	Lertwanakarn T et al. (2020) J Venom Anim Toxins Incl Trop Dis 26 e20200005
[74]	Ligabue-Braun R, Carlini CR (2015) Toxicon 99 102
[75]	Lind P, Eaker D (1982) Eur J Biochem 124(3): 441
[76]	Lu Q et al. (2005) J Thromb Haemost 3(8): 1791
[77]	Mackessy SP, Baxter LM (2006) Zoologischer Anzeiger 245(3–4): 147
[78]	Mackessy SP, Saviola AJ (2016) Integr Comp Biol 56(5): 1004
[79]	Manuwar A et al. (2020) Toxins 12(11): 669
[80]	Marsh N, Williams V (2005) Toxicon 45(8): 1171
[81]	Matsui T, Hamako J (2005) Toxicon 45(8): 1075
[82]	McMullen BA et al. (1989) Biochemistry 28(2): 674
[83]	Mebs D (1989) Endeavour 13(4): 157
[84]	Mebs D (2002) Venomous and Poisonous Animals: A Handbook for Biologists, Toxicologists and Toxinologists, Physicians and Pharmacists. CRC Press, Boca Raton, London, New York, Washington, USA
[85]	Moise L et al. (2002) J Biol Chem 277(14): 12406
[86]	Moreau ME et al. (2005) J Pharmacol Sci 99(1): 6
[87]	Morrissette J et al. (1995) Biophys J 68(6): 2280
[88]	Nakamura M et al. (1995) Toxicon 33(9): 1201
[89]	Nelsen DR et al. (2014) Biol Rev 89(2): 450
[90]	Nikai T et al. (2000) Arch Biochem Biophys 378(1): 6
[91]	Nirthanan S, Gwee MC (2004) J Pharmacol Sci 94(1): 1
[92]	Oshima G et al. (1969) Toxicon 7(3): 229
[93]	Osipov AV et al. (2005) Biochem Biophys Res Commun 328(1): 177
[94]	Osipov A, Utkin Y (2012) Cent Nerv Syst Agents Med Chem 12(4): 315
[95]	Pahari S et al. (2007) BMC Ecol Evol 7 175
[96]	Panagides N et al. (2017) Toxins 9(3): 103
[97]	Pethő G, Reeh PW (2012) Physiol Rev 92(4): 1699
[98]	Petras D et al. (2015) J Proteome Res 14(6): 2539
[99]	Pintor AFV et al. (2011) Toxicon 57(1): 68
[100]	Pirkle H (1998) Thrombosis Haemostasis 79(3): 675
[101]	Pohanka M (2012) Int J Mol Sci 13(2): 2219
[102]	Post Y et al. (2020) Cell 180(2): 233
[103]	Pung YF et al. (2005) J Biol Chem 280(13): 13137
[104]	Rajagopalan N et al. (2007) FASEB J 21(13): 3685
[105]	Ranawaka UK et al. (2013) PLoS Negl Trop Dis 7(10): e2302
[106]	Raufman J-P (1996) Regulat Pept 61(1): 1
[107]	Revis HC, Waller DA (2004) The Auk 121(4): 1262
[108]	Rigoni M et al. (2008) J Cell Sci 117(16): 3561
[109]	Robberecht P et al. (1984) FEBS Lett 166(2): 277
[110]	Rouault M et al. (2006) Biochemistry 45(18): 5800
[111]	Saitou M et al. (2020) Cell Rep 33(7): 108402

[112] Sanchez EF et al. (2017) Toxins 9(12): 392

[113] Schieck A et al. (1972) Naunyn-Schmiedebergs Arch Pharmacol 274(1): 7

[114] Schonour RB et al. (2020) Toxins 12(10): 659

[115] Schweitz H et al. (1994) Proc Natl Acad Sci U S A 91(3): 878

[116] Schweitz H, Moinier D (1999) Perspect Drug Discovery Des 15–16 83

[117] Servent D et al. (2000) Eur J Pharmacol 393(1–3): 197

[118] Servent D et al. (1998) Toxicol Lett 103 199

[119] Shan LL et al. (2016) J Proteom 138 83

[120] Smith LA et al. (1995) Toxicon 33(4): 459

[121] Smith MT et al. (2014) Toxicon 86 1

[122] Sheumack DD et al. (1990) Comp Biochem Physiol B 95(1): 45

[123] Simmons KEL (1966) J Zool 149(2): 145

[124] Šribar J et al. (2019) Sci Rep 9(1): 283

[125] Straight R et al. (1976) Nature 261(5557): 259

[126] Sunagar K et al. (2013) Toxins 5(11): 2172

[127] Sunagawa M et al. (1996) Toxicon 34(6): 691

[128] Suryamohan K et al. (2020) Nat Genet 52(1): 106

[129] Sutcliffe MJ et al. (1994) Nat Struct Biol 1(11): 802

[130] Swenson S, Markland FS (2005) Toxicon 45(8): 1021

[131] Takacs Z et al. (2001) Mol Biol Evol 18(9): 1800

[132] Takasaki C et al. (1988) Toxicon 26(6): 543

[133] Tan CH et al. (2021) J Venom Anim Toxins Incl Trop Dis 27 e20210051

[134] Tan CH et al. (2015) BMC Genom 16 687

[135] Utaisincharoen P et al. (1993) J Biol Chem 268(29): 21975

[136] van den Kerkhof DL et al. (2021) Int J Mol Sci 22(7): 3366

[137] Vandermeers-Piret MC et al. (2000) Eur J Biochem 267(14): 4556

[138] Vonk FJ et al. (2008) Nature 454(7204): 630

[139] Watanabe TX et al. (1995) Jpn J Pharmacol 68(3): 305

[140] Wattam B et al. (2001) Biochem J 356(1): 11

[141] Weldon PJ (2000) Proc Natl Acad Sci U S A 97(24): 12948

[142] Williams JA et al. (1993) Biochem Soc Trans 21(1): 73S

[143] Yang CC (1974) Toxicon 12(1): 1

[144] Yang CC, Chang LS (1991) Biochem J 280(3): 739

[145] Yasuda O et al. (1994) Artery 21(5): 287

[146] Yin X et al. (2020) Sci Rep 10(1): 14142

[147] Zhang ZY et al. (2022) Cell Rep 40(2): 111079

[148] Zingali RB et al. (1993) Biochemistry 32(40): 10794

[149] Zuliani JP et al. (2009) Protein Pept Lett 16(8): 908

3.2.14 Mammals (Mammalia)

Many mammalian species have developed behaviors to avoid being exposed to plant or animal toxins or physiological traits that render them resistant against such toxins. Chimpanzees (*Pan* sp.) are frequently observed to eat soil to bind and neutralize plant poisons that they take up with their food. Possums (*Didelphis virginiana*), the Indian grey mongoose (*Urva edwardsii*), or the scorpion mouse (*Onchomys torridus*) are examples of animals which can tolerate bites of highly venomous snakes or scorpion stings, respectively, likely due to the competences of their immune systems with respect to these toxins.

Given this diversity in evolutionary adaptations in mammals to deal with toxins, it is somewhat surprising that there are only very few mammalian species which use venoms or poisons on their own right for efficiently making prey or for defensive purposes. Toxins are present in three orders of mammals, the Insectivora, the Monotremata, and the Chiroptera [15].

Interestingly, the only mammalian animals which are truely venomous belong to a group of small or even tiny animals, the Insectivora or Eulipotyphla. Other than the Cuban solenodon (*Atopogale cubana*) which has a body mass of approximately 600 g, the shrews belonging to this taxon are generally tiny animals. The largest species in this group is the Asian house shrew (*Suncus murinus)* native to the tropical regions of Asia. It reaches a body length of 15 cm and weighs approximately 100 g. On the other hand, the Etruscan shrew (*Suncus etruscus*) has a maximum body length of 3.5 cm and weighs only 1.8 g. It is actually the smallest terrestrial mammal known. Only somewhat larger are the northern short-tailed shrew (*Blarina brevicauda*; Ph. 3.107) native to the northwestern United States and the Eurasian water shrew (*Neomys fodiens*; Ph. 3.108) with 15–30 g BW. These animals are mainly insectivorous and have to eat almost constantly to sustain their high metabolic rates which have to be kept at maximum levels for the production of enough heat to maintain body temperatures of 38.0–38.5 C. The foraging success of these animals is high due to the presence of toxic proteins in the saliva. Toxic saliva is secreted from submaxillary glands and transferred to the bases of the median incisors in the lower jaw. These teeth form a groove which conducts the saliva into the prey organisms. Just scratching prey animals (insects, amphibians, lizards, birds, or other small mammals) is sufficient to stun them by the induction of severe pain and a rapid fall in blood pressure.

Ph. 3.107: The northern short-tailed shrew, *Blarina brevicauda*, from North America (Source: NajaShots/ iStock/Getty Images Plus).

Ph. 3.108: The European water shrew, *Neomys fodiens* (Source: Oxford Scientific/iStock/Getty Images Plus).

The presence of toxins in the saliva of mammals may have been invented early in evolution (approx. 60 million years ago) as fossils of extinct shrew-like mammals, i.e., *Bisonalveus browni*, show grooved canines. These teeth do not have corresponding surfaces in the lower jaw indicating that they may have been used for stabbing prey [8].

The saliva of the modern northern short-tailed shrew, *Blarina brevicauda*, contains kallikrein-like proteases, the blarinatoxins. One of these proteins has 253 amino acids and a molecular mass of 28 kDa [11], the other, blarinasin, has a molecular mass of 32 kDa [12]. While blarinasin is efficiently able to generate bradykinin form kininogens which supposedly induces pain sensations, the 28 kDa protein may even kill prey animals (LD_{50} (mouse, i.p.) = 0.1 mg/kg BW).

Like other hematophagous animals, the South American vampire bat (*Desmodus rotundus*), which feeds on blood of resting or sleeping wild and domestic mammalian animals, produces saliva that contains vasodilators, platelet aggregation inhibitors, and anticoagulants (inhibitors of coagulation factors and of thrombin). Draculin is a salivary glycoprotein of vampire bats. In *Desmodus rotundus* (Ph. 3.109) it is a single-chain protein composed of 708 amino acids with a molecular mass of approximately 88 kDa. It is an anticoagulant and efficiently inhibits the coagulation factors IXa and Xa [7, 16]. In addition, saliva of *D. rotundus* contains a 2-Kunitz-type domain protein, desmolaris, which specifically inhibits coagulation factor XIa in the intrinsic pathway of the blood coagulation cascade [17]. Other Kunitz-type domain proteins may also inhibit coagulation factor VIIa and thrombin [16]. Moreover, highly active plasminogen activators, desmokinases, have been isolated from the saliva of the vampire bat [2, 13]. These agents induce the dissolution of already formed fibrin clots and keep the host blood in a liquid state during feeding.

Australian platypuses (*Ornithorhynchus anatinus*; Ph. 3.110) and other monotremes represent an early offshoot from the evolutionary lineage leading from reptiles to mammals.

Ph. 3.109: The South American vampire bat, *Desmodus rotundus* (Source: belizar73/iStock/Getty Images Plus).

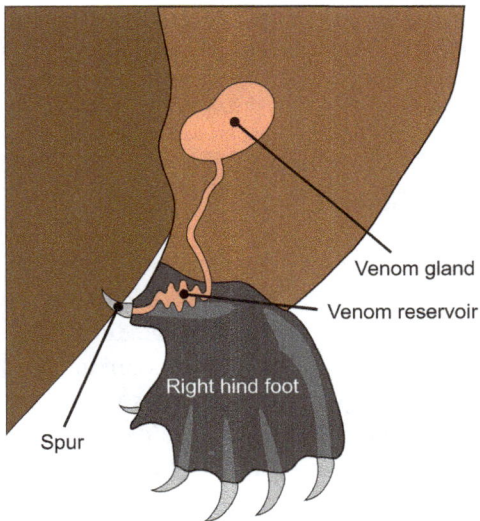

Fig. 3.42: Venom apparatus of the Australian platypus, *Ornithorhynchus anatinus*.

The male platypus carries an erectile venomous spur at its webbed hindfeet (Fig. 3.42). The associated venom gland (crural gland) is only active during the mating season indicating that the biological significance of the venomous spurs lies mainly in intraspecific territorial aggression rather than in general defense against predators [9]. Nevertheless, a frightened platypus may use its spurs against aggressors of other species as well.

Ph. 3.110: The Australian platypus, *Ornithorhynchus anatinus* (Source: JohnCarnemolla/iStock/Getty Images Plus).

⚑ A case of platypus envenomation in a human was reported by Fenner et al. [6]: 'A 57-year-old Australian was envenomated via two spur wounds to the right hand from each hind leg of a male platypus.
Pain was immediate, sustained, and devastating; traditional first aid analgesic methods were ineffective. On admission to hospital, narcotics administered intravenously, both intermittently and by infusion, provided inadequate analgesia. A right wrist block was dramatically effective. After the blockade
narcotic analgesic support was required for several days. The patient spent six days in hospital, and
the envenomated area remained painful, swollen and with little movement for three weeks. Significant
functional impairment of the hand persisted for three months, the cause of which is uncertain.'

O. anatinus venom glands contain more than 5,000 proteins of which more than 1,800
are upregulated during the mating season. Only 10 of these proteins are identified as
components of the venom with growth differentiation factor 15, nucleobindin-2, CD55,
a CXC-chemokine, and corticotropin-releasing factor-binding protein occurring only
in platypus venom [24]. Other, previously identified proteins in platypus venom are
defensin-like peptides [21], C-type natriuretic peptide [20], hyaluronidase [3], and
nerve growth factor [22]. C-type natriuretic peptide causes mast cell degranulation
and histamine release which may contribute to the pain sensations experienced by
sting victims [4]. However, platypus venom has other components that may act directly on nociceptors. Application of platypus venom to isolated rat DRG cells resulted
in an inward electrical current that was dependent on the release of calcium ions
from intracellular stores. This indicates that as yet unidentified ingredients of the
venom activate G protein-coupled receptors in neurons [5].

Among the few venomous mammals are even primates like the nocturnal slow
loris (*Nycticebus* sp.; Ph. 3.111) from Indonesia. Slow loris have a toxic bite which is a
unique feature in primates. The toxins are produced in the brachial organ, a naked,
gland-laden area of skin situated on the flexor surface of the arm. The toxin is obtained

Ph. 3.111: The Indonesian slow lori, *Nycticebus coucang* (Source: Anub Shah/iStock/Getty Images Plus).

by licking the glandular area during grooming. The toxins seem to be activated only when mixed with saliva. Bites are not used for obtaining prey, but for defensive purposes [1]. They may be directed against predators or competing conspecifics. In addition, licking the fur spreads the activated toxins over the entire body surface which may protect adults and offspring of these slow moving animals from predators.

The secretions from the skin glands on the arms contain several proteins and other factors which may be, at least in part, derived from the diet. The major endogenous component is an approximately 18 kDa protein composed of two chains of 70–90 amino acid residues, each, which are linked by disulfide bonds [14]. The N-termini of these peptides exhibit approximately 70% sequence similarity with the two chains of Fel d 1, which is produced in salivary and sebaceous skin glands in cats and is the major allergen of the domestic cat (*Felis catus*) [10].

Slow lori venom may cause death in small mammals and severe systemic symptoms including anaphylactic shock in humans [18, 19, 23].

References

[1]   Barrett M et al. (2021) Toxins 13(5): 318
[2]   Cartwright T (1974) Blood 43(3): 317
[3]   de Plater G et al. (1995) Toxicon 33(2): 157
[4]   de Plater GM et al. (1998) Toxicon 36(6): 847
[5]   de Plater GM et al. (2001) J Neurophysiol 85(3): 1340
[6]   Fenner PJ et al. (1992) Med J Aust 157(11–12): 829
[7]   Fernandez AZ et al. (1998) Biochim Biophys Acta – Gen Subj 1425(2): 291
[8]   Fox RC, Scott CS (2005) Nature 435(7045): 1091
[9]   Grant TR, Temple-Smith PD (1998) Philos Trans Royal Soc B – Biol Sci 353(1372): 1081
[10]   Kaiser L et al. (2003) J Biol Chem 278(39): 37730
[11]   Kita M et al. (2004) Proc Natl Acad Sci U S A 101(20): 7542
[12]   Kita M et al. (2005) Biol Chem 386(2): 177
[13]   Krätzschmar J et al. (1991) Gene 105(2): 229
[14]   Krane S et al. (2003) Naturwissenschaften 90(2): 60
[15]   Ligabue-Braun R et al. (2012) Toxicon 59(7): 680
[16]   Low DH et al. (2013) J Proteom 89 95
[17]   Ma D et al. (2013) Blood 122(25): 4094
[18]   Madani G, Nekaris KA (2014) J Venom Anim Toxins Incl Trop Dis 20(1): 43
[19]   Nekaris KA et al. (2013) J Venom Anim Toxins Incl Trop Dis 19(1): 21
[20]   Torres AM et al. (2002) Toxicon 40(6): 711
[21]   Torres AM et al. (1999) Biochem J 341(3): 785
[22]   Whittington C, Belov K (2007) Aust Mammal 29(1): 57
[23]   Wilde H (1972) Am J Trop Med Hyg 21(5): 592
[24]   Wong ES et al. (2012) Mol Cell Proteomics 11(11): 1354

Poison Information Centers

(Selection of PI numbers below, without guarantee of correctness)

Dial the emergency number of the country you are in (EN), numbers see https://www.taschenhirn.de/allgemeinbildung-reisen/internationale-notrufnummern) **or from abroad** (**PI**, see international dialing code in the internet file above) and you will be connected with the relevant poison information center in this country

Worldwide, PI: (+45) 82 12 12 12
All European countries. EN: 112
Austria, PI: +44(1) 406 43 43
Australia, PI: 131 126 or EN: 000
Belgium, PI: +32(70) 245 245
Brazil, EN: 192
Bulgaria, PI: +359(2) 515 32
Denmark, PI: +45 (35) 316-060
Egypt, EN: 123
England, PI: +44 (171) 635 91 91, 999 works in England and in all form British colonies
Finland, PI: +358(9) 472 977
France, PI: +33 (3) 883 737 37
Germany, EN: 112, this number also works in India, Great Britain, and in all European countries
Greece, PI: +30 (1) 799 37 77
Hungaria, PI: +36 (1) 215 215
India, PI: 1800 345 033
Indonesia, EN: 112
Italy, PI: +39 (6) 490 663
Norway, PI: +47(22) 591 300
People's Republic of China, EN: 120, works also in Japan
Poland, PI: +48 (42) 657 99 00
Russian Federation, PI: +7 (95) 928 16 47
South Africa, EN: 10 177
Spain, PI: +34 (91) 562 84 69
Sweden, PI: +46 (8) 736 03 84
Switzerland, PI: +41(1) 251 51 51
The Netherlands, PI: +31 (39) 274 88 88
Turkey, PI: +90(312)433 70 01
USA, PI: 1-800-222-1222 or 11911 or EN: 911, works also for Philippines
Other Numbers use EN, see above.

https://doi.org/10.1515/9783110728552-004

Index

https://doi.org/10.1515/9783110728552-005

www.ingramcontent.com/pod-product-compliance
Lightning Source LLC
Chambersburg PA
CBHW080911220326
41598CB00034B/5540